Environmental Health Science

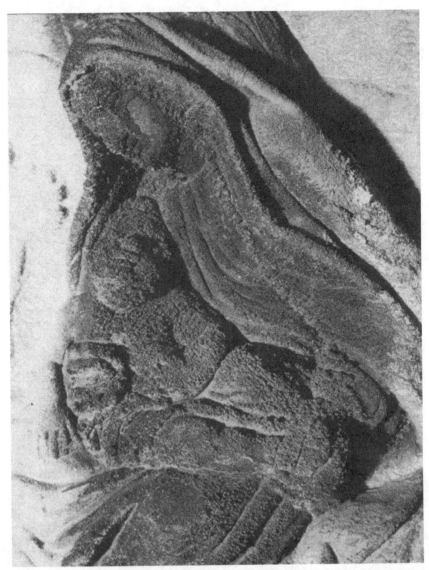

FRONTISPIECE. "This photograph shows the potential for air pollution to damage building materials including art objects. The marble Madonna in this photograph was carved about 1650 beneath a buttress of the Cathedral of Milan. Decades of combustion of high-sulfur oil and coal in this industrial area of Italy resulted in the generation of sulfur dioxide. This combined with rain water to produce sulfuric acid which, in turn, reacted with the marble, eventually disintegrating it with prolonged exposure. The lighter areas show less severe damage."

ENVIRONMENTAL HEALTH SCIENCE

Recognition, Evaluation, and Control of Chemical Health Hazards

SECOND EDITION

MORTON LIPPMANN, Ph.D.
New York University
School of Medicine

RICHARD B. SCHLESINGER, Ph.D.
Pace University
Dyson College of Arts and Sciences

OXFORD
UNIVERSITY PRESS

OXFORD
UNIVERSITY PRESS

Oxford University Press is a department of the University of Oxford. It furthers
the University's objective of excellence in research, scholarship, and education
by publishing worldwide. Oxford is a registered trade mark of Oxford University
Press in the UK and certain other countries.

Published in the United States of America by Oxford University Press
198 Madison Avenue, New York, NY 10016, United States of America.

© Oxford University Press 2018

First Edition published in 2003
Second Edition published in 2018

Library of Congress Cataloging-in-Publication Data
Names: Lippmann, Morton, author. | Schlesinger, Richard B., author.
Title: Environmental health science : recognition, evaluation, and control of
chemical health hazards / Morton Lippmann, Richard B. Schlesinger.
Description: Second edition. | Oxford ; New York : Oxford University Press, [2018] |
Includes bibliographical references and index.
Identifiers: LCCN 2017015383 (print) | LCCN 2017016140 (ebook) |
ISBN 9780190688639 (updf) | ISBN 9780190688646 (epub) |
ISBN 9780190688622 (hardcover : alk. paper)
Subjects: | MESH: Environmental Health |
Environmental Pollution—prevention & control
Classification: LCC RA566 (ebook) | LCC RA566 (print) | NLM WA 30.5 |
DDC 615.9/02—dc23
LC record available at https://lccn.loc.gov/2017015383

9 8 7 6 5 4 3 2 1
Printed by Sheridan Books, Inc., United States of America

This book is dedicated to our wives, Janet Gurian Lippmann and JoAnn Rossi Schlesinger, who through their love and support have made our home environments of the highest quality in all respects.

Preface

This book is a major revision of *Environmental Health Science*, the previous edition of which was published in 2003. While much of the basics from that book have been retained, some chapters have been reorganized and others completely rewritten. The chapters in the prior edition on physical contaminants, namely ionizing and nonionizing radiation and noise, have been removed, so as to better concentrate on aspects of chemical contamination.

The book is designed to accompany an introductory graduate level or advanced undergraduate course and thus uses scientific notation and nomenclature familiar to those who have completed introductory courses in the basic sciences. Where a greater degree of sophistication is needed in order to adequately discuss a specific issue, it is provided herein.

The broad goal of this edition, as it was for the prior one, is to provide a cohesive, up-to-date overview of concepts in environmental health science, focusing on the extent and significance of chemical contamination in our environment as it relates to human health and welfare. Information is presented in terms of the underlying physical, chemical, and biological processes that determine the behavior, fate, and ultimate impact upon health and welfare.

Our concentration on human impact is not meant to belittle additional aspects of the overall environmental stress that are also of great health concern, such as disease transmission by infectious agents through air, water, and animate vectors. The human environment is also clearly shaped by social and economic factors, and the health effects of chemical contaminants may be greatly influenced, or even overwhelmed, by the effects of tobacco smoke, alcohol, and drugs. Finally, there is currently great concern about the possible environmental impact of genetic engineering and nanomaterials. Each of these areas is sufficiently different and complex to warrant separate treatments and is brought into this book where it may relate to the effects of chemical contaminants.

Since our main focus is on the potential for effects on human health and welfare, relatively little space is devoted to disturbances within the environment that have minor impacts on humans. This is not to say that such disturbances may not eventually be shown to have such effects or that they are not important in themselves. It is simply not possible to spread the focus of this book to include them while still providing the depth of coverage desired on each topic. A supplementary bibliography is provided for those readers who wish to delve into specific topics in greater depth.

In fields as broad and often controversial as environmental contamination and environmental health, the perspective offered here will undoubtedly disappoint some people. We have attempted to avoid presenting attractive hypotheses and plausible assumptions as proven facts. This may lead to conclusions that differ from some current consensus views on particular issues, and it may upset those who approach specific issues with an advocacy perspective. The current state of knowledge in our field is almost always inadequate in scope and depth, and available data are often unreliable. The uncritical or selective use of these data by individuals with different perspectives accounts for the frequent incidence of scientists coming to diametrically opposed conclusions on the same issue. Thus, our major objective is to provide the reader with the background needed to intelligently evaluate scientific issues and thereby be able to develop informed judgments when sufficient reliable information is available.

New York, NY
February 2017
M.L. and R.B.S.

Contents

1

Introduction and Historical Perspective

INTRODUCTION

Environmental health science is a multidisciplinary field that applies basic science and engineering knowledge to the recognition, evaluation, and control of physical, chemical, and biological processes influencing human health and welfare. It is based on the premise that exposure to environmental stressors that adversely affect people can be recognized through observation of environmental quality parameters and prevented or ameliorated by the application of source controls or, where possible, through pollution prevention.

In designating air, water, and soil as "clean," "dirty," "contaminated," or "polluted," we are using terms that are subjective or, in specific contexts, have arbitrary meanings. Thus, a human environment can never be absolutely clean because, by definition, it includes at least one contaminant source, a person. An environment passes from being clean to being contaminated when the source of contamination, relative to its rate of elimination, is sufficiently large, or where there are enough sources whose aggregate output is sufficiently large, to exceed some sensory or predetermined physical concentration limit that is considered acceptable. Thus, practices that are environmentally acceptable in rural areas, such as using wood fires for domestic heating or discharging sanitary waste liquids into septic tanks

1

and underground drainage fields, can become unacceptable in densely populated urban regions. Although the term "contamination" is often used interchangeably with "pollution," the latter is better defined as contamination to a degree that renders some resource unfit for its desired use.

The degree of environmental contamination that a society finds acceptable has been highly variable both spatially and temporally and has depended more on the rate of change in contaminant levels than on their absolute amounts. It has also depended on whether there appear to be realistic alternatives. For example, very high levels of soot from inefficient coal combustion used for heating were tolerated for centuries, as long as cleaner combustion alternatives were not available at prices deemed reasonable. With the large-scale conversion of heating and electric utility boilers from coal to oil and gas combustion in recent decades, there has been a major reduction in sulfur dioxide (SO_2) and suspended airborne particulate matter (PM). As we continue to reduce our utilization of "dirty" fossil fuel resources, it is likely that we will see increased utilization of natural gas, with its lower levels of carbon dioxide (CO_2), sulfur (S), trace metals, and organic products of combustion. We can also expect to see increasing reliance on energy sources that do not emit combustion products, such as wind and solar radiation, whose costs continue to decrease. We therefore will benefit from the applications of these technologies that minimize or eliminate the release of air contaminants resulting from fossil fuel combustion.

Public acceptance of environmental contamination also depends on whether it is perceived to be "natural." Fire and its effluents have been part of the human environment throughout recorded history. Thus, until recently, the potential health effects of inhaled combustion products created relatively little concern, even though these effluents contain many toxicants. As another example, the nuclear-power industry currently generates approximately 20% of the electricity in the United States (US), while keeping most discharges of radioactive waste products to a very low level and without producing any of the chemical toxicants emitted by coal or heavy oil combustion. However, many people prefer less reliance on nuclear-power plants because they associate nuclear power with catastrophic releases of radionuclides, as from the Chernobyl and Fukushima reactor failures, and equate radioactivity with cancer. Finally, similar considerations in relation to the acceptability of chemical exposure can be applied to food safety. Under the Delaney clause of the Food and Drug Act, which was adopted in 1958 and eventually repealed in 1996, a variety of food additives, dyes, and packaging materials were banned because they were demonstrated to have the potential to cause cancer in laboratory animals when administered in high doses. On the other hand, many foods contain naturally occurring carcinogens and other toxicants, yet neither the Food and Drug Administration (FDA) nor the general public has shown much interest in banning their distribution.

If the reader is often confused about reports concerning the effects of chemical contaminants on environmental quality and human health, it is not surprising, since many such reports in the popular media tend to be selective with respect to content and emphasis. Often, information is directed for the purpose of advancing a particular viewpoint. In our incomplete state of knowledge about many environmental problems, it is relatively easy to provide plausible documentation for either the pro or the con side of almost every issue.

HISTORICAL PERSPECTIVE

Chemical contamination of the human environment did not originate at any particular time. Natural processes have been generating chemical contaminants throughout most of earth's history. Some natural waters that are remote from human activities have levels of dissolved chemicals that make them unsafe by current standards, and the air above swamps and near geothermal springs is contaminated by S gases. There have always been some poisonings of humans and grazing animals by natural chemical toxicants in foods.

However, most current problems of chemical contamination have arisen from anthropogenic sources, that is, those attributable either directly or indirectly to human activity. In shaping our environment to our needs and convenience, we have used the earth's resources to feed and clothe us and to power our industries, motor vehicles, and homes. Our ability to mold our environment has enabled us to greatly increase our numbers and improve our standard of living. But through our often careless use of natural resources, and our generation and discharge of waste products, we have created and dispersed chemical contaminants.

When early humans discovered how to control and use fire, they must have found that one of its undesirable side effects was inhalation exposure to smoke. When fire was brought indoors to heat caves or shelters, the problem of smoke exposure became more severe, and at least some additional ventilation usually had to be provided in order to enjoy the benefits of the fireplace. However, success in this regard was only partial; mummified human lungs from the preindustrial age show considerable carbonaceous pigmentation. Dwellers in heated caves were undoubtedly also exposed to significant levels of carbon monoxide (CO) and polycyclic aromatic hydrocarbon (PAH) carcinogens.

One of the earliest written discussions of the relationship between environment and health was the Hippocratic essay *On Airs, Waters and Places,* written around 460 BC. It advised physicians to consider the winds, seasons, and sources of water when evaluating the health of their patients. Hippocratic works also described lead (Pb) colic in miners, as well as diseases occurring in other occupational groups.

Other authors in ancient Greece and Rome recognized that materials used in metallurgy were toxic. Pliny the Elder discussed the dangers in handling S and zinc (Zn), and Galen recognized the dangers of acid mists among copper (Cu) miners. Although the ancients were not aware of the possibility, subclinical chronic Pb poisoning may have been widespread, as Pb containing glazes were commonly used on kitchen pottery, and acidic wines and foods extracted some of the metal. The Romans also used Pb pipes for the delivery of drinking water, and some of the Pb was slowly dissolved into the water. The exposures would have been greatest among the more prosperous Romans, who more often had running water and glazed vessels. The decline of the Roman Empire has been partially attributed by some to chronic Pb intoxication, on the basis that the recorded decline in fertility among the upper classes was consistent with the effects of ingested Pb.

The Workplace and Health

Some of the earliest reports relating an apparent association between specific chemical contaminants and human health effects involved occupational exposures, where the levels of exposure were generally much higher than that for the general population. Treatises on occupational diseases began to appear in Europe in the Middle Ages. In 1472, Ulrich Ellenbog of Augsburg wrote an eight-page booklet that discussed the toxic actions of CO, nitric acid (HNO$_3$) vapors, Pb, mercury (Hg), and other metals.

A classic description of mining technology and its hazards, *De Re Metallica* was published in 1556 by the heirs of Georg Bauer, a native of Saxony who was more commonly known by his Latin name, Georgius Agricola. From 1526 until his death in 1555, Agricola had been the official physician of the Bohemian mining town of Joachimstal, a major source of European silver, and more recently of radium (Ra) and uranium (U). The silver (Ag) coins of Joachimstal were known as Thalers, which in English later became dollars. *De Re Metallica* is a scholarly work of 12 books that were translated from Latin into English in 1912 by an American mining engineer and his wife. (The engineer, Herbert C. Hoover, eventually gave up engineering for public service and was elected US President in 1928.) In the last part of the sixth book, Agricola described diseases of the lungs, joints, and eyes that were common among miners. It appears from descriptions of the diseases that the men had silicosis, tuberculosis, lung cancer, and combinations thereof. The book also contained numerous woodcut illustrations. Figures 1–1 and 1–2 show samples that illustrate means that were used to limit chemical exposures.

In 1567, a posthumous work appeared with the title *Von der Bergsucht und anderen Bergkrankheiten* (On the Miners' Sickness and Other Diseases of Miners). It was written by Theophrastus Bombastus von Hohenheim, better known as Paracelsus, who was an itinerant Swiss physician and alchemist. This monograph

A—Furnace. B—Sticks of wood. C—Litharge. D—Plate. E—The foreman WHEN HUNGRY EATS BUTTER, THAT THE POISON WHICH THE CRUCIBLE EXHALES MAY NOT HARM HIM, FOR THIS IS A SPECIAL REMEDY AGAINST THAT POISON.

FIGURE 1–1. Woodcut from Agricola, Book X. The technology of lead smelting was considerably more advanced than the recommended prophylaxis for lead poisoning. Note the barrier plate, D, which protects against splatter burns. (*Source*: Agricola, G. *De Re Metallica*, Basel, 1556. Translated by H.C. Hoover and L.H. Hoover for the *Mining Magazine*, London, 1912. Reprinted by Dover Press.)

was devoted to the occupational diseases of mine and smelter workers. Paracelsus did not consider dust exposure to be the causative factor in the lung diseases he observed in miners but rather explained them in terms of alchemy and the stars. He was, however, considerably more astute in his description of diseases among smelter workers and differentiated between acute and chronic poisonings. His detailed descriptions of mercurialism covered most of the currently recognized symptoms. Paracelsus is also well known for his enunciation of a basic tenet of toxicology, "All substances are poisons; there is none which is not a poison. The right dose differentiates a poison and a remedy."

The most comprehensive description of occupational diseases of its time, and for well over a century thereafter, was a book of 40 chapters titled *De Morbis Artificum* (Diseases of Workers), published in 1700 by Bernardino Ramazzini, a professor of medicine at the University of Modena in Italy and, after 1700, at the University of Padua. Ramazzini is the generally acknowledged "father of occupational medicine." His descriptions of diseases covered most of the trades practiced in his time, including those of dirty and humble trades, such as corpse carriers, porters, and laundresses. He stated, "When a doctor visits a working-class home,

FIGURE 1–2. Woodcut from Agricola, Book IX. Note respiratory protection of the furnace worker. (*Source*: Agricola, G. *De Re Metallica,* Basel, 1556. Translated by H.C. Hoover and L.H. Hoover for the *Mining Magazine,* London, 1912. Reprinted by Dover Press.)

A—HEARTH. B—HEAP. C—SLAG-VENT. D—IRON MASS. E—WOODEN MALLETS. F—HAMMER. G—ANVIL.

he should be content to sit on a three-legged stool, if there isn't a gilded chair, and he should take time for his examination; and to the questions recommended by Hippocrates, he should add one more—What is your occupation?" Unfortunately, there is still a great deal of unrecognized occupational disease today because too many physicians still neglect to ask that important question.

In 1775, an English physician, Sir Percival Pott, provided the first description of occupationally induced cancer, that of scrotal cancer in chimney sweeps. In 1831, Charles Turner Thackrah made a special contribution to this era by the publication of his 200-page book, *The Effects of Arts, Trades and Professions and All Civic States and Habits of Living on Life and Longevity,* based mainly on his experience in the manufacturing district of Leeds, England.

Hazardous working conditions and occupational diseases were also common in the more technologically advanced countries in Asia. Extensive descriptions of the operations involved in the mining and refining of metals were provided in the *Atlas of Important Products in Mountains and Sea of Japan* (1754) and *Atlas of*

FIGURE 1–3. The refining of copper in the Besshi Copper Mine in Japan, circa 1800. (*Source*: "Atlas of Mining and Refining of Copper, 1801," Reprinted in: Miura, T.A. Short History of Occupational Health in Japan [Part I]. *The J. of Science of Labour* 53:509–25, 1977.)

Mining and Refining of Copper (1801) and are illustrated by woodcuts such as the one reproduced in Figure 1–3.

Scattered reports on occupational diseases appeared in the British, French, German, and American literature through the balance of the nineteenth century. Before the end of that century, it was clear that it was desirable to anticipate problems associated with industrial exposures to toxic chemicals before they happened rather than after their effects were apparent in workers and that critical knowledge relating exposures and their effects could usually be gained through the systematic exposure of laboratory animals. Pioneering work along these lines began in the 1880s under K. L. Lehmann in Wurzburg, and by 1884 he had published data on the results of toxicological studies with 35 gases and vapors.

With the rapid growth in industrialization in the nineteenth century, more and more workers were being exposed to a broadening spectrum of chemicals. The result was an increase in occupational disease and disability. This was first apparent in England, and, by 1833, the first of the English Factory Acts was passed by Parliament, establishing the principle that people injured at work are entitled to compensation. While there was no requirement to prevent the conditions that led to the need for compensation, it became more profitable for many businesses to reduce their costs through preventive measures rather than through paying claims. The need for more proactive preventive measures was recognized later, and the English Factory Act of 1878 created a centralized Factory Inspectorate. Most of the major European countries followed the British lead, but it was not until 1911 that Wisconsin became the first US state to establish workmen's compensation and not until 1948 that the last state did so. The first state programs to inspect industry for occupational exposures began in 1913 in New York and Ohio, but nationwide coverage was not achieved until the passage of the federal Occupational Safety and Health Act of 1970.

The US federal government's involvement in the recognition and characterization of occupational health hazards began with the creation of the Bureau of Mines in 1910 and the establishment of an Office of Industrial Hygiene and Sanitation within the Public Health Service in 1914. In that year, these federal agencies jointly conducted the first of a series of comprehensive studies of certain lung disorders in the dusty trades, under the direction of Dr. Anthony J. Lanza. This was a major activity of the Hygienic Laboratory of the Public Health Service's Office of Industrial Hygiene and Sanitation. With the creation of the National Institutes of Health (NIH) in 1937, the Hygienic Laboratory became the Division of Industrial Hygiene (DIH). In 1946, with the increased NIH focus on research, the DIH was transferred to the Bureau of State Services. In 1970, both the National Institute for Occupational Safety and Health (NIOSH) and the Occupational Safety and Health Administration (OSHA) were created to evaluate, regulate, and control occupational health and safety hazards.

Pioneering textbooks began to appear in 1914 with W. Gilman Thompson's, *The Occupational Diseases*. The first text on industrial toxicology, which appeared in 1925, was *Industrial Poisons* by Alice Hamilton. In 1948, the first edition of Patty's *Industrial Hygiene and Toxicology* was published.

In 1918, Harvard appointed Dr. Hamilton to its faculty of public health and became the first university to establish a graduate program in occupational health, a program later broadened to environmental health. The faculty at Harvard School of Public Health started the first American scientific journal devoted specifically to occupational health, the *Journal of Industrial Hygiene*, which appeared in 1919.

Air Contamination and Health

Community air contamination arising from the combustion of fossil fuels first received official recognition at the end of the thirteenth century, when Edward I of England issued an edict to the effect that, during sessions of Parliament, there should be no burning of sea coal or channel coal, so-called because it was brought from Newcastle to London by sea transport via ports on the English Channel. Despite a succession of further royal edicts, taxes, and even occasional prison confinements and torture, the use of coal for producing heat continued in London, especially as the increase in population led to a depletion in the availability of wood for fuel. The first scholarly report on the problem, "Fumifugium or the Inconvenience of Aer and Smoak of London Dissipated, together with some Remedies Humbly Proposed" by John Evelyn, was published by the royal command of Charles II in 1661. Evelyn, one of the founding members of the Royal Society, recognized and discussed the problem in terms of the sources, effects, and feasibility of controls.

Unfortunately, no effective controls were instituted until after the report of the Royal Commission on the effects of the "killer fog" of December 1952, which was associated with approximately 4,000 to 12,000 more deaths in London in that winter season than in prior years. Retrospective examinations of vital statistics demonstrated that there had been numerous prior episodes involving additional deaths during periods of air stagnation in London and other British cities. The reductions in smoke levels achieved in Britain since 1952 have eliminated readily observable deaths directly attributable to fossil fuel combustion and have led to major beneficial changes in visibility and microclimate as well as health.

Deaths specifically attributable to coal smoke had occurred elsewhere prior to 1952 but involved smaller populations. The most notable of these were those attributable to coal smoke pollution in the Meuse Valley in Belgium in 1930 and to steel industry emissions in Donora, Pennsylvania, in 1948.

In the 1940s it became apparent that southern California had an air pollution problem and that it was a very different kind of pollution from that long known in London and the eastern US. The California variety was characterized by oxidant gases, such as ozone (O_3), rather than by reducing gases, such as SO_2. Furthermore, these oxidants were formed in the atmosphere by photochemical processes.

Air pollution research on the federal level began when an occupational health team was sent to Donora, Pennsylvania, to investigate the 1948 pollution episode. In the mid-1950s, it was decided that a separate federal research program was needed, and a separate laboratory program was set up by a federal Public Health Service laboratory. However, regulatory aspects of air pollution control were still considered to be the responsibilities of state and local agencies. The Clean Air Act of 1970 established federal responsibility for air pollution control and, with the creation of the Environmental Protection Agency (EPA) later the same year, became the purview of the EPA. This represented a recognition that air pollution had changed from being a local community problem to being a regional problem and that effective controls had to be taken by national and international authorities, as well as by local authorities and individuals.

By the 1990s, it became evident that fine particles, that is, those smaller than 2.5 um in aerodynamic diameter ($PM_{2.5}$) in the ambient air, were capable of producing increased mortality and morbidity at contemporary levels of air pollution, and a major research effort began to determine the causal components and biological mechanisms for such effects. Also, by the 1990s the international aspects of air pollution had gained widespread recognition. This included not only pollutant transport between the borders of the US and Canada and the US and Mexico but also stratospheric O_3 depletion in the polar regions caused by fluorocarbon emissions and global climate change caused by rises in CO_2 and methane (CH_4) releases associated with human activities. While international agreements have drastically reduced fluorocarbon emissions, effective actions to reduce global climate change remain uncertain as of this writing.

While the focus of this book is on contaminants that are routinely encountered, mention should be made that occasionally industrial accidents do occur. One of the world's worst industrial accidents occurred on December 2, 1984, at the Union Carbide Plant in Bhopal, India, which involved release of a gas cloud of the highly toxic chemical methylisocyanate (MIC). It was apparently initiated by the introduction of water into the MIC storage tank, resulting in an uncontrollable reaction, with liberation of heat and escape of MIC and other decomposition products. Safety systems were either not functioning or were inadequate to deal with large volumes of the escaping chemicals. While more than 200,000 persons were exposed to the gas, the death toll reached about 3,800 and thousands of others suffered significant morbidity.[1]

Water Contamination and Health

Historically, most contamination problems in water have centered on infectious diseases rather than on chemical contamination. Furthermore, the growth of cities was generally limited by their ability to prevent contamination of their water supplies by fecal wastes. The more innovative ancient cities in the Roman Empire had their drinking water supplied from distant sources via aqueducts and enclosed pipes and channels.

In more modern times, the first association between water contamination and human disease was made by John Snow in his classic epidemiologic investigations of the cholera epidemic in London in 1853. It was at about this same time that water pollution was becoming a matter of serious concern in England for aesthetic reasons. The rapid growth of London in the first half of the nineteenth century, and the adoption of the practice of discharging the effluent from newly installed water closets into sewers constructed for carrying away storm drainage, led to such an overpowering stench from the Thames at Westminster that Parliament found it difficult to meet in 1858. The solution for that immediate problem was to extend the sewer system, so as to transfer the wastes far enough downstream from Parliament to alleviate the nuisance locally.

The development of the field of bacteriology in the latter half of the nineteenth century provided a scientific basis for understanding the role of water in the transmission of typhoid and other enteric bacterial diseases. Water-filtration processes capable of reducing bacterial concentrations by one or two orders of magnitude were developed. By the turn of the twentieth century, these processes were starting to come into widespread use. But in cities such as Pittsburgh and Cincinnati, which still used unfiltered river water, annual death rates from typhoid fever remained around 100 per 100,000 in 1900, and in the US as a whole, the typhoid death rate was about 35 per 100,000. The dramatic effects of filtration and chlorination of a public water supply on the incidence of water-borne disease is illustrated in Figure 1–4, which shows the virtual disappearance of typhoid in Philadelphia during the first half of the twentieth century.

Within the first 30 years of the twentieth century, most of the cities and towns using rivers as sources of water had installed water-treatment plants. In 1908, the use of chlorine as a water disinfectant was introduced, and it became a standard treatment. This made possible the production of bacteriologically safe water at very little cost, even when raw water of very poor quality was being treated. However, an epidemic of fatalities associated with *cryptosporidium* contamination in the public water supply of Milwaukee, Wisconsin, in 1993, which resulted in about 100 deaths and severe intestinal disorders in about 400,000 people, shook the complacency of public health authorities and demonstrated that current water purification technology and monitoring systems needed further development to deal with the problem of pathogens resistant to conventional disinfection treatments.

Another method for water purification that has been used since the late nineteenth century is ozonation, since ozone is more effective in treating bacteria and viruses than is chlorine. However, this technique is more widely used in Europe and Asia than in the US.

FIGURE 1–4. Reduction of typhoid fever in Philadelphia following treatment of the water supply.

Although much has been done to reduce disease risks from viable agents, very little is known about the possible health effects of the variety of largely unidentified chemical compounds that enter water-supply sources with sewage and industrial wastes. Many of these compounds are not effectively removed by today's water-treatment plants. Recent reports indicating the presence of trace amounts of known carcinogens in many drinking-water supplies may lead to a change in the design and performance requirements of water treatment and waste-disposal facilities.

Food Contamination and Health

Maintaining the safety of foods has always been important, and the intentional use of chemical preservation techniques, such as salting and smoking, can be traced back to earliest recorded history. Recognition of natural toxicants was often incorporated into dietary laws, which were generally enforced as religious taboos.

Within the past century, modern preservation techniques have made it possible for foods to be processed and distributed for mass markets. It became economically advantageous to make the processed foods attractive in appearance and to keep them that way for extended periods of time. The separation in time and space between production and consumption also created temptations to adulterate the products with fillers and less than wholesome raw materials.

Although the federal government did not begin to enforce food safety regulations until 1906, some individual states recognized their responsibility to provide a safe food supply. As early as 1764, the Massachusetts Bay Colony established a sanitary code for slaughterhouses. California passed a pure food and drug law in 1850, the same year it became a state. In 1856, Massachusetts prohibited the adulteration of milk. Following the enactment of the British Pure Food and Drug Law in 1875, a number of states in the US passed laws regulating the handling and production of food. By 1900, most states had regulations designed to protect the consumer from unsanitary and adulterated foods.

National recognition of the need for federal regulation in the US essentially began with the appointment in 1883 of Dr. Harvey W. Wiley as the chemist for the US Department of Agriculture. He campaigned vigorously against misbranded and adulterated foods. Finally, in 1906, the Sherman Act, regulating interstate transportation of food, was passed.

Most manufacturers observed the Sherman Act, and adulteration with known harmful substances was rare. Economic cheating was widespread, however, and adulterants, sometimes toxic, were common in some foods. Unfortunately, the act did not provide for legal standards or identification of potentially harmful foods. In 1938, the US Congress passed the Food, Drug, and Cosmetic Act. This law, with its various amendments, is administered by the US FDA also created in 1938. Some principal amendments relating to foods were the Pesticide Amendment Act of 1954 (revised 1972), the Food Additive Amendment of 1958, and the Color Additive Amendments of 1960 and 1972. These regulations affected about 60% of the food produced in the US; the remaining 40% remained under state regulation, which in many cases paralleled federal legislation.

In the 1958 Food Additive Amendment exceptions were made for all additives that, because of years of widespread use in foods, were "generally recognized as safe (GRAS) by experts qualified by scientific training and experience." These exceptions, numbering more than 600 items, comprise the so-called GRAS list, which has been and continues to be a subject of much controversy.

The Food Additive Amendment of 1958 also included the cancer or Delaney amendment, which stated: "No additive shall be deemed to be safe if it is found to induce cancer when ingested by man or animal, or if it is found, after tests which are appropriate for the evaluation of the safety of food additives, to induce cancer in man or animal." Although the aim of the Delaney amendment was laudable, serious problems arose in the evaluation of data, protocols for testing for carcinogenicity, and the extrapolation of the data to humans. These problems polarized scientific and legislative authorities, as well as the consuming public, into proponents and opponents of the measure. The controversy re-erupted with each application of the Delaney amendment, such as when cyclamates were banned in 1969, when red dye #2 was prohibited in 1976, and when a proposed ban on saccharin was announced in 1977. The lobby for the diet food industry, and widespread concern about the loss of the only approved nonsugar sweetener among those concerned with obesity as a public health problem, led Congress to approve a specific exemption for saccharin from the provisions of the Delaney amendment. In 1996, the Food Quality Protection Act repealed the Delaney clause and set up an objective limit on lifetime cancer risks at one in a million.

Recently emerging concerns about food supplies include crops based on genetically modified organisms, contamination of meat products by ingestion of animal by-products in animal feed, such as prions that cause brain pathology (mad cow disease), residues of pesticide and growth hormones in animal feeds, and contamination of meat products by persistent organic chemicals produced by combustion, such as dioxin and related compounds. In most cases, the extent of the risks associated with these newer concerns remain unresolved.

REFERENCE

1. Broughton, E. The Bhopal disaster and its aftermath: a review. *Environ. Health.* 2005; 4: 6. doi:10.1186/1476-069X-4-6

2

Characterization of Contaminants and Environments

CHARACTERIZATION OF CONTAMINANTS

Levels of chemicals in the environment that we inhabit are expressed in terms of concentration, but confusion often arises from the use of the same or similar sounding terms that can have different meanings in different contexts. This is especially true when considering concentrations of air and water contaminants. These concentrations are both frequently expressed in terms of a ratio, such as parts per million (ppm) or parts per billion (ppb). When used for air contaminants the units used are conventionally expressed in terms of molar or volume fractions, while when used for water contaminants they are weight fractions. In order to avoid confusion, the concentration units used in this book have appropriate subscripts notations: ppm_v to indicate parts per million parts of air, a volume ratio, and ppm_w, to indicate parts per million parts of water, a weight ratio. Problems can be avoided altogether by expressing all fluid contaminant concentrations as the weight of contaminant per unit volume, for example, cubic meter (m^3) or liter (L) of fluid. In air, the units generally used are mg/m^3 or $\mu g/m^3$, while in water they are most often mg/L or $\mu g/L$.

Air Contaminants

Chemical contaminants can be dispersed into air as gases, liquid droplets, or solid particles. The latter two were given the generic term "aerosols" by Gibbs[1] on the basis of analogy to the term "hydrosol," a term already in use to describe disperse systems in water. On the other hand, gases and vapors, which are present as discrete molecules, form true solutions in air. Particles consisting of moderate to high vapor-pressure materials tend to evaporate rapidly, since those small enough ($< 10 \mu m$) to remain suspended in air for more than a few minutes have large surface-to-volume ratios. Evaporation is also enhanced by the Kelvin effect arising from the interaction of surface tension with the curvature of the droplet surface. Some materials with relatively low vapor pressures can have appreciable fractions in both the vapor and aerosol forms simultaneously and are sometimes referred to as semivolatiles.

Gases and vapors

Gases and vapors, once dispersed in air, generally form mixtures so dilute that their physical properties, such as density, viscosity, enthalpy, and so on, are indistinguishable from those of clean air. Such mixtures may be considered to follow ideal gas-law relationships. There is no practical difference between a gas and a vapor, except that the latter is generally considered to be the gaseous phase of a substance that is normally a solid or liquid at room temperature. While dispersed in air, all gaseous molecules of a given compound are essentially equivalent in their size and capture probabilities by ambient surfaces, respiratory tract surfaces following inhalation, and contaminant collectors or samplers.

Aerosols

Aerosols, as dispersions of solid particles and/or droplets in air, have a very significant additional variable of particle size. Size affects particle motion and, hence, the probabilities for physical phenomena, such as coagulation, dispersion, sedimentation, impaction onto surfaces, interfacial phenomena, and light-scattering. However, it is not possible to characterize a given particle by a single size parameter. For example, a particle's aerodynamic properties depend on density and shape, as well as on linear dimensions, while the effective size for maximal light-scattering is dependent on refractive index and shape.

In some cases, all of the particles within an aerosol are essentially the same size. Such aerosols are considered to be monodisperse. Examples are natural pollens and some laboratory-generated aerosols. More typically, aerosols are composed of particles of many different sizes and hence are called heterodisperse or polydisperse. Since different aerosols have different degrees of size dispersion, it is necessary to specify at least two parameters in characterizing aerosol size: a measure of central tendency, such as a mean or median, and a measure of size dispersion, such as an arithmetic or geometric standard deviation.

Particles generated by a single source or process generally have diameters that follow a log-normal distribution; that is, the logarithms of their individual diameters have a Gaussian distribution. In this case, the most appropriate measure of dispersion is the geometric standard deviation, which is the ratio of the 84.1 percentile size to the 50 percentile size (Fig. 2–1). When more than one source of particles is significant, the resulting mixed aerosol will usually not follow a single log-normal distribution, and it may be necessary to describe it by the sum of several log-normal distributions.

Aerosols have integral properties that depend upon the concentration and size distribution of the constituent particles. In mathematical terms, these properties can be expressed in terms of certain constants or "moments" of the size distribution. Some integral properties, such as light-scattering ability or electrical charge, depend on other particle parameters as well. Some of these important properties are as follows.

Number Concentration. The total number of airborne particles per unit volume of air, without distinction regarding their sizes, is the zeroth moment of the size distribution. In current practice, instruments are available that can count the numbers of particles of all sizes from about 0.005 to 50 μm. In many specific applications, such as fiber counting for airborne asbestos fibers, a more restricted size range may be specified.

Surface Concentration. The total external surface area of all the particles in the aerosol, which is the second moment of the size distribution, may be of interest when surface catalysis or gas adsorption processes are of concern. Aerosol surface is one factor affecting light-scatter and atmospheric-visibility reductions.

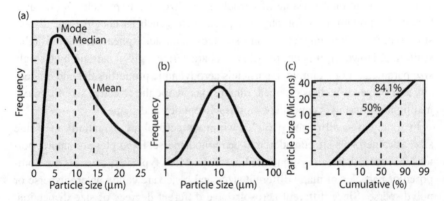

FIGURE 2–1. Particle size distribution data: (a) Plotted on linear coordinates; (b) Plotted on a logarithmic-size scale; (c) In actual practice, logarithmic probability coordinates are used to display the cumulative percentage of particles less than a specific size. The geometric standard deviation (sg) of the distribution is equal to the ratio of the 84.1% size to the 50% size.

Volume Concentration.   The total volume of all the particles, which is the third moment of the size distribution, is of little intrinsic interest in itself. It is, however, closely related to the mass concentration, which for many environmental effects is the primary parameter of interest.

Mass Concentration.   The total mass of all the particles in the aerosol is frequently of interest. The mass of a single particle is the product of its volume and density. If all of the particles have the same density, the total mass concentration is simply the volume concentration times the density. In some cases, such as "size-selective" dust sampling for health hazard assessments, the parameter of interest is the mass concentration over a restricted range of aerodynamic particle sizes. In this application, particles too large to deposit in the target region of concern in the human respiratory tract following their inhalation are excluded from the integral.

Dustfall.   The mass of particles depositing from an aerosol onto a unit surface per unit of time is proportional to the fifth moment of the size distribution. Dustfall has long been of interest in air pollution control because it provides an indication of the soiling properties of the aerosol.

Particles have a number of properties other than linear size that can greatly influence their airborne behavior and their effects on the environment and health. These include the following.

Surface.   For spherical particles, the surface area varies as the square of the diameter. However, for an aerosol of given mass concentration, the total aerosol surface increases with decreasing constituent particle size. Airborne particles have much greater ratios of external surface to volume than do bulk materials, and, therefore, the particles can dissolve on surfaces or participate in surface reactions to a much greater extent than massive samples of the same materials. Furthermore, for nonspherical solid particles or aggregate particles, the ratio of surface to volume is increased, and for particles with internal cracks or pores, the internal surface area can even be greater than the external area.

Volume.   Particle volume varies as the cube of diameter; therefore, the largest particles in a polydisperse aerosol tend to dominate its volume or mass concentration.

Shape.   An airborne particle's shape affects its aerodynamic drag, as well as its surface area, and therefore its motion and deposition probabilities onto surfaces.

Density.   A particle's velocity due to gravitational or inertial forces increases as the square root of its density.

Aerodynamic Diameter. The diameter of a unit-density sphere having the same terminal settling velocity as the particle under consideration is equal to its aerodynamic diameter. Terminal settling velocity is the steady-state velocity of a particle that is falling under the influence of gravity and fluid resistance. Aerodynamic diameter is determined by the actual particle size, the particle density, and an aerodynamic shape factor that is determined by its drag (fluid resistance).

Aerosols are generally classified in terms of their processes of formation. Some of the terminologies used for aerosols of various sizes from a variety of sources are indicated in Figure 2–2a, and their concentration ranges are indicated in Figure 2–2b. While this classification is neither precise nor comprehensive, it is commonly used and accepted in the industrial hygiene and air pollution fields.

Dust. This is an aerosol formed by mechanical subdivision of bulk material into airborne particles having the same chemical composition. A general term for the process of mechanical subdivision is comminution, and it occurs in operations such as crushing, grinding, drilling, and blasting. Dust particles are generally solid and irregular in shape and have diameters greater than 1 μm.

Fume. Fume is an aerosol of solid particles formed by condensation of vapors released into air at elevated temperatures by combustion or sublimation.

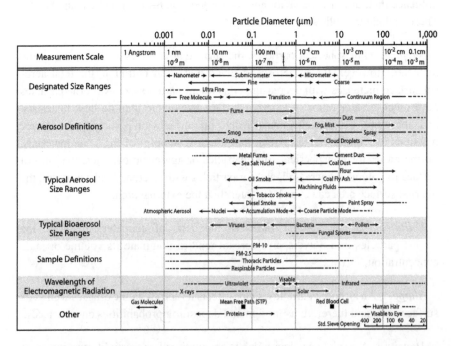

FIGURE 2–2a. Particle size ranges and definitions for aerosols. (*Source*: Hinds, W.C. *Aerosol Technology*, 2nd Ed., New York: Wiley, 1999, p. 9.)

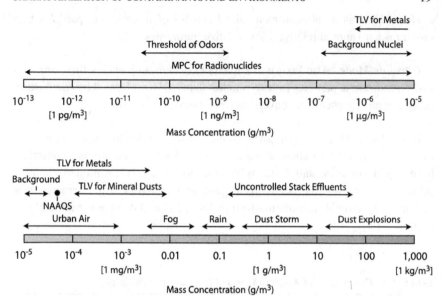

FIGURE 2–2b. Concentration ranges for specific types of aerosols (*Source*: Hinds, W.C. *Aerosol Technology*, 2nd Ed., New York: Wiley, 1999, p. 11.)

The primary particles are generally very small (less than 0.1 μm) and have spherical or characteristic crystalline shapes. They may be chemically identical to the parent material, or they may be composed of an oxidation product, such as a metal oxide. Since they may be formed in high number concentrations, they often rapidly coagulate, forming aggregate clusters of low overall density.

Smoke. Smoke is an aerosol formed by combustion of organic materials. The particles are generally less than 0.5 μm.

Mist. Mist is a droplet aerosol formed by condensation or by mechanical shearing of a bulk liquid, for example, atomization, nebulization, bubbling, or spraying. The droplet size can cover a very large range, usually from about 2 μm to greater than 50 μm.

Fog. Fog is a water aerosol formed by condensation of water vapor onto atmospheric nuclei at high relative humidities. The droplet sizes are generally greater than 1 μm.

Smog. This is a popular term for a pollution aerosol derived from a combination of smoke and fog. The term is also commonly applied to light-scattering aerosols formed by photochemical reactions.

Haze. Haze is a submicrometer-sized aerosol of hygroscopic particles that take up water vapor at relatively low relative humidities.

Ultrafine Mode. Also known as Aitken or condensation nuclei, these are very small atmospheric particles (mostly smaller than 0.05 μm) formed by combustion processes and by chemical conversion from gaseous precursors.

Fine Mode. This is a term given to the particles in the ambient atmosphere that range from 0.1 to about 2.5 μm. These particles generally are spherical, have liquid surfaces, and form by coagulation and condensation of smaller (ultra-fine) particles that derive from gaseous precursors. The characteristics of fine mode aerosols are summarized in Table 2–1 and depicted graphically in Figure 2–3.

TABLE 2–1. Comparison of Ambient Fine and Coarse Mode Particles

	FINE MODE (PM$_{2.5}$)	COARSE MODE (PM$_{10\text{-}2.5}$)
Formed from	Gases	Large solids/droplets
Formed by	Chemical reaction; nucleation; condensation; coagulation; evaporation of fog and cloud droplets in which gases have dissolved and reacted.	Mechanical disruption (e.g., crushing, grinding, abrasion of surfaces); evaporation of sprays; suspension of dusts.
Composed of	Sulfate, SO$_4^=$; nitrate, NO$_3^-$; ammonium, NH$_4^+$; hydrogen ion, H$^+$; elemental carbon; organic compounds (e.g., PAHs, PNAs); metals (e.g., Pb, Cd, V, Ni, Cu, Zn, Mn, Fe); particle-bound water.	Resuspended dusts (e.g., soil dust, street dust); coal and oil fly ash; metal oxides of crustal elements (Si, Al, Ti, Fe); CaCO$_3$, NaCl, sea salt; pollen, mold spores; plant/animal fragments; tire wear debris.
Solubility	Largely soluble, hygroscopic and deliquescent.	Largely insoluble and nonhygroscopic.
Sources	Combustion of coal, oil, gasoline, diesel, wood; atmospheric transformation products of NO$_x$, SO$_2$, and organic compounds includ- ing biogenic species (e.g., terpenes); high temperature processes, smelters, steel mills, etc.	Resuspension of industrial dust and soil tracked onto roads; suspension from disturbed soil (e.g., farming, mining, unpaved roads); biological sources; construction and demolition; coal and oil combustion; ocean spray.
Lifetimes	Days to weeks	Minutes to hours
Travel Distance	100s to 1000s of kilometers	< 1 to 10s of kilometers

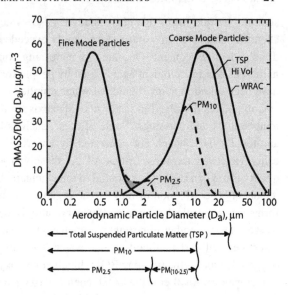

FIGURE 2–3. A multi-modal mass distribution of ambient air PM and the components that are collected by size-selective aerosol samplers.

Coarse Mode. Ambient air particles larger than about 2.5 μm and generally generated by mechanical processes and/or surface dust resuspension are considered coarse mode. The characteristics of coarse mode aerosols are described in Table 2–1 and Figure 2–3.

Water Contaminants

Chemical contaminants can be found in water in solution or as hydrosols; the latter are small immiscible solid or liquid particles that have unipolar charges in a stable suspension. An aqueous suspension of lipid droplets is generally called an emulsion. Many materials with relatively low aqueous solubility can be found in both dissolved and suspended forms.

Dissolved contaminants

Water is known as the universal solvent. While many compounds are not completely soluble in water, there are few that do not have some measurable solubility. In fact, the number of chemical contaminants that can be detected in natural waters is primarily a function of the sensitivity of the analyses. For organic compounds in rivers and lakes, as the limits of detection decrease by an order of magnitude, the numbers of compounds detected increase by an order of magnitude, so that one might expect to find at least 10^{-12} gm/L (approximately 10^{10} molecules per liter) of many organic compounds reported in the literature. Similar considerations undoubtedly apply to inorganic chemicals as well.

Water-quality criteria generally include a nonspecific parameter called "dissolved solids." However, it is customary to exclude natural mineral salts, such as sodium chloride, from this classification. Also, water criteria for specific toxic chemicals dissolved in water are frequently exceeded without an excessive total dissolved-solids content. One area of increasing concern relates to residues of numerous pharmacological agents used by people for therapeutic reasons, which have been found in many municipal water supplies.

Compounds dissolved in water may also exist in the gaseous phase at normal temperatures and pressures. Some of these, such as hydrogen sulfide (H_2S) and ammonia (NH_3), which are generated by anaerobic decay processes, are toxicants. However, the most critical of the dissolved gases with respect to water quality is oxygen (O_2). It is essential to many aquatic life forms and is involved in the chemical conversion of many toxic organic contaminants to more innocuous forms. Thus, a critical parameter of water quality is the concentration of dissolved oxygen (DO). Another important parameter is the extent of the oxygen "demand" associated with contaminants in the water. The most commonly used index of oxygen demand is the five-day BOD (biochemical oxygen demand after five days of incubation). Another is the COD (chemical oxygen demand). The basic oxygen demand parameters are defined in Table 2–2.

Suspended particles

A nonspecific water-quality parameter that is widely used is "suspended solids." The stability of an aqueous suspension depends on particle size, density, and charge distribution. The fate of suspended particles depends on a number of factors, and particles can dissolve, grow, coagulate, or be ingested by various life forms in the water. They can become "floating solids" or part of a surface oil film, or they can fall to the bottom to become part of the sediments.

TABLE 2–2. Dissolved Oxygen and Oxygen Demand Parameters

BOD	Biochemical or biological oxygen demand. The oxygen consumed by a waste through bacterial action.
COD	Chemical oxygen demand. The oxygen consumed by a waste through chemical oxidation.
TOD	The theoretical oxygen demand required to completely oxidize a compound to CO_2, H_2O, PO_4^{-3}, SO_4^{-2}, and NO_3^-
5-day BOD	The BOD consumed by a waste in five days.
Ult, BOD	The ultimate BOD consumed by a waste in an infinite time.
IOD	The "immediate" oxygen demand consumed in 15 minutes (without using chemical oxidizers or bacteria).
DO	Dissolved oxygen (a "negative" DO is a positive IOD).

Note: BOD, biochemical oxygen demand; COD, chemical oxygen demand; TOD, theoretical oxygen demand; Ult, ultimate; IOD, immediate oxygen demand; DO, dissolved oxygen.

There are many kinds of suspended particles in natural waters, and not all of them are contaminants. Any moving water will have currents that cause bottom sediments to become resuspended. Also, natural runoff will carry soil and organic debris into lakes and streams. In any industrialized area, such sediment and surface debris will always contain some chemicals considered to be contaminants. However, a large proportion of the mass of such suspended solids would usually be considered as "natural" and not as contaminants.

Suspended particles can have densities that are less than, equal to, or greater than that of the water, so that the particles can rise as well as fall. Furthermore, the effective density of particles can be reduced by the attachment of gas bubbles. Such bubbles form in water when the water becomes saturated and cannot hold any more of the gas in solution. The solubility of gases in water varies inversely with temperature. For example, oxygen saturation of fresh water is 14.2 mg/L at 0°C and 7.5 mg/L at 30°C, while in sea water the corresponding values are 11.2 and 6.1 mg/L.

One specific class of suspended particles is colloids. Colloids are extremely stable suspensions. A colloid has a very small, uniform particle size and possesses unipolar charges that cause the individual particles to repel one another. The particles are so small, on the order of 0.01 μm, that they penetrate through most filters. For practical purposes, colloids behave like true solutions and are usually not measured as suspended solids in most assays.

Oil, grease, and other organic immiscible liquids can exist as discrete droplets or globules or can coalesce on the surface as a film. In extreme cases, they can produce a fire hazard. In lesser amounts, such films can block the absorption of atmospheric O_2; retard photosynthesis by aquatic plants and thereby reduce O_2 production; coat and destroy algae and other plankton; and make the waters unfit for fish life, swimming, and other recreational uses. Floating solids can simply be an aesthetic blight or, if they contain O_2-consuming or toxic materials, can contribute to the degradation of water quality.

Particles that fall to the bottom of a body of water remain accessible to the water for partial dissolution and periodic resuspension during storms and flow surges. If there is enough sedimentary fallout relative to the depth of the overlying water, the sediments can gradually reduce the depth of the water and thereby affect its surface velocity and flow patterns. Finally, the sediment layer can become anaerobic and a source of toxic compounds that can diffuse into the overlying water.

Food Contaminants

Chemical contaminants of almost every conceivable kind can be found in most types of food. Food can acquire these contaminants at any of several stages in its production, harvesting, processing, packaging, transportation, storage, cooking, and serving. In addition, there are many naturally occurring toxicants in foods, as well as compounds that can become toxicants upon conversion by chemical reactions with other constituents or additives, or by thermal or microbiological conversion reactions during processing, storage, or handling.

Each food product has its own natural history. Most foods are formed by selective metabolic processes of plants and animals. In forming tissue, these processes can act to either enrich or diminish incorporation of specific toxicants that are present in the environment. For animal products, where the flesh of interest in foods was derived from the consumption of other life forms, there are likely to be several stages of biological discrimination and, therefore, large differences between contaminant concentrations in the ambient air and/or water and the concentrations within the flesh of the animals.

Natural toxicants

No segment of the environment to which humans are exposed is as chemically complex as food. Food products contain both nutrient and nonnutrient components, and, until recently, relatively little attention was paid to the latter. The potato, a food staple for many millions of people, contains more than 100 nonnutrient chemical substances, including solanine alkaloids, oxalic acid, arsenic, tannins, and nitrates. Table 2–3 lists some of the toxic compounds present naturally in common plant foods.

Despite the presence of a multitude of toxic chemicals in the diets of normal humans, there is usually little effect because the consumption of each is relatively low. Problems generally arise from an excessive dependence on a limited number of foods during an extended period of time. For example, people consuming large amounts of cabbage have developed goiter, and people with extended daily consumption of a half-gallon of tomato juice developed lycopenia, a skin discoloration. While a balanced diet may help keep the level of each natural toxicant below harmful levels, it may also provide opportunities for interactions among toxicants. For example, the toxic effects of cadmium (Cd) are reduced by an accompanying elevated level of zinc (Zn), and copper (Cu) reduces the toxic effects of molybdenum (Mo). On the other hand, not all such interactions may be beneficial. Chemical contaminant interactions are discussed in more detail in chapter 6.

Natural contaminants

Foods can be contaminated by a variety of natural processes not involving either direct or indirect human intervention. Such processes can result in contamination by-products of decay and decomposition that are generated between the growth of the food and its consumption; microbiological and animal pests and/or their residues, wastes, and metabolites; and chemical element congeners of normal nutrient materials, which become incorporated into the food during growth, such

TABLE 2–3. Some Intrinsic Components of Food Having Known Toxicity

COMPOUNDS	FOOD SOURCES	SUSPECTED OR KNOWN TOXIC ENDPOINTS
Solanine, chaconine	White potato[a]	Nervous system
HCN (hydrogen cyanide)	Many plants, as adducts, released when plant tissue is damaged	Hemoglobin, cyanosis
Vasoactive amines	Pineapple, banana, plum	Cardiovascular system
Xanthines (caffeine, theophylline, theobromine)	Coffee[b], tea, cocoa, kola nut	CNS[d] stimulation, other biochemical changes, cardiac effects
Myristicin	Nutmeg, mace	Nervous system
Carotatoxin	Carrots, celery	Nervous system
Synephrine	Lemons	Vasoconstriction
Hemagglutinins (protein)	Soybeans, other legumes	Agglutination of red blood cells
Norepinephrine	Bananas	Vasoconstriction
Lathyrus toxins	Legumes of genus *Lathyrus*	Lathyrism (neurological disease)
Tannins	Tea, coffee, cocoa	Carcinogenic
Safrole and other methylenedioxy benzenes	Oil of sassafras, cinnamon, nutmeg, anise, parsley, celery, black pepper	Carcinogenic (not all members of the class)
5- and 8- Methoxypsoralen (light activated)	Parsley, parsnip, celery	Carcinogenic, with UV light
Ethyl acrylate	Pineapple	Carcinogenic in animals
Estragole	Basil, fennel	Carcinogenic in animals
Goitrin[c]	Cabbage, turnips	Antithyroid activity

[a]Solanaceous glycoalkaloids are present in other Solanaceae, including eggplant and tomato.
[b]Coffee contains more than 600 compounds in addition to caffeine. This is typical of natural foods. Included are many different classes of organic compounds.
[c]This is not present in the original plant but is formed by enzymatic reactions from nontoxic precursors following harvest, during processing, or following digestion.
[d]CNS, central nervous system.

as excessive levels of nitrates, mercury (Hg), selenium (Se), and so on, taken up from soils having high concentrations due to geochemical anomalies. As an example of natural contaminants, Table 2–4 lists some mycotoxins that are produced by molds growing on common plant foods.

TABLE 2–4. Some Mycotoxin Contaminants of Food

MYCOTOXIN	GENUS OF PRODUCING MOLD	TYPICAL SUBSTRATE	TOXIC EFFECT
Aflatoxin B$_1$	*Aspergillus*	Peanuts, oil seeds, corn	Carcinogenesis
Ochratoxin A	*Aspergillus*	Grains	Kidney toxicity
Sterigmatocystin	*Aspergillus*	Grains	Carcinogenesis
Patulin	*Penicillium*	Apples	Carcinogenesis
Cyclopiazonic acid	*Penicillium*	Grains	Tremors, paralysis
Luteoskyrin	*Penicillium*	Rice	Liver toxicity
Islandotoxin	*Penicillium*	Rice	Carcinogenesis

Anthropogenic contaminants

Human activities can greatly increase concentrations of the contaminants already present naturally in food and can also introduce entirely new ones. Some of these latter materials, such as pesticides, are applied intentionally during the growth of the food and become contaminants to the extent that they persist as residues within or on the surfaces of foods long after their intended function has been completed. Others, such as polychlorinated biphenyls (PCBs), are completely inadvertent food contaminants, since they were never intentionally applied. However, they have been widely used in consumer products and electrical system components, have become ubiquitous and persistent environmental contaminants, and have reached excessive levels in some fish and birds through biological concentration processes. Table 2–5 shows some food contaminants resulting from industrial processes.

Air contaminants

Contaminants in the ambient air may produce pervasive contamination of foods, but they are not likely to result in acute intoxications. The atmosphere is a continuous envelope reaching essentially all components of the human environment, and it is capable of rapidly diluting the contaminants discharged into it.

Water contaminants

Contamination of food via water takes place along several distinct pathways. One is by the consumption of fish and shellfish that have taken up chemicals from the water they inhabit or from lower aquatic life forms. Another is via the consumption of fruits and grains that took up contaminants from irrigation waters. A more indirect path is the consumption of animal products from livestock consuming contaminated irrigation water and from the crops grown with these waters. Finally, residual contaminants in drinking waters can reach us either directly or through transfers from water used in food processing and/or cooking.

TABLE 2–5. Some Food Contaminants of Industrial Origin[a]

CHEMICAL	MAJOR SOURCES	FOODS SUBJECT TO CONTAMINATION
Arsenic	Smelting, mining	Many, including fish[b]
Cadmium	Smelting, sewage sludge	Grains, vegetables, meat
Lead	Smelting, mining, solder in can seams, lead glazed pottery and ceramic ware, auto-mobile exhaust	Several, including acidic foods coming into contact with lead ceramic ware and pottery
Mercury, alkyl mercurials	Chlorine, soda lye manufacturing	Fish
Aldrin, dieldrin, DDT, mirex	Pesticide usage[c]	Fish, milk, eggs
Polychlorinated biphenyls	Electrical industry	Fish, human milk
Polychlorinated benzodioxins	Impurities in certain chemicals; incineration; bleached paper manufacturing	Fish, milk, beef fat

[a]Note that the metals listed are also present in foods because of natural occurrence.
[b]Most of the arsenic in food appears to be organically bound. These forms are substantially less toxic than inorganic arsenic. The EPA considers the latter form to be carcinogenic by ingestion.
[c]Pesticide residues in foods for which no official tolerance was ever granted, or was rescinded, are considered contaminants.

Pesticide residues

Chemical pesticides are applied to crops and soil to increase the quantity and quality of the harvest. They may also be applied to the harvested food during transportation and/or storage to minimize spoilage and losses. The US Food and Drug Administration (FDA) sets specific residue limits, termed tolerances, for major pesticides. When these limits are exceeded, the foods involved are subject to seizure and withdrawal from the market. Pesticide residues in and on foods have resulted in readily measurable human body burdens. Since most of the chlorinated hydrocarbon pesticides are known or suspected carcinogens, there is a great concern and effort devoted to the monitoring and control of their residues in the food supply, and there are continuing efforts to ban more of them from agricultural usage.

Residues of drugs and growth stimulants

Veterinary drugs, such as antibiotics, and growth stimulants are given to domestic animals to increase the quantity and/or quality of the meat and to control epizootic disease. They may be applied by injection, implantation, or ingestion, and some fraction may remain as a detectable residue in the flesh or in the eggs or milk products produced by the animals. Antibiotic residues may be capable of causing adverse reactions among allergic individuals. Also, continued ingestion of low levels in foods may lead to a reduced potency when the antibiotics are used therapeutically in people at a later time and/or to the development of drug-resistant strains of animal and human pathogens.

Packaging material residues

Foods can take up chemicals from the materials used to package or contain them. Problems may arise in canned foods; for example, condensed milk was, historically, contaminated by lead in the solder used to seal the cans. More recently, most of the focus of concern has centered on plastic packaging materials, especially the polymeric materials such as polyvinyl chloride (PVC), acrylonitrile, and bisphenol A. The polymers themselves have long interlocking-chain structures and are not considered toxic. However, the polymers are always contaminated, to some extent, with unpolymerized monomers, and some of these are known to be toxic.

Accidents and misuse

Acute chemical intoxications via food are generally attributable to accidental cross-contamination of foods and industrial chemicals or pesticides in transportation or storage, mislabeling by manufacturers, or a failure to understand or follow label instructions. There have been many cases of cross-contamination between bread flour and pesticides, which have led to acute poisonings and some deaths. A notable example of mislabeling occurred in Michigan in 1975, when a fire retardant containing polybrominated biphenyls (PBBs) was mistakenly bagged in containers intended for a cattle feed supplement. The insidious poisoning of the cattle that resulted was not immediately recognized, and a large number of farmers, their families, and people in the general population had severe PBB contamination and suffered health effects from consuming the contaminated meat and milk from these animals.

There have been numerous cases of Hg intoxication resulting from the direct consumption of bread made from seed grains treated with mercurial pesticides. Some occurred because the grain was not properly labeled. Others have occurred when the labels were not read, sometimes because they were in the wrong language or because the people involved were illiterate.

Finally, contamination sometimes has resulted from a major release due to an explosion or other industrial accident. An example is the factory explosion in Seveso, Italy, in July 1976, which spread 2 to 3 kg of the highly toxic chemical dioxin downwind over a considerable distance. The extent of the contamination was not disclosed to the farmers affected and, therefore, domestic animals grazed on contaminated ground, and contaminated milk and eggs reached many people.

Food additives

Food additives are not contaminants, in the sense that they are intentionally applied to food products for a particular purpose. At the time their usages were approved, there presumably was no evidence that they would produce significant adverse effects. However, in recent years, many food additives have been banned by the FDA or are being reevaluated because of evidence that they are capable of producing cancer in laboratory animals. Notable examples that attracted considerable public attention in the past were the food coloring dye known as red dye #2, the artificial sweetener saccharin, and the nitrites used to preserve red meat. The latter can combine with secondary amines in the digestive tract to form carcinogenic nitrosamines.

The controversy about food additive safety, and the occasional banning of food additives on the basis of evidence for carcinogenic potential, has led to understandable confusion and concern among the general public about the safety of food additives that are still approved and has resulted, in part, in the proliferation of "natural food" retail stores, cooperatives, and restaurants. It has also led to an increased effort by the FDA to more thoroughly evaluate the safety of commonly used food additives.

Excessive use of food additives may cause increases in diseases other than cancer. For example, excessive ingestion of sodium chloride ($NaCl$) has been associated with human hypertension, while an excessive daily intake of phosphates can lead to premature cessation of bone growth in children.

Many processed foods are fortified with vitamins. When these are added to the vitamins naturally occurring in foods and to the vitamins ingested in tablets or as liquids, abnormally large daily intakes may occur. Excessive intakes of vitamins A and D in particular have caused a variety of acute and chronic health effects. However, it is important to note that vitamin deficiencies cause many more health problems than does excessive ingestion.

Special issues

Genetic manipulations of plants to introduce more desirable properties and to increase yields may result in alterations in their levels of essential nutrients and natural toxicants. Common plants, such as tomatoes and potatoes, which have toxic foliage, need special attention with respect to potentially altered toxicant levels. On the other hand, selective breeding can be employed to reduce the concentrations of natural toxic substances in foods. For example, lima beans that have low cyanogenetic glycoside content have been developed in order to minimize their cyanide-generating capacity.

While toxicants in food may represent public health problems, the more important dangers associated with food consumption arise from other considerations. The major problem is undoubtedly overeating, especially of foods rich in fats and sugars, resulting in obesity, cardiovascular disease, hypertension, diabetes, and dental caries. Problems also arise frequently in special populations, such as teen-agers, who may consume large amounts of "junk" food, resulting in unbalanced diets that permit the development of deficiency diseases.

CHARACTERIZATION OF ENVIRONMENTS

The environment may be divided into two overall components, the abiotic, or nonliving, and the biotic, or living. The abiotic environment, in turn, can be divided into three components: the atmosphere (i.e., air), the hydrosphere (i.e., water), and the lithosphere (i.e., earth's crust). The aggregate of all of the life forms within all these three components is termed the biosphere. Although these components are often separated in terms of many discussions in this book, it is important to realize that transport of contaminants takes place within and among all components. Thus, release of a contaminant into one environmental compo-nent does not insure that it will remain there; contamination must be considered in terms of the environment as a whole.

The Atmosphere

The earth's atmosphere is simple in some respects and complex in others. It is relatively uniform in composition with respect to its major mass components, namely O_2 and nitrogen (N_2), yet extremely variable in some minor components, such as water vapor and carbon dioxide (CO_2), which play major roles in its heat and radiation fluxes. It has a complex structure based on temperature gradients. This structure governs its mixing characteristics and the buildup of contaminants yet is usually invisible to us except when light-scattering particles suspended in the air make it visible as haze.

The specific structure of the atmosphere is of major importance to the dilution and dispersion of contaminants. It is governed by the lapse, which is the rate of change of air temperature with height above the ground. The changes in tempera-ture and pressure with height are shown in Figure 2–4.

The lowest of the atmospheric layers is the known as the troposphere. It con-tains about 75% of the mass of the atmosphere and almost all of its moisture. It extends to a height that varies from about 9 km at the poles to about 15 km at the equator, and it has an average lapse rate of about −6.5°C/km. The boundary between the troposphere and the next layer, the stratosphere, is known as the tropopause. The stratosphere contains essentially all of the remainder of the mass of the atmosphere; it is nearly isothermal (i.e., the temperature does not change with altitude) in lower regions and shows a temperature increase with height in the upper regions.

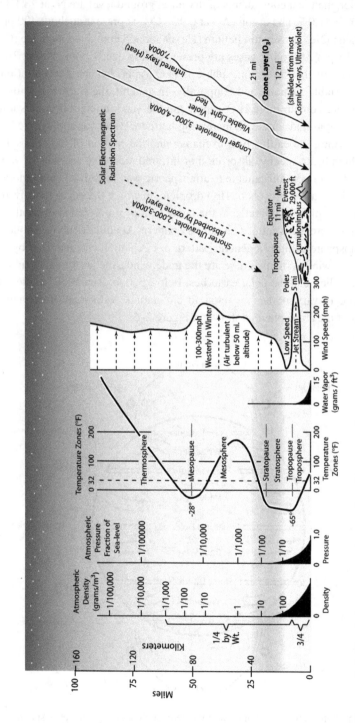

Figure 2-4. Physical structure and properties of the atmosphere.

The major chemical constituents of dry air at ground level are N_2 at 78.1% by volume, O_2 at 21.0%, and argon (Ar) at 0.9%. CO_2 is present at about 400 ppm_v (0.04%), neon (Ne) at 18 ppm_v, helium (He) at about 5 ppm_v, and methane (CH_4) at about 1.8 ppm_v. All other gases are present at less than 1 ppm_v. About 2% of the total mass of the lower atmosphere is H_2O vapor, but the concentration is extremely variable spatially and temporally. In general, the warmer portions of the atmosphere contain more H_2O vapor, and the vapor content becomes lower with increasing altitude and with increasing latitude. Water vapor plays a critical role in governing the earth's heat exchange and the motion of the atmosphere due to its high heat capacity, absorption of infrared radiation, and heat of vaporization. Further effects attributable to atmospheric water result when air motions create clouds, that is, aerosols of H_2O droplets, in which the energy received as sunshine in one place is liberated as the latent heat of vaporization in another.

The atmosphere has a defined circulatory pattern. At altitudes above about 500 m, the atmosphere exhibits a general pattern of circulation characterized by several dominant wind systems. These are the trade winds, the jet stream or midlatitudinal westerlies, and the polar easterlies. Below 500 m, the surface of the earth exerts effects of varying degrees upon air circulation. A diagram of the general circulation of the atmosphere is shown in Figure 2–5.

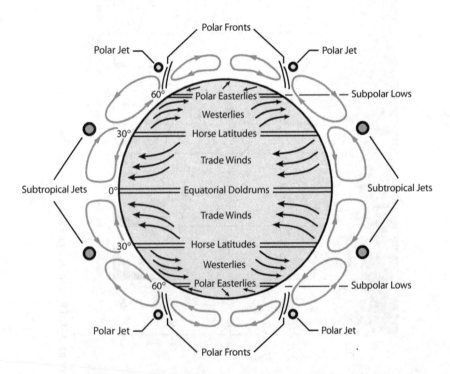

FIGURE 2–5. Schematic representation of the general circulation of the atmosphere.

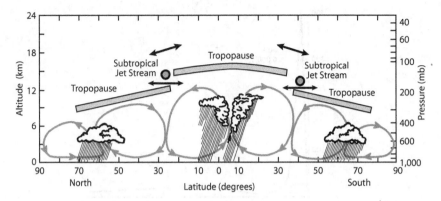

FIGURE 2–6. Schematic representation of the circulatory system in the troposphere during summer in the Northern Hemisphere, showing the discontinuity in the troposphere at the subtropical jet stream.

Throughout the troposphere, there is a rapid decrease in temperature toward the poles. One result of this is a high-speed wind known as the subtropical jet stream. As seen in Figure 2–6, the tropopause is discontinuous in the region of the jet stream; it is through these "gaps" in the mid-latitudes that much of the circulation occurs between the stratosphere and the troposphere. In addition to this general circulation, the atmosphere has secondary and small-scale circulation patterns. Secondary circulations are exemplified by the migratory high- and low-pressure areas seen in daily weather charts. Examples of small-scale circulations are land-sea breezes, mountain-valley winds, and thunderstorms.

Earth-atmosphere Energy Balance

The sun is the source of essentially all of the energy that reaches the earth. Radiant energy from the sun covers the entire electromagnetic spectrum. However, most of it occurs in and near the visible portion, that is, wave lengths from 0.4 μm to 0.7 μm (Fig. 2–7).

Of the incoming radiant energy, about 30% to 50% is scattered back toward space, reflected by the atmosphere due primarily to clouds and, to some extent, by solid particles or by the earth's surface. On a global basis, the average reflectivity, termed "albedo," of the earth's surface and atmosphere is about 35%. The actual albedo of any specific surface is highly dependent upon the particular area and its characteristics; ice- and snow-covered polar regions have high reflectivity, while the reflectivity of oceans is relatively low, with most incident energy being absorbed.

About 20% of the incident radiant energy is absorbed as it passes through the atmosphere. Stratospheric O_3 absorbs about 1% to 3%, primarily in the short-wave ultraviolet (UV) portion of the spectrum; this effectively limits further penetration

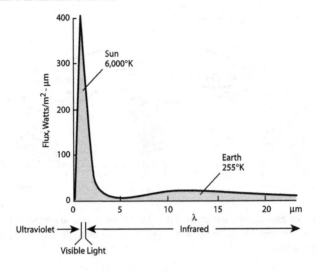

FIGURE 2–7. Emission spectra to and from the earth. The sun and earth radiate as "black bodies" having temperatures of 6000°K and 250°K respectively.

to those wavelengths greater than 0.3 μm. In the troposphere, 17% to 19% of the incoming radiation is absorbed, due primarily to H_2O vapor and secondarily to CO_2. The total atmospheric absorption for radiant energy with wavelengths of 0.3 to 0.7 μm is not very large, and these effectively penetrate an essentially "transparent" atmospheric window.

In total, about 50% of the incoming solar radiation reaches the earth's surface and is absorbed. The surface reradiates energy back into the atmosphere through a broad range of wavelengths, but with a flat maximum in the long-wave, infrared (IR) portion of the spectrum at about 10 to 12 μm (Fig. 2–7). The atmosphere is nearly opaque to this radiation; most is absorbed by H_2O vapor and droplets and by CO_2. Some is then reradiated back to earth or out into space. The earth-atmosphere energy balance is summarized in Figure 2–8. By allowing effective penetration of short-wave solar radiation, yet retaining a large fraction of the reradiated long-wave radiation, the atmosphere acts as an insulator, keeping heat near the surface of the earth. This phenomenon is known as the greenhouse effect.

The solar flux, which is the intensity of radiation as measured by the amount of energy transferred per unit area per unit time, decreases with increasing latitude. For example, the average mid-winter solar flux is over 800 cal/cm²/day at the equator and less than 200 cal/cm²/day at 50°N. latitude. On the other hand, the flux of IR radiation from the atmosphere to space only drops from about 470 cal/cm²/day at the equator to about 400 at 50°N. The average radiation into space must equal that absorbed from the sun. Thus, it follows that a substantial amount of energy must flow from the tropics toward the poles within the troposphere. This flow of

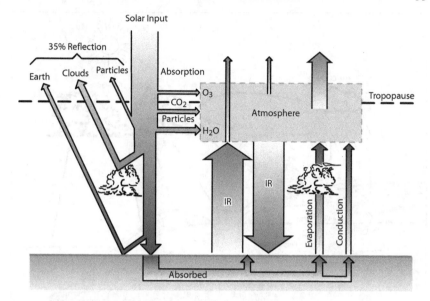

FIGURE 2–8. Earth-atmosphere energy balance.

energy is accomplished primarily by systems of warm air currents toward the poles and cool currents toward the tropics and partially by corresponding ocean currents.

The Hydrosphere

Hydrologic cycle

The hydrosphere includes a variety of distinctly different aquatic environments that are essential to human life and economic productivity. It is a dynamic system due to the hydrologic cycle, in which liquid H_2O evaporates and is transported through the atmosphere as water vapor. This cycle is illustrated schematically in Figure 2–9. While atmospheric H_2O vapor mass constitutes only a small fraction of the total mass of the hydrosphere, it is critical both to the heat balance of the globe, as previously discussed, and to the replenishment of fresh H_2O needed for drinking and for agriculture purposes.

As indicated in Table 2–6, most of the hydrosphere is contained in the oceans, which are saline. Furthermore, most of the fresh H_2O is relatively inaccessible, present in groundwater or polar ice. Its inaccessibility is indicated by its rate of turnover, that is, by the time required for each part to be exchanged in the hydrologic cycle.

Fresh surface waters

Chemically pure waters, or those containing only H_2O, are not found in nature. Water's great power as a "universal" solvent insures that natural waters contain

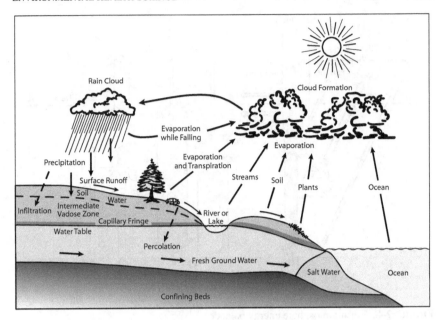

FIGURE 2–9. Schematic representation of the hydrologic cycle.

many other substances largely derived from the lithosphere. Table 2–7 presents a listing of some major chemicals found in natural, fresh waterways in the United States. The table applies to those waters that support a varied and profuse fish fauna and are considered nonpolluted. The concentrations of minerals present at

TABLE 2–6. The Hydrosphere

COMPONENT	APPROXIMATE VOLUME (10^3 KM^3)	TURNOVER TIME (YEARS)
Oceans	1,370,000	3,000
Groundwater	60,000	5,000[a]
(Groundwater zones with active turnover)	(4000)	(300)[b]
Polar ice caps	24,000	8,000
Surface waters of land	280	7
Rivers	1.2	0.031
Soil moisture	80	1
Water vapor in atmosphere	14	0.027
Total hydrosphere:	1,454,000	2800

[a]Inclusive of groundwater runoff to oceans, thus bypassing rivers—4,200 years.
[b]Inclusive of groundwater runoff to oceans, thus bypassing rivers—280 years.

TABLE 2–7. Some Average Properties of Natural Waterways in the United States[a]

	5% OF THE WATERS CONTAIN LESS THAN	95% OF THE WATERS CONTAIN LESS THAN
Total dissolved solids	72	400
Bicarbonate (HCO_3^-)	40	180
Sulfate (SO_4^{-2})	11	90
Nitrate (NO_3^-)	0.2	4.2
Calcium (Ca^+) } Magnesium (Mg^+)	18.5	66.0
Sodium and potassium ($Na^+ + K^+$)	6	85
Free carbon dioxide (CO_2)	0.1	5.0
Ammonia (NH_3)	0.5	2.5
Chloride (Cl^-)	3	170
Iron (Fe)	0.1	0.7

[a]Numbers are in ppm_w (parts per million).

lower concentrations, including many of the toxic chemicals of interest, are more variable and cannot readily be summarized. Table 2–8 list some trace metals commonly found in nonpolluted river waters.

Surface runoff into lakes and rivers is used extensively for drinking H_2O supplies. Sometimes it is used with minimal treatment, as in New York City and Boston, which collect their H_2O in distant watersheds that are largely isolated from contaminated surface runoff. In other cities located on rivers or lakes with extensive commercial traffic and upstream sources of contamination, the H_2O requires physical and chemical treatment and disinfection before it is safe to use in water-supply systems.

TABLE 2–8. Trace-Level Metals Commonly Found in River Water

Cd
Co
Cr (VI)
Cu
Hg
Mo
Pb
V
Zn

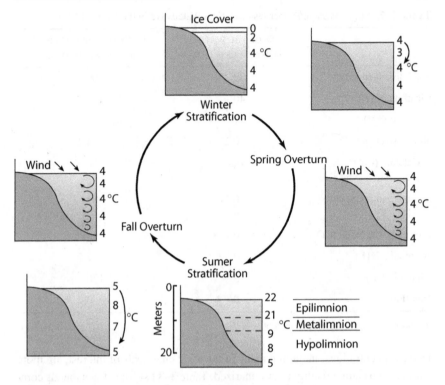

FIGURE 2–10. The seasonal cycle of temperature in a temperate lake.

Lakes and other enclosed bodies of H_2O, especially those in temperate zones, show a thermal stratification during different seasons of the year. This pattern is due to the interplay of temperature, H_2O density, and wind and is shown in Figure 2–10. The maximum density of fresh H_2O water occurs at a temperature of about 4°C; density decreases as the temperature is further reduced to the freezing point, 0°C. Thus, in winter, a lake surface will be covered by a layer of ice, while waters slightly warmer than 0°C will remain at various depths below the surface. This ice layer prevents wind-induced circulation, suppressing vertical and horizontal movements and further loss of heat to the atmosphere. This situation is known as winter stagnation.

As the weather gets warmer in the spring, and the winter ice begins to break up, the surface water starts to warm. As its temperature increases toward 4°C, and it becomes denser, this surface water sinks below the colder, and less dense, layer beneath it. This overturning of the water, which may last for several weeks, results in a temperature profile that is essentially uniform at all depths of the lake; as the ice melts, vertical circulation is further aided by wind action. As the weather gets progressively warmer toward summer, the surface water becomes warmer still and thus less dense than the underlying colder water. The water is directly stratified, in terms of temperature, in the condition known as summer stratification or stagnation. This condition generally persists from April to November in northern latitudes of the Northern Hemisphere. The upper water layer is termed the epilimnion and has a relatively uniform warm temperature. The bottom layer, the hypolimnion, is characterized by a relatively uniform cold temperature. Between these two layers is the metalimnion, a zone of maximum temperature change. The rapid temperature change itself is termed a thermocline. As autumn approaches, the surface water layer once again cools and sinks. The water becomes mixed and stirred to increasing depths, and equality of surface and bottom temperatures occur once again, as it did in the spring, in a process termed fall overturn. When the surface freezes, winter stagnation is reestablished.

This temperature cycle has important implications in terms of essential nutrient circulation and contaminant levels in temperate aquatic environments. Thermal gradients are also gradients for concentrations of dissolved gases. The O_2 absorbed at the surface layers is distributed throughout the lake by water circulating within the epilimnion. Waste gases produced during the decomposition of bottom deposits are released by contact with air over surface waters. Within the thermocline, where mixing with surface water is minimal, there is a sharp drop in DO and a rise in concentration of gases of decomposition, until DO reaches a minimum below the level of the thermocline. Because the stability of the waters during summer stratification reduces mixing, the hypolimnion tends to become depleted of O_2. This effect may be enhanced by the discharge of O_2-demanding contaminants.

As part of the semiannual overturn and mixing of the entire lake, nutrients are mixed and redistributed. Those in lower waters are brought to surface water and made available for use in photosynthesis. Thus, overturns may be periods of decreased water quality, as sudden blooms of aquatic plants, such as algae, occur. Lakes, therefore, often exhibit seasonal differences in water quality. In some inland lakes, water in the lower depths remains unmixed with the main water mass during turnovers, contributing to the stagnation of nutrient cycling. The biological productivity of surface water depends heavily on temperature, dissolved and suspended nutrients and contaminants, and O_2 concentration. There is continual interchange of nutrients and contaminants between the H_2O, bottom sediments, suspended particles, and various life forms.

Groundwater

While most fresh water is underground (Table 2–6), surface waters supply over 80% of current usage. As surface water supplies become inadequate, groundwater usage is bound to increase. Groundwater is generally free of significant contamination by bacteria and suspended solids but will usually contain significant amounts of dissolved solids. These characteristics result from the intimate and extended contact with minerals that constitute the lithosphere. The soil particles

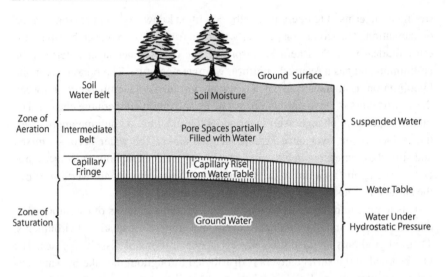

FIGURE 2–11. Diagram of water zones. Groundwater is the part of subsurface water within the zone of saturation.

filter out the bacteria and suspended particles and increase the mineral content by dissolution to the point of chemical equilibrium.

As shown in Figure 2–11, groundwater lies in the zone of saturation, where it fills essentially all the voids in the rock stratum. Its upper limit is known as the water table. When usable volumes of water can be extracted from a saturated zone, it is called an aquifer. When the groundwater is under considerable pressure, the water will flow upward without mechanical pumping through a well pipe drilled into it. This is known as an artesian well (Fig. 2–12).

The depth of the water table can be quite variable in time, depending on the rates of water extraction and recharge. Natural recharge through precipitation and infiltration into the ground may be inadequate in areas with heavy use of groundwater, and the resulting fall in the water table may limit the supply and/or cause significant subsidence of the land surface. These problems can be partially or completely overcome in some areas by installing recharge basins or wells. Recharge basins are designed to catch storm water, which percolates through their base to the water table. Recharge wells are used for industrial waste waters and cooling waters, which generally are pretreated to an acceptable quality before being pumped into the groundwater.

Coastal waters and estuaries

The relatively shallow waters of the continental shelves and the estuaries, where tidal action results in mixing of fresh H_2O runoff with saline ocean water, are of particular importance because of their very high biological productivity. They are the major harvest grounds for many commercial shellfish and finned fish. In addition, many ocean fish migrate into estuaries to spawn. Their life cycles can be interrupted by changes in DO, contaminant concentrations, or the benthic environment.

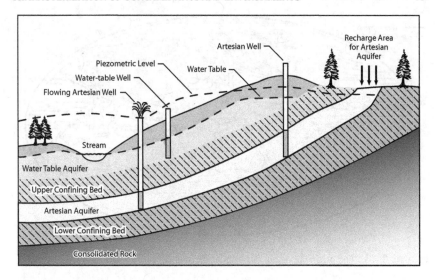

FIGURE 2–12. Subsurface and groundwater phase of the hydrologic cycle.

Estuaries are heavily used for marine commerce and are located within major population centers. New York, Philadelphia, and Washington, D.C., are situated on the estuaries of the Hudson, Delaware, and Potomac Rivers, respectively. The flow in an estuary is cyclic, rising and falling for 6.2 hours from ebb slack (sea-water efflux) to flood slack (sea-water influx) and back to ebb slack. A full tidal cycle is completed in 12.4 hours.

Oceans

The surface waters of the oceans, that is, those up to 200 m deep, are relatively uniform in temperature, density, and salinity because of mixing due to surface winds. There are characteristic surface currents of the oceans (Fig. 2–13), which largely coincide with surface wind patterns. The intermediate zone, down to a depth of about 1000 m, is characterized by decreasing temperature and increasing density and salinity with depth. This zone has a very low vertical diffusivity and therefore acts in effect as a barrier to contaminant transfer between the surface and deep waters. The deep ocean waters, with temperatures of 1° to 4°C, occupy about 75% of the total volume. Thus, most of the ocean waters contribute little to the hydrologic cycle and, in fact, have been relatively inaccessible to human activity.

The Lithosphere

In terms of contamination, soil is the most significant part of the lithosphere. The characteristics of soil determine the rate of spread of chemical contaminants placed on or in it and therefore their access to ground water. Soil, which is derived from the weathering of rocks, is a very complex system, with solid, liquid, and gaseous phases. Its solids consist of particles of different chemical

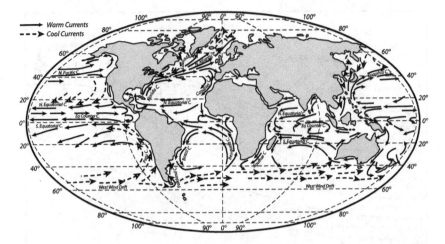

FIGURE 2–13. The principal ocean currents of the world.

and mineralogical composition, which also vary in particle size, shape, and packing characteristics. The particle configuration determines the characteristics of the channels (pores) that contain air and/or water. The composition of both fluid phases is also variable, both temporally and spatially.

Soils consist of layers, termed horizons, which are roughly parallel to the ground surface, with each differing in properties, such as color and texture, from its adjoining layers. A succession of horizons is termed the soil profile. Transfer of minerals and organic material in solution or suspension occurs between horizons via water moving through the pores.

The solid phase of soil may be separated into inorganic and organic components. The inorganic particles are generally classified into fractions, depending on their size. In the US Department of Agriculture classification, particles smaller than 2 μm in diameter are clay, those between 2 and 50 μm are silt, those between 50 μm and 2 mm are sand, and those larger than 2 mm in diameter are gravel. The overall textural designation of a soil depends on the ratio of the masses of the less than 2 mm fractions, and these are generally expressed in terms of a textural triangle, as illustrated in Figure 2–14.

The fraction that largely governs the physical properties and chemical retentions of soil is the colloidal clay. While sand and silt are composed mainly of quartz and other primary mineral particles, clay is composed of a large group of minerals. Some are amorphous, but many are highly structured microcrystals. The most prevalent of these are the layered aluminosilicates. The clay particles have the highest specific surface. They adsorb water, causing the soil to swell upon wetting and shrink upon drying. Most clays have negative electrical charges and form electrostatic double layers with exchangeable cations in the water phase.

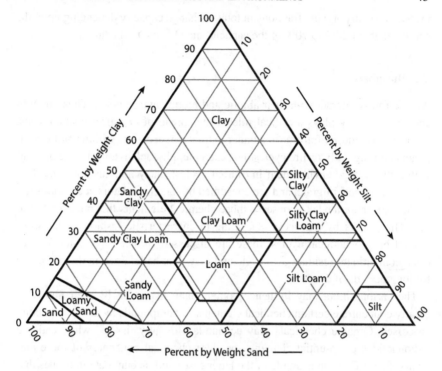

FIGURE 2–14. Textural soil triangle, showing the percentages of clay, silt, and sand in the basic soil textural classes.

The quantity of cations adsorbed per unit mass of soil is known as the cation exchange capacity, which may range as high as 0.60 mEq per gram. The attraction of a cation to a negatively charged clay particle increases with increasing chemical valence, while more highly hydrated cations are more easily displaced than less hydrated ones. Thus, the order of preference of the common soil cations is $Al^{+3} > Ca^{+2} > Mg^{+2} > K^+ Na^+ > Li^+$. When lime is added to an acid soil, Ca^{+2} ions displace many of the hydrogen ions on the surface of the clay particles. Soils with little colloidal materials, such as those consisting largely of sand, have low natural ion exchange capacities. Although ion exchange capacities of soils are usually measured in terms of cations, positively charged soil particles also occur, and soils also have some degree of anion exchange capacity. The very large surface area of soil and its great ion exchange capacity permit it to retain nutrients to a far greater extent than are retained in the soil moisture.

Aside from life forms, the organic fraction of soil consists of decaying plant and animals, plus a relatively chemically stable colloidal fraction called humus. The humic component also plays a part in ion exchange, with the relative role of inorganic versus organic fractions in total exchange capacity depending upon the soil type; about 20% to 70% of the capacity may be due to the humus.

The Biosphere

All life forms interact with their abiotic and biotic environments. Thus, the biosphere frequently plays a critical role in the spread of chemical contaminants, especially via those lifeforms that undergo transport across or within abiotic components during daily or life cycles. For example, phytoplankton can rise and fall within the aquatic environment in diurnal cycles; many animals consume their prey in one place, migrate, and are consumed by larger predators at another site.

The biosphere tends to conserve those chemical elements that are essential to life. This occurs by a process of cycling in characteristic fashion in pathways termed biogeochemical cycles; these are discussed further in Chapter 5. Very often contaminant residues also cycle and may thus be retained in the biosphere for long periods of time.

The various biota may influence contaminant dispersion by their ability to greatly concentrate certain chemical elements or compounds by active metabolic processes. Organic chemicals may accumulate in fatty tissues, while cationic radionuclides of essential elements, or close chemical congeners of these elements (Sr for Ca is an example of the latter case), will accumulate in tissues that store those elements. Metabolic discrimination can work both ways; some chemicals will become concentrated in tissues relative to their environmental levels, while others will be found at much lower levels in biological tissue.

REFERENCE

1. Gibbs, W.E. *Clouds and Smokes*. New York, NY: Blakiston; 1924.

3

Sources of Contaminants

Chemical contaminants in the environment can be classified in several ways: natural or anthropogenic; according to the medium in which they are dispersed, namely air, water, soil, food; by specific elemental and/or compound class; or by state of matter, namely solid, liquid, gas, or dispersion (colloid, suspension, etc.). Contaminants can also be classified as primary or secondary. The former are directly released by some source, while the latter are formed within the environment via reaction of precursor chemicals, which can include primary contaminants.

In this chapter, contaminants are classified according to their sources, and, therefore, it becomes necessary to consider all physical and chemical forms and all media simultaneously. This approach provides the best opportunity to associate specific contaminants with the activities generating them. Furthermore, it permits an appreciation of the total impact of a given sphere of activity and of the possible implications of alternative control strategies in those cases where current environmental contamination levels are excessive.

PRIMARY CONTAMINANTS

Natural Contaminants

Chemically pure air and water are not found in nature because there is always an abundance of sources of both chemically and biologically active agents. The life cycles of plants generate pollens, the decomposition of vegetable and animal matter releases various organic and inorganic compounds, and the hydrologic cycle leaches metals from the earth's crust. Thus, unless the agent of interest is unknown in nature, it is generally difficult to apportion measured levels at any particular location to their natural and anthropogenic sources. Table 3–1 shows various natural and anthropogenic sources for aerosol emissions, while Table 3–2 shows that, on a global basis, emissions from natural sources often exceed those from anthropogenic sources. However, in any given region, the local

TABLE 3–1. Natural and Anthropogenic Sources of Aerosols

AEROSOL TYPE	ANTHROPOGENIC	NATURAL
	MAIN SOURCE ACTIVITIES	MAIN SOURCES
Sulfates (as HSO_4^-)	Fossil fuels and smelting	Dimethylsulfide and H_2S from oceans, land biota, and soils; Volcanic SO_2
Organic carbon	Fossil fuels, outdoor cooking	Photochemical conversion of terpenes to condensable products and primary biogenics
Black carbon	Fossil fuels, outdoor cooking	
Smoke	Biomass burning; smoke is largely composed of organic and black carbon and is therefore often included in those categories	Natural fires
Nitrates (as NO_3^-)	NO_x from biomass burning, fossil fuel, and aircraft; agricultural soil NO_x	Lightning, natural soil, and stratospheric NO_x
Ammonium (as NH_4^+)	Enhanced soil emissions from application of nitrogen fertilizer; domestic animals; human emissions; biomass burning; fossil fuel and industry	Natural soils, wild animals, and oceans
Sea salt		Formation of jets and bubbles from wind
Dust	Agriculturally disturbed lands and increased desertification	Wind-blown dust from deserts and other arid areas

TABLE 3–2. Particulate Atmospheric Emissions on a Global Scale

PRIMARY EMISSIONS	
NATURAL SOURCES	Emission Range (10^6 metric tons/year)[a]
Wildfires	5–150
Sea salt Spray	1,000–10,000
Volcanic Activity	5–10,000
Wind Erosion (Soil Dust)	1,000–3,000
Botanical Debris	26–80
ANTHROPOGENIC SOURCES	
Industrial Activity	40–130
Combustion By Products	10–30
Biomass Burning	50–190
Wind Erosion	850
SECONDARY PRODUCTION	
NATURAL SOURCES	
Gas to particle conversion	100–260
Photochemical reactions	40–200
ANTHROPOGENIC SOURCES	
Gas to particle conversion	260–460
Photochemical reactions	5–25
TOTALS	
Natural Sources	2,200–24,000
Anthropogenic Sources	320–640

[a] The wide range is a reflection of uncertainty in the various assumptions upon which the various estimates are based.

Data from Kreidenweis et al, 1999. Aersols and Clouds. In: *Atmospheric Chemistry and Global Change*, G Brasseur, J Orlando, G. Tyndall, eds. Oxford Univ. Press, NY; WC Hinds, *Aerosol Technology*, Wiley Interscience, NY, 1999; IPCC 2001: Climate Change, 2001.

sources often have a much greater influence on concentrations, so the increment from natural sources of, for example, gases such as ammonia (NH_3), nitrogen oxides (NO and NO_2), and hydrogen sulfide (H_2S) may be negligible. In order to appreciate when and to what extent natural sources are important to observed environmental levels, it is always necessary to consider contaminants released by natural processes.

While many "natural" contaminants can be found in areas not significantly influenced by human activity, it does not necessarily follow that the observed levels are independent of human activity. A notable example is air contamination by ragweed pollen. Ragweed grows well in recently disturbed soil but is at a competitive disadvantage in soils with established covers of natural vegetation. In the sense that it grows best on construction sites, abandoned fields, and urban lots and coexists with cereal grains on some agricultural fields, much of it may be considered to be anthropogenic in origin.

Viable particles

A lengthy discussion of viable particles is beyond the scope of this book. However, since their prevalence is often largely a function of human activity, and since they do have a great impact on human health, they are discussed briefly in order to place them in perspective with nonviable chemical contaminants. Viable particulates and their size ranges are are shown in Table 3–3.

Airborne microorganisms cause a substantial share of infectious disease incidence, especially in indoor environments. Fungal spores can become airborne by active processes or by the action of wind, rain, or insects. Bacteria and viruses become airborne in aqueous aerosols formed by a variety of processes, including sneezing and coughing, spraying of sewage in treatment and disposal, and wind action on standing water in ponds and lakes. Viable contaminants can also be ingested in drinking water or with contaminated food products and may produce similar effects following either inhalation or ingestion. The differences will be in the degree of respiratory tract and gastrointestinal tract involvement, the overall pathogenic potential, and the geographic pattern of disease incidence.

TABLE 3–3. Viable Particulates

PARTICULATE	STOKES' DIAMETER[a] (μM)
Viruses	0.015–0.45
Bacteria	0.3–15
Fungi	3–100
Algae	0.5
Protozoa	2–10,000
Moss spores	6–30
Fern spores	20–60
Pollen grains (wind-borne)	10–100
Plant fragments, seeds, insects, other microfauna	100

[a]The Stokes' diameter for a particle is the diameter of a sphere having the same bulk density and same terminal settling velocity as the particle.

Products of metabolism

All living organisms create metabolic waste products that are discharged into their environments and normally recycled or neutralized in the immediate vicinity without significant impact on their own or the larger human environment. Difficulties usually arise when human activities alter the natural patterns of life cycles or the density of biota populations. Examples are poultry farms and animal feed lots, where the population density is so great that the volume of waste products often cannot be effectively disposed of or recycled on-site by natural processes. These problems are discussed further under the topic of agricultural wastes.

The respiratory activity of animals and plants and the photosynthetic activity of green plants results in the exchange of oxygen (O_2) and carbon dioxide (CO_2). While these processes are in approximate balance on a global basis, on a local scale there is a considerable diurnal variation in CO_2 concentration. CO_2 is taken in by plants during daylight at a rate dependent upon the type and density of vegetation, the amount of solar radiation, and the CO_2 concentration. It is released to the atmosphere from decomposing organic material, largely via bacterial activity in the soil. The release rate is dependent on the type of soil, its moisture content, and its temperature. Much of the CO_2 released is taken up by plants just above the ground and is thus effectively short-circuited. The concentration of CO_2 in soil air can be two orders of magnitude greater than that in the surface air.

Vegetation is also responsible for the release of substantial quantities of hydrocarbons. Perhaps the most significant of these is a class known as terpenes. Although released as vapors, they readily condense in the atmosphere to form fine particles. A fairly concentrated light-scattering aerosol may be formed over areas where terpenes are released in significant quantities, such as coniferous forests. These aerosols can be the most characteristic feature of a region, as in the Great Smoky Mountains of North Carolina.

Products of organic decomposition

When living things die, their tissue constituents are recycled to the environment. Some decomposition takes place by physical processes such as evaporation or incineration. However, in the natural state, most organic matter is utilized as an energy and materials source by a succession of microorganisms and is eventually broken down into simple and stable inorganic chemical entities, such as water and CO_2.

When organic matter is broken down by aerobic microorganisms, the process is known as decay. Oxidation continues until stable end products are produced and there are no odorous gases. The nitrogen in the protein undergoes conversion to ammonium, nitrite, and, finally, nitrate. The sulfur compounds are converted to sulfate. On the other hand, in anaerobic decomposition, or putrefaction, the process of oxidation is incomplete, and the breakdown products include many unstable and odorous gases. The initial products are organic acids, acid carbonates, CO_2, and H_2S. Intermediate products include NH_3, acid carbonates, and sulfides, including mercaptans. The final products include CO_2, NH_3, CH_4, and H_2S. NH_3, H_2S, and the mercaptans have strong and unpleasant odors. H_2S is toxic to many animals, including humans. CH_4 is odorless and nontoxic but is combustible and can build up to explosive concentrations in underground air spaces.

Erosion

Wind action on land can resuspend settled dust and can erode materials from surfaces. Similarly, wave action and flowing waters can erode the surfaces over which they pass, increasing the dissolved and suspended particulate burdens of the water. While these are natural processes, they can be greatly enhanced by human activities. It is much easier to suspend tilled or trampled soil than soil with a natural cover of vegetation, and tilled soil may also be much more susceptible to erosion by storm water. Erosion can also take place with manmade wind or water motion, such as that created by high-speed vehicles, like cars, trucks, and motorboats. Mining and construction activities can also result in heavy local concentrations of dust.

Fire

Lightning strikes and spontaneous combustion of organic matter have always caused and continue to cause fires in forest and brush lands. Such fires generate carbon monoxide (CO), CO_2, and a wide variety of organic materials that are products of incomplete combustion and/or pyrosynthesis. While many forest fires are attributable to anthropogenic activities, human intervention has also limited the spread of fires started by natural causes.

Sea salt spray

The breaking of waves in the oceans creates aqueous aerosols. Most of the droplets fall back into the water, but many of the smaller ones remain airborne long enough for the water to evaporate. A large proportion of the resulting salt particles are sufficiently small to remain airborne for considerable times and distances.

Natural radioactivity

Radioactive isotopes (radionuclides) are continually being released to the atmosphere from two sources. One is the interaction of cosmic rays with atmospheric gases to produce tritium (^3H), carbon-14 (^{14}C), and a variety of other nuclides. Carbon-14, with a half-life ($T_{1/2}$) of 5,730 years, is produced from ^{14}N and is continually incorporated into the biota. The proportion of ^{14}C to ^{12}C, therefore, indicates the elapsed time since the carbon was incorporated into living tissue and has been used as an index of the age of archeological and biota specimens.

In addition there are a variety of radionuclides that have been in the lithosphere since it was formed. The ones of greatest interest to humans are ^{235}U ($T_{1/2} = 7.1$

$\times 10^{10}$ years), ^{238}U ($T_{1/2} = 4.5 \times 10^9$ years), ^{232}Th ($T_{1/2} = 1.4 \times 10^{10}$ years), and ^{40}K ($T_{1/2} = 1.26 \times 10^9$ years). Some radionuclides, for example, ^{40}K, decay directly to stable daughter isotopes. Others, for example, ^{235}U, ^{238}U, and ^{232}Th, decay through complicated chains and have numerous daughters with varying half-lives and emissions. Certain radionuclides are large contributors to natural atmospheric radioactivity via formation of radioactive noble gases that diffuse from the earth's crust.

When radium-226 (^{226}Ra), a member of the ^{238}U chain, decays, it becomes radon gas (^{222}Rn). Similarly, radium-224 (^{224}Ra), a member of the Th chain, decays to thoron gas (^{220}Rn), another isotope of the element radon. These gases can diffuse into the atmosphere out of the solid matrix holding the parent isotopes. The ^{222}Rn, with a half-life of 3.8 days, is released to a greater extent than is the ^{220}Rn, with its much shorter half-life of 54 seconds. The "daughter" products of radon and thoron decay, which are also radioactive, are molecular-sized particles, and they rapidly diffuse onto available surfaces, primarily other airborne particles. Since most of these airborne particles are very small and remain in the atmosphere for a relatively long time, these daughter products can stay airborne for significant time intervals.

The concentration of radon daughters in the atmosphere is quite variable. It depends on the strength of the local sources, for example, the concentration of radium in the soil and/or building materials, the porosity of the matrix containing the radium, and meteorological factors. Very high concentrations can be found in enclosed spaces, such as mines, caves, and the basements of homes, schools, and commercial buildings where the underlying soil, building materials, or both have relatively high radium contents. On a population basis, radon and daughters can account for 50% to 90% of the total background radiation exposure to humans depending upon the region.

Vulcanism

Volcanic eruptions occur relatively infrequently and at irregular intervals; those with significant releases to the troposphere occur, on average, only a few times each year. Even less frequently, perhaps once or twice per century, there is a major eruption with sufficient energy to inject large amounts of ash and SO_2 into the stratosphere. The eruptions of Krakatoa (1883) and Mt. Agung (1963) in present-day Indonesia injected enough light-scattering particles into the stratosphere to cause readily observable optical effects that persisted for several years. The sulfur released by volcanoes represents a significant fraction of the total emissions to the atmosphere.

Anthropogenic Contaminants

Fossil fuel combustion

Total energy consumption in the United States (US) can be apportioned to various sectors, namely transportation (27% of total consumption), industrial (22%) residential/commercial (12%), and electric power generation (39%). Approximately 81%

TABLE 3–4. Fossil Fuel Utilization

FUEL	SECTOR[a]		
	POWER GENERATION	TRANSPORTATION	SPACE HEATING
Coal	42	0	<1
Petroleum	1	92	3
Natural Gas	22	3	75

[a] Each value represents the relative percentage of total energy use supplied by the indicated fossil fuel source for each sector. Numbers may not add up to 100% due to use of renewal fuels or nuclear power in some sectors.
Data from: US Department of Energy, US Energy Information Administration (2015).

of this energy demand is derived from the combustion of fossil fuels, that is, natural gas, oil, and coal; the rest is supplied by nuclear energy and renewable sources, such as solar, geothermal, wind, and biomass. Table 3–4 shows the relative percentage of total energy use by various sectors that can be attributed solely to fossil fuels.

Electric power production and space heating

Fossil fuels contain varying mixtures of hydrocarbons and minerals. Natural gas, as it is commercially distributed, has negligible amounts of sulfur, nitrogen, and noncombustible mineral ash but may contain noble gases, including radon, and radon daughters. Components of petroleum are classified according to their range of boiling temperatures at atmospheric pressure. The lighter fractions of oil (those having lower boiling points) will also be low in nitrogen, sulfur, and ash. Coal and the heavier fractions of the oil have mineral contents depending on their source and the amount of pretreatment, if any, they receive. Coal typically contains about 10% mineral ash. While it is possible to approach complete combustion of fossil fuels in stationary furnaces, this becomes increasingly difficult with the increasing viscosity of fuel oil and the increasing content of volatiles (decreasing fixed carbon content) in coal. The difficulty lies in achieving a sufficiently prolonged contact between the oil droplet or coal particle surface and the oxygen in the flame.

If combustion is not completed within the high temperature zone of the flame, the effluent will contain unburned and/or incomplete products of combustion. In an efficient flame, most of the carbon in the fuel will be oxidized to CO_2, the hydrogen to water vapor and the sulfur to SO_2. The source strength depends on the sulfur content of the fuel, and power plants can reduce their SO_2 emissions by switching to desulfurized fuels with naturally lower sulfur content or by scrubbing the combustion effluents before their discharge to the ambient air. In 1990, the federal government mandated phased reductions in sulfur emissions. By 1995, emissions were substantially below the target, and further substantial reductions have occurred since then.

The nitrogen in the fuel and the N_2 and O_2 in the air react within the combustion zone, and are released as nitric oxide (NO), and nitrogen dioxide (NO_2). The amounts of NO and NO_2 are quite variable, and their summed concentration is usually expressed as nitrogen oxides (NO_x). There is generally much more NO_x in the furnace effluent than could be accounted for by the N_2 content of the fuel. The excess is formed in the flame by fixation of the N_2 in the combustion zone, a process whose rate increases rapidly with rising flame temperature. The reaction between atmospheric N_2 and O_2 to form NO is reversible; if the products of combustion were allowed to cool gradually, the NO would decompose. However, since most large furnaces are designed to permit a rapid extraction of heat for useful work, the gases are rapidly quenched, and some of the NO does not decompose. Beginning in the furnace and stack, and continuing in the atmosphere, the NO undergoes conversion to NO_2, a brownish-colored and much more toxic gas that plays a critical role in the photochemical formation of ozone (O_3) in the troposphere.

The mineral content of the fuel is converted to an ash composed of mixed mineral oxides. The more volatile materials, such as oxides of lead (Pb), cadmium (Cd), and mercury (Hg), and the radium and polycyclic aromatic hydrocarbons, are released as vapors, but condense as, or on, fine particulates in the stack. Most of the mass of the ash falls to the bottom of the furnace (bottom ash), while the fine particles (fly ash) rise up the stack with the combustion gases. Coal-fired utility furnaces generally use various air-cleaning devices to collect most of the fly ash.

The fate and effects of the effluents from fossil fuel combustion depend on how they are released. Large utility furnaces may discharge effluents through high stacks, diminishing their maximum ground-level concentrations but increasing their residence times in the atmosphere and increasing their spatial dispersion. Space-heating furnaces usually discharge their effluents at or near roof level. The maximum ground-level concentrations can, therefore, be higher, and thus any effects they produce are more likely to be localized.

A potential surface water contaminant produced by electric power generating plants is heat. Spent steam leaving the turbine passes through a condenser, where it is cooled; the condensed water is then returned to the boiler. The simplest cooling method is to allow cold water from a river or stream to pass through pipes in the condenser. This allows heat exchange between the steam and this cooling water. In a once-through system using river water, where the water passes through the condenser only once before returning to the waterway, the discharge water can be much warmer than the intake, affecting the ecology of the river. In some cases, there is inadequate river flow, and structures known as cooling towers are used to discharge the heat to the atmosphere via evaporative cooling. The mists generated by such towers may produce fogs, affecting visibility and traffic safety. These fog droplets will also contain traces of the toxic chemicals that are used to prevent algae growth within the tower. The potential for thermal pollution is greater in nuclear-power electric generating plants since they release approximately 15% more heat per unit of electricity generated than do fossil-fuel facilities.

Transportation

In a highly mobile society such as ours, a great deal of time and energy is consumed in transporting both goods and people. As a result, a major portion of our total energy consumption can be attributed to that required for movement. In consuming energy, there is always waste. For transportation, or mobile sources, these are heat, fuel spillage, exhaust products, and various by-products. Fuel combustion that occurs in motor vehicles is generally incomplete, and the exhaust wastes include CO_2, H_2O, SO_2, NO, CO, and a host of organic compounds. They may also include the oxidation products of any fuel additives, such as manganese compounds.

The use of transportation systems creates additional environmental contamination. Vehicular motion resuspends settled dust particles. The vehicles rust and wear, spreading iron and copper oxides, rubber particles from tire wear, and asbestos fibers from clutch and brake linings along their paths. They leak or vaporize engine oil and other working fluids, elevate noise levels, and contribute to environmental litter, either through the actions of their occupants or by the abandonment of the vehicles or parts thereof. Finally, vehicles alter the environment in very significant but less obvious ways. Land transport vehicles need roads, parking areas, and service stations. Major fractions of our urban land surfaces are covered with concrete and asphalt to accommodate the needs of wheeled transport. The mobility that autos provide has led to the rapid growth of the suburbs that, by and large, are less energy-efficient than cities and has thereby increased the demand and costs of energy for heat and power usage.

Motor vehicle exhaust. A major source of air contamination in the United States is motor vehicle exhaust. Before 1968, there was no intentional control of tail pipe emissions, except in California. With the then constantly increasing numbers of cars in use, the levels of CO, hydrocarbons (HC), NO_x, and Pb in urban air were high enough to cause concern among health authorities. This resulted in the setting by the federal government of performance goals for automobile engine emissions, with the objective of reducing emissions of CO, HC, and NO_x. The CO and hydrocarbon emissions limits were met by the use of catalytic converters in the exhaust system. Since catalytic converters were poisoned by Pb, which was used as an antiknock additive, they could only be used with unleaded fuel. Thus, Pb additives were eliminated from the fuel supply. While catalytic converters increased the extent of oxidation of hydrocarbons and CO in the exhaust, they also increased the oxidation state of any sulfur in fuel. Historically, the concentration of sulfur (S) in gasoline was relatively low, about 0.03% by weight, and essentially all of the S was emitted from the tailpipe as sulfur dioxide (SO_2). The SO_2 that was released was not generally considered to be a significant problem, since considerably larger amounts of SO_2 were released by fossil fuel combustion in stationary (i.e., non-moving) sources. However, in passing through a catalytic converter, some of the SO_2 in the engine exhaust was oxidized to sulfur trioxide (SO_3), which is rapidly hydrolyzed by water vapor in the exhaust to form sulfuric acid mist (H_2SO_4). Thus, the catalytic converter became a source of H_2SO_4, a more highly irritating chemical than SO_2. This problem turned out to be self-limiting, at least in part. The initially high rate of formation of H_2SO_4 dropped off markedly as the vehicles aged, and subsequent efforts to control the concentration of ambient air fine particles led to substantial reductions in the S content of motor vehicle fuels.

Other consequences of the reliance on catalytic converters and the modification of engine designs to reduce CO and HC emissions in the early 1970s were reductions in performance and fuel economy. To meet mandates for improved fuel economy, the industry developed onboard microchip-based fuel supply and ignition timing controls that optimized emissions, performance, and economy. It also stimulated a minor boom in the use of diesel (compression-ignition) engines in automobiles, based on their greater efficiency. However, relatively few diesel-powered cars were sold in the US because of their generally somewhat sluggish performance, their black smoke emissions, and the characteristic odor of the exhaust. Thus, the conventional Otto-cycle (spark ignition) internal-combustion engine continued to be used in most US automobiles, while diesel engines were dominant for heavy-duty trucks. In the US, stringent diesel engine emission standards for black carbon and NOx were established, while in Europe, where diesel engines had gradually become the dominant type for autos as well as for trucks, less stringent emission limits were established.

While both Otto-cycle and diesel-cycle engines operate at maximum efficiency and with minimal emissions when running at constant power, emissions increase greatly during other operating modes, such as idling, acceleration, and deceleration. At the turn of the twenty-first century, the use of hybrid designs became a viable option. The hybrids combine a smaller internal combustion engine with electric motor driven flywheels. The internal combustion engine runs at a nearly constant rate, at which it is quite efficient and generates a minimum of pollutant effluents, and the batteries provide peaking power for acceleration. The battery packs used in these hybrids are recharged by the alternator during cruising and deceleration. Some hybrids are using Otto-cycle engines, while others use diesel-cycle engines of more modern designs. Both of these relatively small diesel engines, and the larger diesel engines used in trucks and buses made in recent years, produce much cleaner exhaust than those made in earlier years, especially in terms of black carbon particles.

Aircraft exhaust. The black plume exhausts of the early commercial jetliners have been essentially eliminated in the newer engines, as well as retrofitted older engines. However, engine emissions during idling can be a serious problem at airports. All combustion engines, including aircraft jets, generate NO. The release of NO by jets in the stratosphere has raised the possibility of significant destruction of stratospheric O_3 by interaction with the NO. The fear of a possible reduction in stratospheric O_3, with a projected subsequent increase in human skin cancer, was a significant factor in the decision of the federal government not to subsidize the production of a US supersonic commercial transport. In 2003, the only supersonic transports in regular use, the Anglo-French Concorde, were taken out of service.

Water contamination by boats and ships. A major problem in aquatic environments is oil contamination. There has been considerable attention paid to oceanic oil spills resulting from the breakup of deep-draft oil tankers. However, the oil contamination of the oceans by more mundane sources, such as bilge flushing and leakage, may have greater overall effects. Within coastal and inland waterways, leakage and inadvertent spills create most of the oil-related problems. Another major problem in inland waters is sanitary waste, which is discharged directly into the water by commercial and pleasure boats. This problem should diminish through the enforcement of holding tank regulations by the US Coast Guard.

Water contamination by disasters. Two major incidents resulted in release of millions of gallons of oil into the environment. In 2010, an explosion on the offshore drilling rig Deep Water Horizon resulted in the release of approximately 210 million gallons of crude oil into the Gulf of Mexico off the coast of Houston, Texas, in what was the largest oil spill in US history. In 1989, the Exxon Valdez oil tanker ran aground in Prince William Sound, Alaska, releasing between 11 million and 38 million gallons of crude oil into the water, which was the second largest oil spill in US history.

Contaminants resulting from unregulated activities

In the US, the Environmental Protection Agency regulates major sources of anthropogenic pollution, but other significant sources remain unregulated. An example is particulate matter (PM) having aerodynamic diameters >10 μm in wind-blown soil, products of forests and plant growth, and smoke from forest fires. Emission inventories have traditionally been focused on three categories of anthropogenic activities: fuel combustion, industrial processes, and transportation. Furthermore, a significant fraction of ambient air PM can be attributed to gaseous precursor emissions that combine to form particles downwind of the sources (e.g., sulfate, nitrate, and organic fine particles), which can account for as much as 90% of PM_{10} in the eastern US and ~30% of PM_{10} in some western US locations. By contrast with $PM_{2.5}$, which remain suspended in ambient air for many days, coarser particles in fugitive dust, and dust from wind erosion and agricultural operations, tend to settle out within a few hours.

Agriculture

Modern agriculture is similar to mass-production industry in many respects. It relies heavily on highly specialized automated equipment and the economics of high-volume production. In the interests of efficiency, it utilizes manufactured fertilizers and pesticides and concentrates waste products to the point where they often create disposal problems. The problems of chemical contamination of the environment by agricultural operations may appear to be remote to urban populations, but the key role of agriculture in the national economy makes it important to understand its total impact on the environment.

Agricultural use of fire. Controlled burning of stubble on harvested fields is a cost-effective means of preparing the soil for the next crop while simultaneously enriching the soil in nutrients. It also produces large amounts of smoke, which may be objectionable to downwind populations. In many areas, the controlled burning of agricultural fields is allowed only after a permit has been issued and only under specified meteorological conditions.

Soil tillage. Mechanical soil tillage, especially when the soil is dry, can result in substantial quantities of dust immediately or can make it possible for subsequent wind storms to pick up dust from the tilled surfaces. Marginal dry land agriculture has resulted in major dust storms and extensive soil erosion.

Fertilizers and soil conditioners. One classic definition of contamination is a resource out of place. Too much inorganic plant nutrient in a waterway is a prime example, since it can cause an undesirable proliferation of rooted aquatic plants and algae and lead to oxygen depletion and anaerobic decay. The drainage of fertilizers and other soil conditioners from agricultural fields to streams can cause such effects in downstream waters. The main inorganic nutrients used as fertilizers are nitrogen (as nitrate) and phosphorous (as phosphate). The application of fertilizers, manures, and lime as powders or sprays can also create local air contamination problems, which may extend beyond the confines of the farm.

Pesticides. Pesticides are used because they reduce or eliminate pest species that consume, destroy, or compete with the efficient production of the product of interest. The ideal pesticide is specific and biodegradable. A pesticide is specific when it destroys the pest without affecting nontarget organisms. A pesticide is biodegradable when it is destroyed or degraded by microorganisms within a reasonably short time after it has fulfilled its intended function. The time frame during which the pesticide should persist without biodegradation depends upon the pest. In some cases, such as rat control, instantaneous killing is adequate, while for insect control, the pesticide may not be effective unless it persists for the full life cycle of the insect. Persistent pesticides, such as DDT and other organochlorines, had practical advantages, which led to their widespread use in the mid-twentieth century because of their extremely low order of acute toxicity, and them being relatively inexpensive. Part of their economic advantage rested in their persistence, since they did not require as frequent reapplication as the more degradable pesticides.

Unfortunately, the ability to develop pesticides of sufficient specificity is limited. Frequently, the pesticides used are toxic to species other than the target organisms. The unintended toxicity may be immediate and apparent, such as a fish kill in a farm pond. On the other hand, the toxicity may take a long time to develop, and even longer to recognize, such as the effects of the persistent organochlorine pesticides on the reproductive capacities of wild birds in areas removed from the direct applications of the pesticides.

Another problem is the potential for human health effects from the ingestion of pesticide residues on or in foods. Thus, agricultural operations with pesticides are sources of environmental effects off the farm in several ways: the reduction in populations of desirable mobile nontarget species of insects, birds, and mammals because of exposure on the farm; the effects on aquatic organisms resulting from the transport in surface storm drainage; the effects on biota in near and remote downwind regions because of the drift of aerosol sprays or from surface deposits that are volatilized by solar heating and are transported long distances as vapors or condensed fine particles; and the effects of residues remaining in or on agricultural produce.

Waste products.   In some agricultural operations, such as fruit growing, only marketable products are harvested and there is relatively little waste to dispose of. In other operations, such as the production of rooted crops (i.e., potatoes, beets, carrots, etc.), all of the foliage is waste. In animal husbandry, especially on feedlots, the animals generate large amounts of fecal wastes during their growth. In the slaughterhouse, large amounts of very rich organic wastes are generated. Such wastes, if not properly processed, can cause significant problems in terms of excessive organic nutrient levels in receiving waters and odor problems in downwind areas.

Domestic activities

The wastes generated in our homes also contribute to chemical contamination of the environment. Each person disposes of about 500 L per day of wastewater and approximately 1.6 kg per day of domestic refuse. About half of the refuse is paper and about 10% metal. If buried, the solid wastes can contaminate groundwater with dissolved chemicals and oxygen demand. If burned in older municipal incinerators, some of the metals, volatile organics, and products of incomplete combustion will be released into the ambient air. It is extremely difficult to maintain efficient combustion with a material as variable in heating value and moisture content as domestic refuse. Many households minimize the amounts of garbage disposed of as solid wastes by using a garbage disposal on the drain of the kitchen sink. This merely transfers the oxygen demand burden to the sewage treatment facility and/or the receiving waters. The treatment facilities must also cope with sanitary and laundry wastes from individual residences, along with the storm waters that pick up some of their waste burdens in passing over the surfaces of residential properties.

Municipal services

Municipal governments are generally responsible for disposing of solid and liquid wastes generated by their residents and to reduce the volume and strength of the wastes they collect. For example, an efficient sanitary landfill disposes of the solid wastes without significant reentry to the general environment. In sanitary landfilling, the wastes are spread in thin layers, compacted, and covered with soil each day. In some cases, where wetlands are used as fill sites, there is a negative impact, in the sense that the wetlands environment is destroyed in the landfill process.

Some municipalities use large central incinerators to reduce the mass and volume of solid wastes by approximately 90%. The residual ash is a relatively clean fill material and can be used as a cover for the refuse in the landfill. Unfortunately, as discussed, the incinerators put partially burned hydrocarbons, ash, and trace metals into the atmosphere. Community opposition to new incinerators has frequently been sufficiently intense to prevent their construction. However, incinerators can be operated as energy recovery systems, utilizing the heating value of the refuse.

Municipal sewage treatment facilities, are, at best, only partially successful in removing wastes and contaminants from incoming waters. Some operate only primary treatment facilities, consisting of gravity settlement chambers. These remove the more massive suspended solids but very little of the finer solids and hardly any of the dissolved solids. Most of the rest of the facilities include secondary treatment involving some kind of biological oxidation. Such plants may remove from 50% to 95% of the oxygen demand from the wastes but usually have lower efficiencies for the removal of nitrates, phosphates, and trace metals. The large portion of the oxygen demand removed from the incoming sewage by the treatment plant does not disappear. Some is oxidized to harmless end products in the process, but the rest is simply transferred into sewage sludge, which must itself be disposed of. Further discussion of water treatment technologies is presented in a subsequent chapter.

Sludge disposal remains a major problem for many large cities. For many years, New York City and other coastal cities dumped digested sewage sludge into designated dumping grounds within coastal waters. In New York's case, this was a site approximately 12 miles out to sea. As a result of potential hazards to marine life and public health, the city terminated such dumping. Land disposal can involve significant problems in odor generation and/or groundwater contamination. Some of the sludge is now being incinerated.

Another municipal activity that generates chemical contamination in some regions is the road and street salting done in the winter to improve traffic safety. The salt corrodes metals; damages plants, shrubs, and trees; and increases the salinity of the receiving waters. However, the environmental contamination from street salting appears generally acceptable in comparison to the alternative (i.e., impassable and unsafe roadways).

Nuclear power reactors

In power reactors, controlled fission is used to generate electricity. The first nuclear central power station began operation in 1957 in Pennsylvania. However, no new nuclear power reactors have been ordered since 1979, when a reactor failure occurred at one of the units at Three Mile Island in Pennsylvania. While the reactor itself was destroyed, there was no significant release of radioactivity or reactor fission products to the environment because of the effectiveness of the containment structure, which is a feature of all US power reactors.

A quite different exposure scenario occurred when a power reactor failed at Chernobyl in Ukraine in 1986. This reactor had no containment structure. As a result, about 1,000 cases of acute radiation poisoning occurred, including 28 early fatalities. There were approximately 95,000 persons within 3 km of the reactor who were exposed to quite high levels of radiation. About 135,000 people living within 30 km were evacuated about 10 days later. A radioactive cloud moved downwind into central and northern Europe, and the fallout from the cloud caused somewhat elevated radiation exposures to many millions. Within the next 10 years, there were several thousand cases of thyroid tumors in downwind populations in Ukraine and Belarus that were attributable to radioactive iodine in dairy products that delivered a concentrated radiation dose to the thyroid. The exposed populations are being monitored to determine the extent of the elevation that will actually occur for tumors at other organs of the body with longer lag times.

While there may be future generations of nuclear power reactors in the US at some time, that time is at least several decades in the future. The fraction of the electrical power supply generated by nuclear fission can be expected to decline as some of the older units reach their economically useful lifetime and are retired, decommissioned, and dismantled. In the meantime, there is a growing inventory of spent fuel rods stored under water in tanks maintained at nuclear power plants. Ultimate disposal of radioactive wastes is a contentious issue. There are low-level (i.e., not very radioactive materials) and high-level (i.e., highly radioactive sub-stances), primarily fission products produced at the power plant and separated during reprocessing, wastes; the former are buried at commercial sites in the United States. The latter are stored in liquid form in underground tanks. In 2002, the US Congress endorsed a plan to transfer high-level power plant wastes to an underground storage facility at Yucca Mountain in Nevada as a secure long-term repository, but such transfers have not occurred.

Nanomaterials

A recent concern related to environmental contamination involves nanoparticles. These are defined as particles having diameters less than 100 nm. Because of their small size, nanoparticles have a very large surface to volume ratio. This results in the potential for high surface reactivity. Furthermore, the physicochemical prop-erties of nanoparticles differ from those of their larger size counterparts.

Like many other contaminants, these too can have natural or anthropogenic origins. Naturally occurring nanoparticles are introduced into the environment via ocean spray, volcanic emissions, forest fires, weathering, and soil erosion and can be transported thousands of miles due to their small size. As such, they are widely distributed throughout the atmosphere, hydrosphere, lithosphere, and even biosphere. In fact, viruses can be considered natural nanoparticles.

Anthropogenic nanoparticles can be produced either incidentally due to certain industrial and mechanical processes, such as unfiltered exhaust of diesel engines, mining operations, or from coal, oil, and gas boilers, or on purpose, in which case they are termed engineered nanoparticles. These latter materials have been increasingly used in industrial applications and occur in thousands of commercial products, including food, cosmetics and personal care products, drug delivery systems, and biomedical imaging systems. Most engineered nanoparticles consist of carbon, silicon, metals, or metal oxides.

SECONDARY CONTAMINANTS

Many of the chemical contaminants in the environment, including some of the most important ones, such as O_3 in urban air, chloroform ($CHCl_3$) in drinking water, and aflatoxin in certain foods, are not directly released into the environment by any primary source. Rather, they are formed within the environment by chemi-cal reactions among their precursor compounds, as in O_3 or $CHCl_3$ formation, or by biological processes, as in the formation of aflatoxin by the action of *Aspergillus flavus* mold on stored peanuts and corn. This section presents some of the impor-tant mechanisms for forming secondary environmental contaminants, citing a few notable examples of transformations within the ambient air and surface waters that have affected public health or attracted considerable public attention.

Chemical Transformations in the Atmosphere

Relatively little attention was paid to chemical reactions within the atmosphere until the 1940s, when the characteristic southern California smog, with its visibility reductions and eye-irritation properties, became persistent. It was very different in character from the classic urban air pollution of London or Pittsburgh, which was characterized by its soot and SO_2 content and reducing properties. In contrast, the Los Angeles smog had oxidizing properties. It was found to contain O_3, which seemed inexplicable initially, since an inventory of known sources did not yield an adequate O_3 source and, at the time, it seemed unlikely that atmospheric chemistry would account for the O_3. The available information on reaction kinetics suggested that the concentrations of potential reactants in the air were too low. Reactions require molecular collisions, and there are many fewer collisions per unit time in air than, for instance, in water, where the effective density is more than 800 times greater. However, chemical transformations within the atmosphere, while they initially appeared unlikely, do lead to the formation of O_3 and a host of other reactive species. Advances in atmospheric chemistry not only helped to develop an understanding of the formation of the oxidizing atmospheres of southern California but also were able to help lead to a better appreciation of the transformations undergone by SO_2 within the atmosphere and, therefore, of our understanding of the modes of action and effects produced by the more classic kinds of reducing contaminants. The processes that determine the composition and characteristics of the ambient air aerosol are summarized in Figure 3–1.

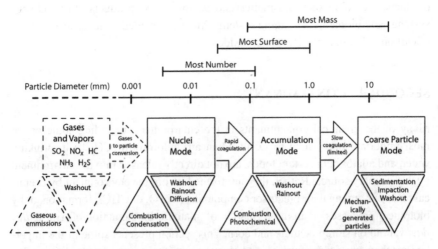

FIGURE 3–1. Schematic representation of the ambient air aerosol and processes that modify it. (*Source*: Hinds, W.C. *Aerosol Technology*, New York: Wiley, 1999.).

Photochemical reactions

An important ingredient in many of the atmospheric transformations is actinic radiation, or sunlight. Reactions depending on light are known as photochemical reactions. Solar radiation consists of a broad spectrum of wavelengths and energies. The higher-energy wavelengths dissociate gas molecules in the upper atmosphere. Photodissociation of O_2 produces a mixture of the highly reactive atomic oxygen (O), the stable diatomic molecular oxygen (O_2), and the relatively reactive triatomic ozone (O_3). There is relatively little O_3 below the stratospheric O_3 layer, except for that formed in urbanized areas. The background tropospheric level of about 0.03 ppm (30 ppb) is attributable to the small amount of mixing that occurs between the stratosphere and troposphere.

In photochemical reactions, solar energy raises the reactivity of the absorbing molecules to a level enabling them to react with other molecules. Only colored substances can absorb the longer wavelength photons that can penetrate the stratospheric O_3 layer. In plants, for example, the key substance is chlorophyll. In a polluted atmosphere, the key substance is the brownish-colored gas, NO_2. Absorption of light quanta having sufficient energy (those with wavelengths up to 0.38 μm) will cause photodissociation:

$$NO_2 + hv \rightarrow NO + O \tag{3-1}$$

where hv is an index of the energy of the light, h = Planck's constant, and v = frequency. Some organic compounds can also undergo photodissociation to form free radicals, for example:

$$R - \overset{\displaystyle O}{\underset{\displaystyle H}{\overset{\|}{C}}} + hv \rightarrow \dot{R} + H\dot{C}O \tag{3-2}$$

The highly reactive atoms and free radicals formed in these dissociations may combine with their original partners but will more likely interact with other molecular species in the atmosphere, utilizing the extra energy received from the solar radiation in their dissociation. Free radicals are so reactive that in aqueous systems their lifetimes are measured in microseconds or nanoseconds. However, in the atmosphere they are present in concentrations of less than 1 part per 100 million and can have half-lives of minutes to hours. Free radicals can also be formed in nonphotochemical reactions, such as the reactions of O_3 with olefins:

$$O_3 + RCH = CHR \rightarrow RCHO + R\dot{O} + H\dot{C}O \tag{3-3}$$

Solar energy quanta, which are not sufficiently energetic to dissociate molecules, can excite them, enabling them to react more readily. Oxygen absorbs visible light poorly, but there is so much oxygen in the atmosphere that a sufficient number of excited molecules are formed to play a role in reactions with aldehydes and possibly other hydrocarbons.

After the initial steps of photodissociation and excitation, an extremely complex series of secondary and chain reactions can be initiated, depending on the mixture of raw materials available in the atmosphere and on the ambient temperature and humidity. There are many possible sequences of reactions, and those presented here are examples of potential ones.

Some free radicals attach to O_2, forming peroxy radicals. These can react with NO_2 to form peroxyacetylnitrates, which are responsible for much of the eye irritation and plant damage produced by smog. The peroxy radicals can also react with other primary and secondary air contaminants to form other organic compounds, for example, alcohols, ethers, and acids.

Many of the secondary products are relatively unstable and undergo further chemical and photochemical reactions, which may generate NO_2, O_3, and a variety of free radicals; these then participate in further chemical and photochemical reactions. This gas-phase photochemistry can continue through many generations of complex chain reactions and has the net effect of converting most of the NO in the atmosphere to NO_2 and of increasing the levels of O_3 and of light-scattering aerosols of sulfur oxides and highly oxidized hydrocarbons.

The net result of the series of complex reactions is that levels of the primary contaminants, that is, the hydrocarbons and NO from auto exhaust, increase through morning rush hours. However, the buildup of NO is stopped by its conversion to NO_2. Aldehydes are partly a primary contaminant, but their continued rise through the morning and the late-morning fall in hydrocarbons indicate that the aldehyde level is augmented by photochemical processes. As the NO_2 levels fall, the O_3 levels rise. It is interesting that the evening rush hour does not produce the same sequence of events. The better ventilation in the evening limits the accumulation of the primary contaminants. Also, the sunlight is not as strong, nor does it last as long as in the morning. Finally, residual O_3 reacts with NO to form more NO_2, accounting for the late-afternoon rise of the latter. Much of what is known about these complex series of chemical reactions has been learned in smog-chamber tests in which primary contaminants are irradiated by ultraviolet light. The experiment illustrated in Figure 3–2 began with initial concentrations of 2 ppm$_v$ of propylene (C_3H_6), 1 ppm$_v$ of NO, and 0.05 ppm$_v$ of NO_2. The decline in C_3H_6 and NO, and the successive rises of NO_2 and O_3, closely parallel those seen in ambient atmospheres.

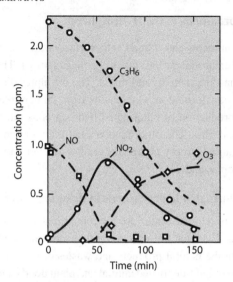

FIGURE 3–2. Concentration changes due to irradiation of a mixture of NO, NO$_2$, and C$_3$H$_6$

Reactions of sulfur dioxide and other gas-phase emissions

SO$_2$ released to the atmosphere can remain in the air for several days. Some of it, especially when emitted at or near ground level, is taken up by vegetation and materials or by water droplets in the atmosphere. However, most of it will undergo a series of chemical transformations before it reaches the earth's surface. The first step in the transformation is oxidation to SO$_3$. This can take place by photochemical oxidation, but the rate is relatively slow and this mechanism does not account for much of the conversion. Most is believed to occur by catalytic oxidation on the surfaces of airborne particles and within droplets. The SO$_3$ vapor is rapidly hydrolyzed by atmospheric water to form small droplets of H$_2$SO$_4$. These droplets take on water of hydration, coagulate with other atmospheric particles, and scavenge gas molecules, including more SO$_2$ and NH$_3$. They thus accumulate trace-metal cations and ammonium ions and provide a medium for conversion of the absorbed SO$_2$ to sulfite and sulfate ions. The end stage of the process is generally an aerosol with a mass median diameter of about 0.5 μm, composed primarily of ammonium sulfate but also containing traces of other sulfate and nitrate salts. This fine-particle aerosol is very stable in dry conditions and can travel for hundreds to thousands of kilometers before it is removed from the atmosphere, primarily via precipitation.

Chemical Transformations in the Hydrosphere

The primary contaminants discharged into surface waters can be transformed within the aquatic environment into other chemical species. The possibilities for such transformations are virtually unlimited. The examples discussed here illustrate two kinds of transformation. One results from human intervention, namely the unintentional production of chlorinated hydrocarbons during the process of disinfection with chlorine. The other involves the intervention of aquatic bacteria, whereby inorganic mercury compounds are converted into much more toxic organic mercury compounds.

Synthesis of chlorinated hydrocarbons in drinking water

Surface waters in the United States are all contaminated, to some degree, with organic compounds. Even natural waters in protected watersheds will have some contamination from the natural products and wastes of aquatic life and from the storm-water pickup of land-deposited animal and plant debris and air-contaminant fallout. Many communities produce their drinking water from river water that has been contaminated upstream by agricultural runoff, municipal sewage discharges, and industrial wastes and spills. The physical and/or chemical treatments given to these waters to remove their contaminants are never 100% efficient. Thus, when the waters are disinfected, they still contain a variety of organic materials. Some waters are disinfected with O_3, but most are chlorinated. In the process, some of the hydrocarbons react with chlorine to produce chloroform and other trihalomethanes, which are classified as carcinogenic compounds. Many drinking-water supplies also have measurable levels of carbon tetrachloride, methylene chloride, and other halogenated hydrocarbons, but some of these do not appear to be formed to any appreciable extent in the treatment process. By-products of drinking water chlorination include various trihalomethanes, haloacids, haloacetonitriles, haloaldehydes, haloketones, halophenols, other aldehydes, and carboxylic acids.

Methylation of mercury by aquatic organisms

In the natural environment, Hg is found in soil, air, and water. The natural level for Hg in water is 0.06 ppb_w in the northeastern US. Hg can enter the environment via a host of sources, either directly or indirectly. There are direct Hg releases in the waste effluents from manufacturing processes. The use of fossil fuels, such as coal (containing approximately 0.5 ppm_w Hg) and the burning of Hg-containing paper constitute indirect means of Hg release. Additionally, accidental misuse must also be considered. The most well-known source of Hg contamination is the chloroalkali industry, which formerly discharged 0.25 to 0.5 lb Hg per ton of caustic soda produced. It is important to consider the various forms of Hg present in water. The methyl form is the most toxic; other alkyl forms are also toxic but to lesser degrees. Phenyl and inorganic Hg derivatives are also less toxic than the methyl form, but when released into water they can be converted to the methyl form by a variety of microorganisms.

Various factors influence the rate of bacterial methylation in water. Methylation occurs rapidly in the winter and early spring but slows during summer and fall. Waters having low pH contain more of the methyl form than do those having high pH. Also, aerobic conditions enhance methylation. When exposed to similar concentrations of the methyl form and inorganic mercury, fish are able to absorb the methyl form from water 100 times as fast as the inorganic Hg and are able to absorb five times as much of the methyl form from food as compared to inorganic Hg. Once absorbed, the methyl form is retained two to five times as long as is inorganic Hg. With increased fish size, both the uptake of the methyl form from the environment and its clearance is decreased. Since the methyl form is strongly bound to muscle, it accumulates with increased muscle mass and increased duration of exposure. Assuming a steady Hg-containing diet, it takes about a year to establish a balance between Hg intake and excretion. Very high concentrations of Hg have been found in salmon caught in lakes contaminated by industrial discharges. While the US Food and Drug Administration sets a limit for Hg in fish, some tuna and swordfish exceed this limit. However, examination of the tissues of museum specimens captured before the advent of large-scale anthropogenic releases of Hg to the environment show similar levels, suggesting that background sources of Hg still exceed anthropogenic sources.

4

Dispersion of Contaminants

The contaminants generated by primary and secondary sources tend to spread continuously within environmental media. This chapter discusses the dispersion of contaminants within and between the atmosphere, the hydrosphere, the lithosphere, and the biosphere. While each component of the environment is discussed separately, it is not possible to treat them in isolation, as there is contaminant transfer among all of them. For example, there is very active interchange of contaminants between the hydrosphere and the biosphere, and atmosphere to hydrosphere exchange can be a major pathway for airborne contaminants to enter waterways. Figure 4–1 depicts potential pathways of movement among the biotic and abiotic environments.

DISPERSION IN THE ATMOSPHERE

Dispersion of contaminants within the atmosphere is generally referred to as diffusion. In still air, the process occurs by molecular diffusion and can be described by Fick's equation developed in the 1880s for physiological applications. This states that the net rate of flow is directly proportional to the product of the concentration gradient and a constant known as the molecular diffusivity, or coefficient

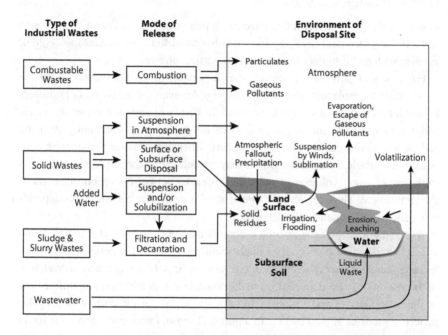

FIGURE 4–1. Sources and mode of release of contaminants into the environment.

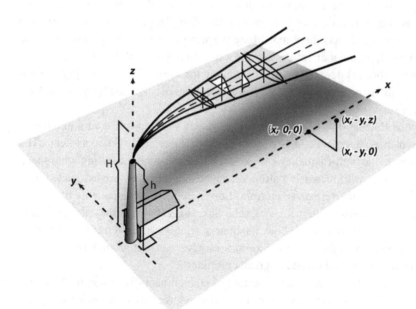

FIGURE 4–2. Conventional coordinate system and basic geometry of plume dispersion. h = geometric (actual) stack height, H = effective stack height.

of diffusion. The coefficient decreases with increasing molecular or particle size since the displacement resulting from a collision of the contaminant molecule or particle with an air molecule decreases as particle size increases.

For practical purposes, however, the dispersion of contaminants within the atmosphere by molecular diffusion is negligible, even for gaseous contaminants of small molecular sizes. This is because the displacements are generally infinitesimal compared to the displacements of the air volumes containing them by turbulence within the air. Thus, atmospheric diffusion is synonymous with convective or turbulent diffusion, which may be defined as the random motion and exchange of small volumes of air, called parcels, between neighboring parts of the atmosphere. Thus, the actual dispersion of atmospheric contaminants depends primarily on the scale of this turbulence.

Atmospheric turbulence is such a complicated phenomenon that it has defied attempts to describe it in rigorous mathematic terms. Even the definition of turbulence remains a matter of debate. For our purpose, it is the almost random variation of the velocity of a fluid, in contrast to the constancy in steady-state streamline flow or the periodicity of wave motion. In turbulent air, parcels exchange in irregular patterned motions termed eddies. In a practical sense, fluctuations about the mean or average wind speed are called turbulence. Such variations are associated with complicated eddy patterns of flow in both the horizontal and vertical directions.

The characteristics of turbulence depend upon whether the wind velocity component of interest is directed along the average horizontal wind direction, the transverse direction, or the vertical direction. The conventional coordinate system (Fig. 4–2) has the x-axis along the wind direction, the y-axis along the transverse horizontal direction, and the z-axis in the vertical direction. The average wind speed is denoted by $\bar{u}$, and is along the x-axis. The fluctuating components of the wind along the x-, y-, and z-directions are denoted by $u,'$ $v,'$ and $w,'$ respectively.

Actual fluctuations in wind velocity represent a superposition of a number of different sized turbulent eddies (Fig. 4–3). Some are related to known effects. The intensity of the high-frequency components increases with increasing wind speed and decreases with increasing altitudes, suggesting that they are associated with the friction of air over surface terrain. Lower frequency components have been observed to increase with increasing solar flux, suggesting that they are associated with thermal effects. Above about 500 meters, frictional effects from level ground have little effect on the motion of air masses. Over rougher ground, the effects of surface friction would extend to greater altitudes.

When considering the extent and rate of contaminant dispersion in the atmosphere, contaminant sources can be divided into three different categories: point sources, such as tall stacks; line sources, such as highways; and area sources, such as whole urban regions. The simplest category is an elevated point source, such as a power-plant smokestack. The light-scattering properties of the aerosol in the

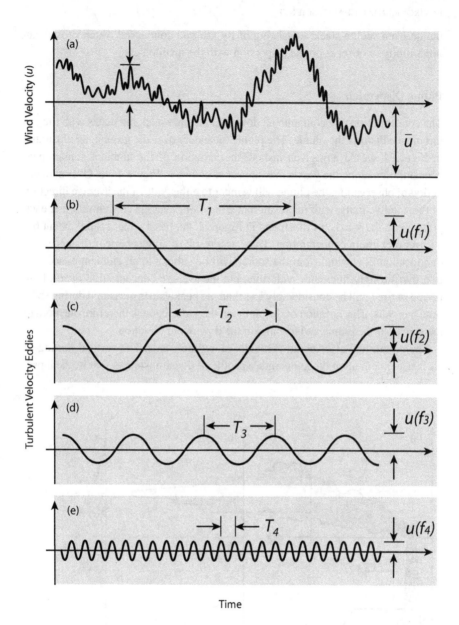

Figure 4–3. (a) Tracing of wind velocity showing fluctuations in velocity. These fluctuations are due to the superposition of eddies, each of which has a different frequency and intensity (b)–(e). The contribution due to four eddies. The intensity of each eddy is labeled $u(f_n)$ for turbulent components at frequency $f_n = T_n^{-1}$. For example, $u(f_2)$ = intensity of component at frequency $f_2 = T_2^{-1}$, etc. The value of u fluctuates about its average.

plume from such a stack, consisting of fly ash and condensed water, provide an opportunity to observe plume dispersion with the unaided eye.

Plume Dispersion

The concentration of a contaminant downwind from a stack fluctuates with time as turbulence distorts the plume. The plume meanders over the ground, as illustrated in Figure 4–4a. At any given instant, the variation of the ambient contaminant concentration along the y-axis may appear as shown in Figure 4–4b. Over a longer period of observation, the plume will wander to either side of the average direction of the wind velocity, with the result that a one-hour average concentration at each point along the y-axis, as illustrated in Figure 4–4c, has a wider spatial extent but a lower maximum concentration. Thus, as a result of atmospheric turbulence, any one location downwind from the source will be subject to an ambient concentration that strongly fluctuates with time, but the average concentration over a long period of time will be considerably less than the peak values measured during short time intervals. The variation of the plume in the y-direction is therefore due both to the spread of the plume and to shifts in the downwind direction.

The plume will also spread vertically along the z-axis. The vertical mixing of air is dependent upon the temperature profile of the atmosphere, that is, the lapse.

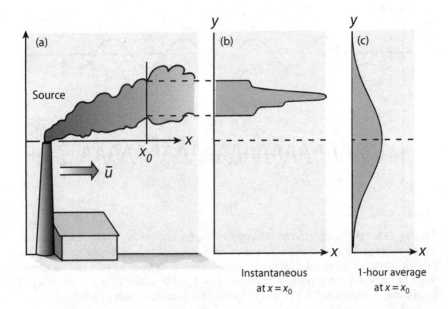

FIGURE 4–4. (a) Instantaneous top view of a plume; (b) instantaneous horizontal profile of plume concentration, x, along a transverse direction at some distance downwind from the source; (c) one-hour average profile for the same downwind distance.

The immediate ground-level concentrations of air contaminants may be reduced by good vertical mixing, since dispersal into higher regions dilutes the contaminant. Poor vertical mixing may allow contaminants released at low altitudes to remain there in relatively concentrated form.

The simplest way to illustrate vertical mixing is in terms of the concept of a parcel of air. This parcel is a small volume of air that has a uniform temperature throughout. Furthermore, if some disturbance causes this parcel to rise or fall in the atmosphere, it generally will do so adiabatically, that is, with no heat exchange between the parcel and its surroundings.

A rising parcel of air expands and therefore cools; a descending parcel contracts and warms. If the actual lapse rate of the atmosphere is identical to that of the adiabatically rising or falling parcel, the parcel will neither increase nor decrease its rate of ascent or descent. In other words, the parcel will be in equilibrium with the surrounding air. In this case, the atmospheric lapse rate is known as the adiabatic lapse rate. Since vertical motion is neither encouraged nor discouraged, the atmospheric condition is said to be neutral. Mixing only occurs if the parcel continues to be buoyant as it cools.

The exact atmospheric lapse rate depends upon the amount of moisture in the atmosphere. The adiabatic lapse rate in dry air is approximately 10°C/km. In moist air, water vapor has a greater heat capacity than does the oxygen, nitrogen, and argon and may also undergo phase changes from vapor to liquid and vice versa, with release or absorption of the latent heat of vaporization. Thus, the actual moist adiabatic lapse rate is always less than the dry rate, ranging as low as 3.5°C/km depending on the degree of moisture. The average tropospheric lapse rate is about 6.5°C/km.

Because of the various energy-transport processes occurring in the atmosphere, the actual lapse rate is very often different from the adiabatic lapse. If the lapse is greater than the adiabatic lapse (i.e., when the rate that ambient temperature decreases with altitude is greater than the adiabatic rate), the rising air parcel will be warmer, and less dense, than the surrounding air. It will, therefore, continue to rise, since upward motion is accelerated. The lapse for this situation is called superadiabatic; the atmosphere is said to be unstable, and vertical mixing is facilitated (Fig. 4–5).

If the actual lapse is less than the adiabatic rate, the parcel of air will begin to rise until it reaches a point where its temperature equals that of the surrounding air; upward movement then ceases, opposed by negative buoyancy. This atmospheric condition leads to a stable condition. The parcel rises or falls only a short distance until it reaches a point of temperature equilibrium (Fig. 4–6).

An extreme case of atmospheric stability occurs when the atmospheric lapse is negative; that is, temperature increases with altitude (Fig. 4–7). This condition is known as a temperature inversion. The atmospheric strata within which the inversion occurs is termed an inversion layer. Vertical air movement within inversion

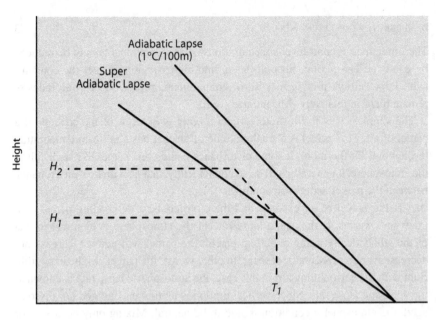

FIGURE 4–5. The instability of the superadiabatic atmosphere. A parcel of air raised in height from H_1 to H_2 cools adiabatically, and its rate of rise is accelerated because it becomes warmer and, therefore, less dense than the ambient atmosphere.

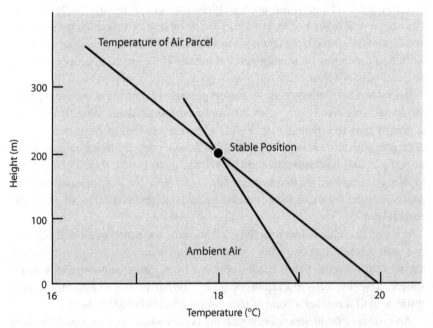

FIGURE 4–6. A rising parcel of air will cease its upward motion if at some point its temperature is identical to the ambient temperature, and the ambient lapse rate is less than the adiabatic rate.

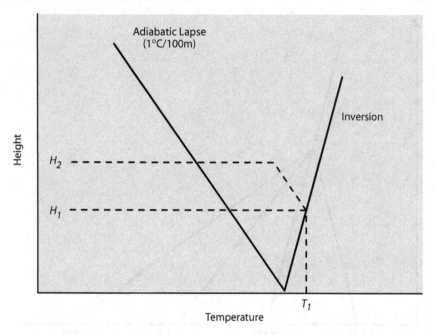

FIGURE 4–7. The inherent stability of the inverted temperature gradient. A parcel of air raised in height from H_1 to H_2 cools adiabatically and sinks to its original position because it becomes more dense than the ambient atmosphere.

layers is essentially nil (low-mixing) and contaminants accumulate within them. These layers are frequently visible because of the light-scattering properties of contaminant particles contained therein.

Depending upon how they are formed, inversion layers may have their base on the ground or they may be elevated. Figure 4–8 shows the development of a ground-based inversion, which occurs because the lapse near the ground is strongly influenced by radiative heat transfer. This property frequently results in a diurnal variation in lapse near ground level.

The effects of lapse on the vertical spread of a plume are illustrated in Figure 4–9. In terms of the potential for adverse effects on the environment and health, we are generally concerned about ground-level concentrations rather than the concentrations aloft within the plume, and since most effects are concentration dependent, we are most concerned with the maximum ground-level concentration. Thus, for stack discharges as illustrated in Figure 4–9, the worst condition is fumigation, where the emissions are all retained within the static inversion layer. At the other extreme, the looping that occurs with a very strong lapse can also bring the plume to the ground close to the stack. However, the time of contact at any one location will be very brief, and the high degree of turbulence will rapidly dilute the contaminant.

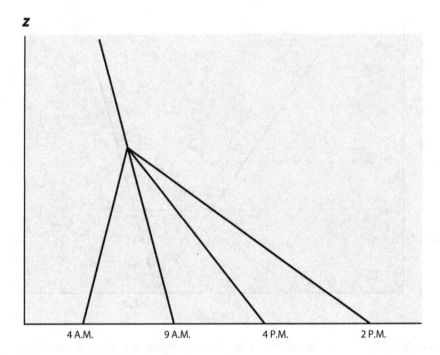

z

4 A.M.   9 A.M.   4 P.M.   2 P.M.

FIGURE 4–8. Typical diurnal variation of temperatures near the ground. *4 A.M.*: Radiation from the earth to the black sky cools the ground lower than the air, producing a ground-based inversion; *9 A.M.*: Ground heated rapidly after sunrise (slightly subadiabatic); *2 P.M.*: Continued heating (superadiabatic); *4 P.M.*: Cooling in the afternoon returns the temperature profile to near adiabatic.

Equations have been developed for predicting the downwind concentrations of contaminants discharged from elevated point sources. These are all empirical, since, as discussed earlier, no adequate theory has yet been developed that can cope with the complexity of atmospheric turbulence. Even those equations based on clearly artificial constraints appear to be formidable to the nonmeteorologist. For example, if we make the simplifying assumption that the average wind speed, $\bar{u}$, does not vary with position, and if contaminants are emitted from an elevated point source, the downwind ground level ($z = 0$) concentration can be approximated by:

$$X(x,y,0) = \frac{Q}{\pi\sigma_y\sigma_z\bar{u}}e^-(H^2/2\sigma_z^2 + y^2/2\sigma_z^2) \qquad (4\text{–}1)$$

where
  χ = concentration of contaminant in mass per cubic meter at $(x,y,0)$
  Q = emission rate from the source in mass per unit time
  σ_y = standard deviation of concentration in y-direction

σ_z = standard deviation of concentration in z-direction
H = effective height of discharge above the ground in meters
y = crosswind distance in meters
x = distance downwind in meters

Power-plant stacks emit hot gases with an upward velocity. These emissions are therefore known as buoyant plumes, since they tend to rise until their momentum is spent and they reach thermal equilibrium with the surrounding air. This buoyancy produces an effective increment to the actual physical stack height. The effective stack height (H) used in dispersion equations includes both the actual height and the increment due to buoyancy (Fig. 4–2).

If one's interest is restricted to the ground-level concentration directly downwind of an elevated source, that is, where $y = 0$, Eq. *(4–1)* reduces to:

$$X(x,0,0) = \frac{Q}{\pi \sigma_y \sigma_z \bar{u}} e^{-(H^2/2\sigma_z^2)} \qquad (4\text{--}2)$$

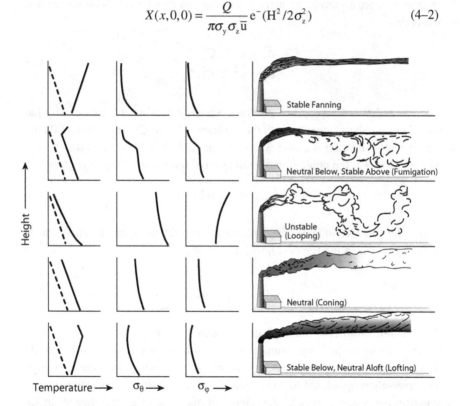

FIGURE 4–9. Various types of smoke-plume patterns observed in the atmosphere. The dashed curves in the left-hand column show the adiabatic lapse, and the solid lines the observed profiles. The abscissas of the columns for the horizontal and vertical wind-direction standard deviations (σ_θ and σ_φ) represent a range of approximately 0° to 25°. [*Source*: Slade, D.H. (Ed.) *Meteorology and Atomic Energy*, 1968. U.S. Atomic Energy Commission, TID-24190, Oak Ridge, TN, 1968.]

A feature of Eqs. *(4–1)* and *(4–2)* is the fact that the maximum concentration depends upon the mass emission rate. For a given release height, H, this is independent of the concentration of the contaminant in the exhaust gas. Thus, emitting a more dilute mixture without decreasing the mass emission rate will not reduce the ground-level concentration. The implicit assumption is that only contaminant, and no air, is emitted from the stack and that dilution in ambient air after emission is the primary means by which the concentration is reduced. If the ambient concentration downwind from an elevated point source is given by Eq. *(4–1)* with empirical values for the x-dependence of (σ_y) and (σ_z), then several statements can be made concerning the magnitude and distribution of the ground-level concentration. The location that receives the maximum ground-level concentration is, of course, on the plume line $(y = z = 0)$. This location depends upon the empirically determined dependence of the standard deviation σ_z on the value of x. For lack of simpler or more accurate alternatives, a power law dependence is often assumed, whereby:

$$\sigma_y^2 = x^{(2-n)} \, C_y^2 / 2$$
$$\sigma_z^2 = x^{(2-m)} \, C_z^2 / 2 \qquad\qquad (4\text{--}3)$$

where n and m are appropriate numbers representing atmospheric stability factors; these numbers, together with the constants Cy and Cz, are called "diffusion parameters." Thus, for the case where $n = m$, and $Cy = Cz = C$, so that σ_y / σ_z is independent of x, the point of maximum ground level concentration is given by:

$$x = (H^2 / C^2) \, (1/2 - n) \qquad\qquad (4\text{--}4)$$

Under the same conditions, the ambient concentration at this location can be obtained by evaluating Eq. *(4–1)*, with the result:

$$X_{max} = \frac{2Q}{e\pi \bar{u} H^2} \qquad\qquad (4\text{--}5)$$

Representative values of n and C^2 are shown in Table 4–1.

Under these simplifying assumptions, it can be seen from Eq. *(4–5)* that the maximum ground-level concentration is directly proportional to source strength and inversely proportional to wind speed. It is also inversely proportional to the square of the effective stack height. Thus, tall stacks, those greater than about 100 m, can greatly reduce local ground-level concentrations due to two factors. One is that the maximum ground-level concentration varies as the inverse square of height. The other is that they generally reach above the morning inversion layer and, as shown in Figure 4–9, this results in the condition known as lofting, where the plume theoretically never reaches the ground.

TABLE 4–1. Values of N and C^2 for Several Lapse Conditions

CONDITION	N	C^2			
		AT 25 M	AT 50 M	AT 75 M	AT 100 M
Unstable	0.20	0.043	0.030	0.024	0.015
Neutral	0.25	0.014	0.010	0.008	0.005
Moderate inversion	0.33	0.006	0.004	0.003	0.002
Large inversion	0.50	0.004	0.003	0.002	0.001

The preceding discussion of dispersion from an elevated point source is limited to the most basic relations and to their application under many simplifying assumptions. They were developed for applications close to the source. The situation becomes much more complex when considering other types of sources. For example, line sources such as roadways are spatially extended, and the wind direction with respect to source geometry influences the downwind dispersion of contaminants. In area sources, contaminants are received not only from sites directly upwind but also from those to the sides as well.

In addition to the imperfections in all of the available empirical relationships for predicting downwind concentrations of airborne effluents, there are a number of other complicating factors that limit their general application. One key assumption is that there is a free flow field around the top of the stack. This is frequently not the case. Figure 4–10 shows a common complication. Many power plants and factories are built within river valleys because of the need for large volumes of cooling water and/or waterborne transportation. In deep valleys, even a tall stack may not extend above the valley walls. Table 4–2 describes some of the effects of natural surface features upon plume dispersion.

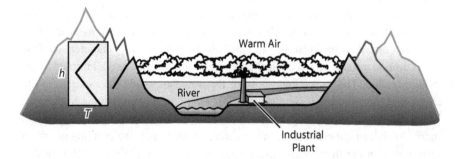

FIGURE 4–10. Fumigation of a valley floor caused by an inversion layer that restricts diffusion from a stack.

TABLE 4–2. Effects of Terrain Features Upon Plume Dispersion

TOPOGRAPHIC FEATURE	EFFECTS
• Elevated regions	a. Increased wind speed (and increased ventilation) over hilltops.
	b. Occasional impacting of elevated plumes on ground level.
• Deep valleys	a. Channeling of wind flow along the valley axis, resulting in higher average concentrations in the valley.
	b. Development of stable drainage winds during calm, nighttime conditions, resulting in higher concentrations along the valley floor.
• Undulating regions	a. Increased atmospheric turbulence near the ground level during times of moderate or strong winds. This results in lower pollutant concentrations at locations near sources.
	b. Accumulation of pollutants in low spots during calm, nighttime conditions (i.e., localized drainage-wind conditions).
• Regions of tree cover	a. Enhanced turbulence near the ground during moderate or strong winds, resulting in lower concentrations for locations near sources.
	b. In fully covered regions, blockage of elevated plumes, resulting in lower concentrations at ground level.
• Bodies of water	a. Increased moisture content in the local atmosphere, favoring fog formation at low-lying spots, and affecting the removal rate of SO_2 and other pollutants from the atmosphere.
	b. For larger bodies of water, formation of local circulation (lake and sea breezes) which can cause ground-level fumigation on the landward side of sources during sunny daytime conditions.

The ability of a stack to disperse contaminants can also be diminished by the presence of a large building in the vicinity. The plume becomes distorted, even if it does not actually contact the building. This effect occurs because the plume is carried in an air stream that accommodates itself to the shape of the building. If the airflow is disturbed locally, that portion of the plume that penetrates the disturbed flow region will also become distorted. Plume distortions near buildings are controlled by local air motions. Figure 4–11 shows characteristic flow zones around a sharp-edged cubical building oriented with one wall normal to the wind direction. The background flow, or the flow that would have existed in the absence of the building, is shown at the left, where the streamlines are horizontal. The mean velocity increases upward from zero at the ground, rapidly at first and more slowly at high elevations. In the center of the figure, the building creates a disturbance in the flow, whose main characteristic is the highly turbulent wake. Within the wake, adjacent to the ground and walls and roof of the building, there exists a region called a cavity; in the cavity, the mean flow is in the direction of

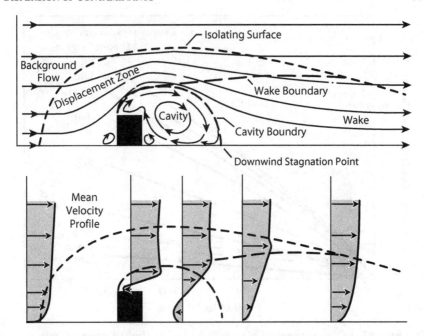

FIGURE 4–11. General arrangement of air flow zones near a sharp-edged building.

the background flow in the outer portion and opposite to the background flow near the axis. Changes in building shape and orientation to the wind affect the cavity dimensions and flow. Rounded buildings have smaller displacement zones and wakes.

All of the preceding discussion is based on the behavior of an inert gas within a plume. It does not take into account removal of a contaminant by precipitation or by contact with the surface of the ground or vegetation. It also does not take the gravitational sedimentation of particles into account, although the sedimentation rate of particles smaller than 20 μm diameter is so low that there is little significant error introduced when applying the empirical equations to the turbulent diffusion of small particles within air.

Empirical correction factors for the diffusion formulas have been developed to account for the capture of particles smaller than 20 μm diameter and reactive gases by ground-level surfaces. They employ the concept of deposition velocity (v_g), which is determined experimentally, and defined as follows:

$$v_g = \frac{\textit{mass deposited per unit area per unit time}}{\textit{mass concentration above the surface}}$$

The product of the mass concentration given by the diffusion equation and v_g in m/sec will be the deposition in mass/m^2/sec.

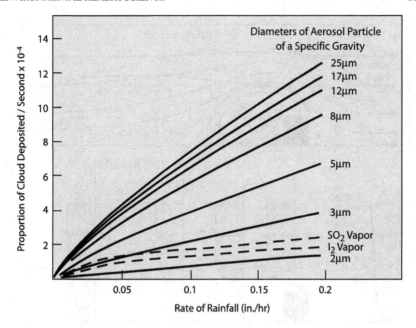

FIGURE 4–12. Percentage removal of particles by particle size and rate of rainfall.

The effect of precipitation on the removal of particles and soluble gases can be accounted for using a washout factor rather than a deposition velocity. The washout factor is defined as the fraction deposited per unit time, and some experimental values are illustrated in Figure 4–12. Table 4–3 provides a summary of the effects of various meteorological factors upon plume dispersion.

Dispersion from Ground-Level Sources

In contrast to elevated sources, most of which can be considered discrete point sources whose impacts on the environment are at distant sites, most ground-level sources have their major effects immediately downwind. Furthermore, since such emissions occur within the air most affected by surface roughness, they rapidly become well mixed into the surface air, a layer of varying height but that normally extends from about 100 m to 2000 m. As illustrated in Figure 4–11, the effluent from a short stack or roof vent will be influenced by the aerodynamic flow pattern around the building(s) with which it is associated. In estimating the downwind concentration patterns caused by this phenomenon, the most conservative approach is to assume that the effluent is trapped in the wake and brought quickly to the ground without being enveloped and mixed in the cavity directly behind the structure.

TABLE 4–3. Summary of Effects of Meteorological Conditions Upon Plume Behavior

CONDITION	EFFECT
Wind speed	Determines initial dilution
Wind direction	Determines downwind geometry
Atmospheric stability category	Determines plume spread associated with turbulent motions in the atmosphere
Mixing layer depth (or height of inversion base)	Determines limit to the vertical spread of the plume-important for downwind flow distances greater than 1000 m
Humidity	High humidity associated with visibility decreases for water-vapor plumes
Surface temperature	Determines possibility of ice formation from water-vapor plumes
Precipitation	Determines possibility of washout of contaminants near the source

If the effluent is trapped in the cavity, it will be mixed rapidly into a volume determined essentially by the cross-sectional area of the structure and wind speed. An appropriate formula for the downwind concentration is:

$$X = \frac{Q}{(\pi \sigma_y \sigma_z + cA)\bar{u}} e^{-}[y^2/2\sigma_y^2] \tag{4–6}$$

where A is the cross-sectional area of the building normal to the wind and c is a shape factor, ranging from 1/2 for a relatively streamlined shape to 2 for a blunt building. This formula gives a maximum concentration directly downwind of the building (σ_y, $\sigma_z = 0$) of $Q/cA\ \bar{u}$, a simple volume approximation.

In urban areas where there are many buildings with short stacks and roof vents, the situation becomes more complex, but the likelihood of thorough mixing of the various effluents is generally quite high. There is usually relatively little variation of concentration with height for, for example, CO emitted by motor vehicles at street level or for sulfur dioxide (SO_2) from fossil fuel combustion from rooftops of private homes or apartment houses

Long-Range Transport

While elevated discharges are an effective means of reducing maximum ground-level concentrations, they obviously do nothing to diminish the total amounts of contaminants emitted into the atmosphere. In fact, by minimizing contact between the ground-level surfaces and the contaminants, they actually act to increase the residence times and total atmospheric burdens of the contaminants. The development of a persistent seasonal haze layer over the eastern third of the continental US is due in large part to sulfate particles in the air that are transported in the prevailing winds over many hundreds of kilometers. These, in turn, were formed in the atmosphere from the airborne oxidation and hydrolysis of SO_2, much of which was discharged at elevated levels by the tall stacks of utility boilers. Even in the vast continental atmosphere, dilution is no longer the "solution to pollution."

The spread of environmental contaminants via the air depends upon their physical and chemical forms. Nonreactive and water-insoluble gases, vapors, and submicrometer particles can remain airborne for weeks within the troposphere and can, therefore, spread throughout the globe. Evidence for such transport of chemical contaminants is the presence of lead, nuclear weapons test debris, and various organic pesticides in samples of polar ice collected at locations that are remote from known sources of such contaminants.

The transport of contaminants from near-surface sources covers time scales measured in hours and days. When airborne contaminants were generated by thermonuclear explosions during the period of above-ground testing, they were spread throughout the atmosphere on a global scale, following injection of some of the debris into the stratosphere. Our knowledge of stratospheric dispersion was largely derived from gas and dust samples obtained by aircraft or balloons at an altitude of about 35 km following these tests. Stewart et al.[1] developed a model of stratospheric-tropospheric exchange based upon the observed pattern of nuclear weapons fallout. According to this model, air enters the stratosphere in the tropical regions, where it is heated and rises to an altitude of about 30 km, at which level it begins to move toward the poles. The tropopause is lower in the polar regions than at the equator, and tropopause discontinuities in the temperate regions facilitate transfer from the stratosphere to the troposphere. The jet streams, with velocities of 100 to 300 km/hr, occur at these discontinuities. The rate of transfer from the lower stratosphere is most rapid in the winter and early spring.

A summary overview of the transport and fate of atmospheric emissions is provided in Figure 4–13.

The atmosphere is a source of chemical contaminants that can deposit on the surface of the earth, for example, by gravitational settlement and via precipitation. Precipitation on the land leads to surface runoff, which picks up additional contamination by dissolution and suspension of contaminants deposited on the surface since the last rain. On the other hand, the land surface also acts as a filter and chemical buffer to the surface runoff, reducing part of its contaminant load. During light precipitation, the runoff may become relatively concentrated with dissolved contaminants that do not react with nor are readily absorbed by surface soil; at the same time, the runoff would be relatively well filtered of suspended

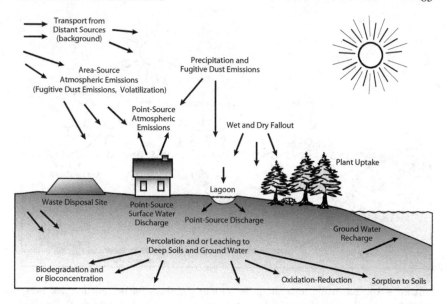

FIGURE 4–13. Schematic illustrating the transport and fate of atmospheric emissions in the environment. (*Source*: Paustenbach, D. The Risk Assessment of Environmental Hazards, New York: Wiley, 1989.)

solids. On the other hand, during heavy precipitation, the soluble contaminants are very much diluted, while the turbulent flow stirs up relatively high concentrations of suspended solids.

DISPERSION IN THE HYDROSPHERE

The aquatic environment is considerably more varied and complex than is the atmosphere with respect to contaminant dispersion. There are large differences in dilution volumes, mixing characteristics, and transport rates between rivers, lakes, estuaries, coastal waters, and oceans, and a generalized approach to the dispersion of contaminants introduced into bodies of water is not possible. Some of the potential routes of contaminant transfer between the hydrosphere, the atmosphere and the biosphere are shown in Figure 4–14.

Streams, Rivers, and Lakes

Rivers and streams vary so much in bed structure, depth, and flow rate that it is not possible to generalize dispersion rates for the contaminants discharged into them. Further complications are added for "managed" rivers. Most of the larger rivers in the United States have been extensively dammed for purposes of flood control,

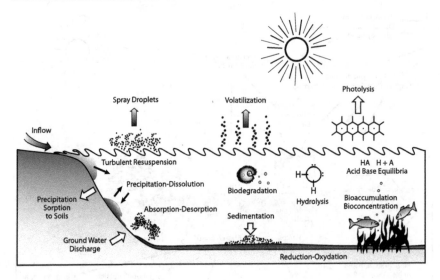

FIGURE 4–14. Schematic illustrating the various pathways for degradation and movement of chemicals and sediments in the hydrosphere. (*Source*: Paustenbach, D. The Risk Assessment of Environmental Hazards, New York: Wiley, 1989.)

hydroelectric power generation, and navigation, and large stretches of them are really a series of shallow lakes rather than a free-flowing river. There are also major withdrawals of river water for drinking-water supplies, industrial and utility usage, and irrigation of agricultural lands. Some of these waters, especially the industrial and utility process and cooling waters, are largely returned to the river, albeit with added contaminant loads. On the other hand, the irrigation waters are largely lost to evaporation, groundwater recharge, and possibly to drainage into another watershed.

Lakes have the additional component of greater depth of water. As discussed in chapter 2, the water can become thermally stratified, with the lower portion (hypolimnion) becoming isolated from atmospheric oxygen. In nutrient-rich lakes, this results in complete oxygen depletion and anaerobic decay within the hypolimnion. The spring and fall turnovers of the waters in the lake then spread the products of this decay throughout the lake.

The natural dispersion of wastewater effluents in lakes is often poor. In the absence of wind and tide, there is little turbulent mixing. Lake waters are normally colder and therefore heavier than wastewater effluents. Effluents discharged at or near the surface of denser receiving waters are likely to overrun them. Discharged at some depth below the surface, they rise like smoke plumes and, on reaching the surface, fan out radially. Because chemical diffusion is slow, natural dispersion or mixing of the unlike fluids is mainly a function of winds and currents.

Estuaries

The spread of contaminants within estuarine waters depends heavily on the structure of the estuary. While in some the sea water forms a distinct wedge beneath the fresh water, more typically there is sufficient turbulent mixing to create a more gradual change from fresh to salt water. At the other extreme, there is so much turbulent mixing that both the horizontal and vertical concentration gradients disappear.

Estuary flow is complicated, and there is no generalized dispersion model; the physical characteristics of each estuary must be studied on an individual basis. For example, in a study of diffusion and convection in the Delaware River Basin, Parker et al.[2] constructed a scale model of the Delaware Basin at the US Army Waterways Experiment Station at Vicksburg, Mississippi. The model was 750 ft long and 130 ft wide, and the expected dilution was studied using dyes. Figure 4–15 illustrates the type of information obtained from a study of the dilution of dye over a period of 58 tidal cycles (approximately 1 month). In this particular study, which involved instantaneous injection of a given dose, the concentration at the end of the 58 cycles remained at approximately 1% of the maximum concentration during the initial tidal cycle. The conditions of the experiment were conservative; there were no losses due to sedimentation or biological uptake. Thus, only the diminution owing to mixing was measured.

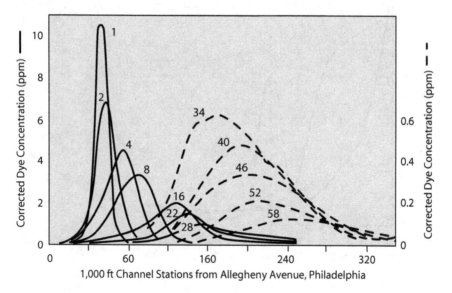

FIGURE 4–15. Longitudinal distribution of contaminants after a designated number of tidal cycles in a scale model of the Delaware River near Philadelphia.

Oceans

Dispersion of contaminants in ocean waters depends primarily on the character-istic physical features discussed in chapter 2. These are temperature, density, and salinity gradients with depth and the surface currents. Vertical mixing is vigor-ous within the surface zone due to wind action but is very poor within the lower parts of the intermediate zone. This is illustrated in Figure 4–16, which shows the distribution of radioactive debris at 6, 28, and 48 hours after it was deposited on the surface of the water. Since material deposited on the ocean surface mixes only very slowly into the deep water, it is possible to find measurable amounts of residual material on the surface at a considerable time following deposition. For example, Figure 4–17 shows the first year's path for weapons test debris from a June 1954 test. Comparison of this figure with Figure 2–13 shows that the debris followed the normal circulation of ocean surface water.[3]

While the rate of movement of the bottom water has not been mapped, rates of vertical diffusion obtained from measurements of radium and other substances dissolved in ocean waters have been modeled (Fig. 4–18).[3] Dissolved ^{238}U gives rise by radioactive decay to ^{230}Th, which precipitates rapidly to the bottom sedi-ments. The decay of ^{230}Th results in the formation of ^{223}Ra, which tends to return into solution at the ocean bottom. The vertical gradient of dissolved radium is a

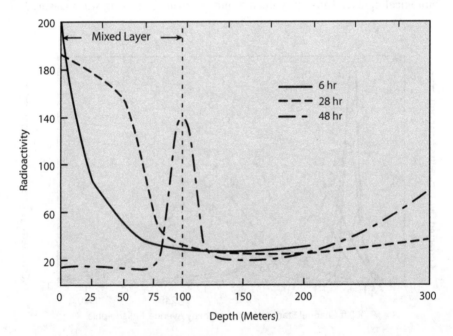

FIGURE 4–16. The vertical distribution of radioactivity in the ocean at selected times after fallout of debris from a nuclear explosion.

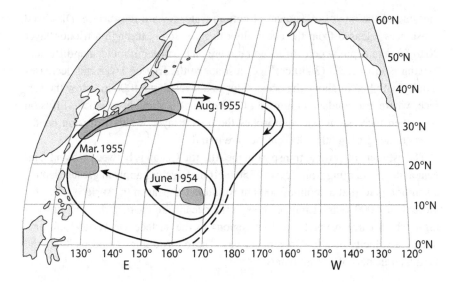

FIGURE 4–17. The horizontal dispersion of nuclear weapons debris in the north Pacific Ocean after tests by the United States in the Marshall Islands, 1951. Striped regions indicate areas of maximum contamination.

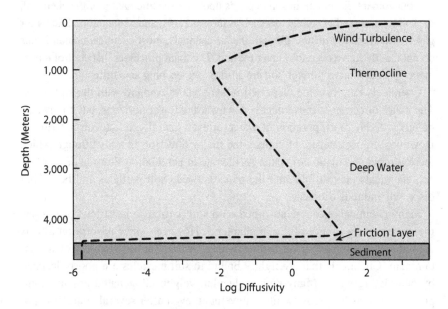

FIGURE 4–18. Vertical diffusion in the oceans.

measure of the rate of movement of bottom water toward the surface. Dissolved substances released from the ocean floor diffuse slowly through a friction layer 20 to 50 m in depth, where diffusion rates are on the order of molecular diffusion. Mixing is most rapid (3–30 cm^2/sec) just above the friction layer and decreases rapidly with height above the ocean floor to a level about 1000 m below the surface, where a secondary minimum (10^{-2} cm^2/sec) is thought to exist. Diffusion rates then increase as one approaches the mixed layer, where diffusion coefficients ranging from 50 to 500 cm^2/sec are found.

The estimate of vertical transport in the Atlantic Ocean is between 0.5 and 2 m/year at depths ranging from 750 to 1750 m. If these values apply at greater depths, a solution placed at a depth of 3000 m would not appear in the surface water for more than 1,000 years. Limited areas of the oceans are characterized by upwellings, which carry water from deep regions to the surface via vertical currents having a velocity of a few meters per day. These upwellings could facilitate the vertical transport of contaminants.

DISPERSION IN THE LITHOSPHERE: SOILS

Like the hydrosphere, the soil environment is quite complex with regard to contaminant dispersion. Different soil types have individual characteristics that affect the movement of chemicals introduced, intentionally or otherwise, into or onto the soil.

Contaminant dispersion in soil depends upon the specific nature of the chemical, the soil type, and other soil factors such as moisture, pH, and temperature. Except for gases, which readily diffuse through the air channels, most chemical contaminants do not readily move once they enter the soil. The main processes that control movement of contaminants through soil are adsorption, leaching, and diffusion.

Chemicals may become dispersed in soils via movement with the soil water. The major direction of movement is downward to lower horizons, but it may also be upward, due to evaporation of water, or even laterally, due to surface runoff. Downward movement is, of course, the major direction in soils through which there is rapid percolation of water. As the rate of percolation slows in deeper layers, chemicals may diffuse into the pores between soil particles and be further dispersed and redistributed.

Many chemicals become adsorbed onto soil particles via various processes; these are discussed more fully in chapter 5. However, since adsorption affects subsequent leaching and diffusion, it may restrict the distribution of a contaminant. Chemicals that are tightly bound to soil particles are poorly leached by percolating water. Many high molecular weight halogenated organic compounds, such as PCBs, show little movement, even after several years of exposure to water percolating through the soil. In this case, their low degree of water solubility is also a factor. As other examples, ammonium and phosphate ions are strongly bound to soil particles, while nitrate is not and readily leaches through the soil. Thus, strongly adsorbed contaminants may be found only in the upper few inches of the soil, while those less strongly bound readily travel through lower horizons. These are, of course, generalizations, since binding and subsequent leaching of any contaminant is highly dependent upon the specific soil type.

DISPERSION IN THE BIOSPHERE

In the biosphere, essential nutrients and energy are transferred from organism to organism along pathways called food chains. Very often, single chains are interconnected, resulting in a more complex interlocking pattern termed a food web.

Solar energy is utilized by organisms, known as producers, for the assimilation of simple, inorganic chemicals (nutrients) into energy-rich compounds. The primary producers are green plants, whose chlorophyll absorbs solar radiation. They then generate organic compounds from CO_2 and H_2O via photosynthesis. The potential food energy and nutrients within the plants are then transferred to the primary consumers, the herbivores, that is, organisms that eat plants. By eating the herbivores, carnivorous secondary consumers obtain energy and nutrients indirectly from plants. Food webs can also involve omnivores, organisms intermediate between the primary and secondary consumers of both plants and animals. Carnivorous tertiary consumers feed on secondary consumers and so on, with the number of levels of carnivores varying with the specific biotic community. The organic compounds in waste products and in dead plant and animal bodies provide the substrate for the decomposers, largely bacteria, which break down these compounds, releasing inorganic nutrients. A generalized food chain is shown in Figure 4–19.

FIGURE 4–19. A generalized food chain.

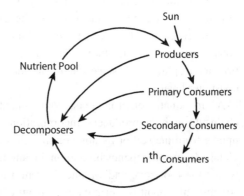

FIGURE 4–20. A trophic pyramid, arranged according to biomass, in an aquatic ecosystem in Silver Springs, FL. The numbers represent gm of dry biomass per m^2. P = Producers; C_1 = Primary Consumers (herbivores); C_2 = Secondary Consumers; C_3 = Tertiary Consumers.

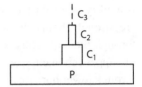

Organisms that obtain their nutrients from plants by the same number of steps along a food chain are said to belong to the same trophic level. Green plants are the first level, herbivores the second, and so on. Trophic structure is often graphically represented as an ecological pyramid, with producers comprising the base of the pyramid and successive trophic levels making up the tiers. Pyramids may be based upon the total number of organisms in each level, their biomass, or their energy content. One such pyramid is shown in Figure 4–20. Humans are generally at the apex of their ecological pyramid, that is, at the end of the food chain. Thus, we are very dependent upon the natural environment for our nutrient needs.

Transport via food chains is an important route by which chemical contaminants may reach us. In addition, damage to the environment may also occur along the transport chain. Contamination of food chains may occur in the aquatic or terrestrial environments. There is, however, never a sharp delineation between these two, and transport of chemicals often occurs between them. For example, carnivorous birds may consume contaminated marine organisms, which then deposit their droppings on land. Humans, while terrestrial organisms, often consume marine creatures directly.

In aquatic environments, contaminants may become adsorbed on solids suspended in the water. The solids may eventually settle to the bottom, deposit on the surfaces of aquatic plants, or be directly taken up by filter feeding marine organisms. Bottom sediments receive the settling remains of dead organisms or deposits of contaminated excreta. Dissolved chemicals may also be directly taken up by plants or animals.

The sediments often provide the organic substrate that supports the benthic organisms. Contaminants taken up by the latter may then be passed to higher trophic levels, and then back to the sediments, or possibly to humans via their consumption of aquatic biota. This chain of contaminant uptake bypasses the primary trophic level, since chemicals may directly enter the food chain at a consumer level.

As in aquatic environments, contaminants generally enter terrestrial food chains through the producers. Incorporation into green plants may occur by root uptake from the soil or by absorption following deposition upon foliar surfaces. Uptake from soil depends both upon the amount of time the contaminant remains within the root zone and whether it is in a form usable by the plant, while the structure of a plant is a prime determinant of its ability to act as an efficient trap for contaminant deposition on surfaces.

In some biota, the concentrations of particular chemicals may be greater than those in the surrounding environment. The process by which this occurs is called biological magnification or bioconcentration. The degree of increase is often indexed by a concentration factor; this is the ratio of the contaminant concentration in the organism to that in its environment. The effects of bioconcentration may lead to reduced populations among members of a food chain. However, when the acute toxicity of a contaminant is low, harmful effects may be delayed until chronic symptoms develop. By this time, damage to the biota may persist long after emissions are curtailed. Bioconcentration poses the greatest threat to those consumers, such as humans, that are at the upper trophic levels. On the other hand, in some cases, a process of biological exclusion may occur; that is, an organism actually has a lower concentration of a contaminant than is present in its environment. The following sections provide some notable examples of contaminant transport in the biosphere.

Radionuclides

Most of the basic chemical elements required by plants are obtained from the soil in terrestrial environments or from the water and sediments in aquatic environments. As these nutrients are taken up, so are contaminants. This is especially true of many radionuclides. The property of radiation itself does not usually affect the uptake of an element, since a biological system will incorporate material based solely upon its chemical properties, without discrimination between radioactive and nonradioactive isotopes. The technology available for isotope identification and quantitation has made it possible to do tracer studies with radionuclides, providing some of the best available information on food-chain transport of contaminants.

Studies in the Pacific Ocean atolls performed following nuclear testing in the 1950s showed that a difference existed between the types of radionuclides that enter aquatic and terrestrial food chains. Marine organisms contained the highest amounts of radionuclides (e.g., ^{54}Mn, ^{59}Fe, ^{65}Zn, ^{30}Co) that either formed strong complexes with organic matter or that occurred in particulate or colloidal form (e.g., ^{95}Zr, ^{106}Rh, ^{144}Ce, ^{144}Pr). Soluble fission products, for example, ^{90}Sr and ^{137}Cs, were found in the highest concentrations in land biota.[4]

In general, the relative proportions of radionuclides that enter food chains, as well as their concentration factors, are greater in nutrient-poor environments. For example, concentration factors tend to be greater in fresh water than in sea water, since the former usually contains lower concentrations of mineral nutrients. Concentration factors in aquatic environments, however, tend to be greater than in terrestrial environments, largely due to more rapid nutrient cycling.

A potential danger to humans is posed by the selective concentration by food crop plants of radionuclides from the soil. In one analysis,[4] for example, relative concentration factor values (ppm_w. in dry plant material/ppm_w in dry soil) of particular elements were examined and found to be 10 to 1000 (strongly concentrated) for K, Rb, P, Na, Li; 1–100 (slightly concentrated) for Mg, Ca, Sr, B, Zn;

TABLE 4–4. Typical Concentration Factors of Elements for Various Classes of Aquatic Organisms

BIOTA	CF (CONCENTRATION ORGANISM/CONCENTRATION WATER)					
	Co	Cs	I	K	Sr	Zn
Fresh-water plants	6760	907	69	—	200	3155
Fresh-water molluscs	32,408	—	320	—	—	33,544
Fresh-water crustacea	—	—	—	—	—	1,800
Fresh-water fish	1615	3,608	9	4,400	14	1,744
Marine plants	553	51	1,065	13	21	900
Marine molluscs	166	15	5,010	8	1.7	47,000
Marine crustacea	1700	18	31	12	0.6	5,300
Marine fish	650	48	10	16	0.43	3,400

Source: Eisenbud, M. *Environmental Radioactivity,* 4th ed. New York: Academic Press, 1997.

and smaller than 0.01 (strongly excluded) for Pb, Pu, Zr, Y. Table 4–4 presents average concentration factors of some elements for the main groups of edible organisms in aquatic environments.

A classic example of food chain concentration involved the Finnish Laplanders and Alaskan Eskimos. The matlike vegetation of the tundra provided a highly efficient collection surface for fallout. Caribou, which feed upon this vegetation, are a staple in the diet of these people. The Eskimos were found to have two to three times the level of ^{137}Cs present in the caribou.[5]

The potential danger to humans of the bioconcentration of any contaminant is dependent upon whether the particular site of sequestration in the plant or animal is eaten. To use radionuclides as the example, ^{90}Sr is concentrated in shells and bones, while ^{60}Co and ^{137}Cs concentrate in the edible tissues of organisms. In addition, other factors also affect the amounts eaten by humans. Food processing and cooking may result in the reduction of the amounts of many contaminants from levels found in raw food.

Pesticides

Together with radionuclides, pesticides have provided good examples of contaminant transport along food chains, with the best examples of pesticide transport provided by DDT. Although pesticide residues may reach humans directly via the water supply, the greatest danger to both humans and the environment is due to bioconcentration along food chains in aquatic environments.

Because it is fat-soluble, DDT is concentrated in the fatty tissue at each link of the food chain. Residues are passed to large game fish and carnivorous birds. For example, a Long Island marsh had been sprayed with DDT for 20 years to control mosquitoes; plankton were found to contain 0.04 ppm_w, minnows 2 ppm_w and carnivorous gulls 75 ppm_w (whole body, wet weight) in their tissues, and eastern US oysters were found to concentrate DDT up to as much as 70,000 times.[6]

Polychlorinated Biphenyls, Dioxin, and Dioxin-Like Chemicals

PCBs and dioxins undergo bioconcentration along the aquatic food chain; they are concentrated in fatty tissue because of their high lipid solubility. Shrimp exposed to 10 ppb_w for only 48 hours accumulated 1300 ppb_w[7]; large concentrations subsequently occurred in fish and carnivorous birds. Peregrine falcons from the coast of California were found to have up to 2000 ppm_w of PCBs in their fatty tissue.[7] With the imposition on controls of PCBs and dioxin and related compounds in recent decades, the levels of these polychlorinated compounds in the environment and in human tissues has declined markedly.[8]

Mercury

Mercury (Hg) compounds can undergo bioconcentration in aquatic food chains by up to as much as 10,000 times the concentration in the sea water. The greatest degree of concentration of Hg occurs for its alkyl compounds, such as methyl mercury. In one stream in central Sweden, for example, water was found to contain 0.13 $\mu g/gm$ H_2O, while a 1-kilogram pike had 300 $\mu g/gm$ body weight. Some degree of bioconcentration may occur in terrestrial food chains. For example, some seed-eating birds were found to have concentration factors of Hg in their livers of only 2 to 3; however, the factors for predatory birds were in the 100s or 1000s.[9]

REFERENCES

1. Stewart, N. G. et al. World-Wide Deposition of Long-Lived Fission Products from Nuclear Test Explosions. U.K. Atomic Energy Auth. Res. Grp. Rep. 1957; MPIR 2354.
2. Parker, F. L., Schmidt, G. D., Cottrell, W. B., and Mann, L. A. Dispersion of radio-contaminants in an estuary. *Health Phys.* 1961; 6:66–85.
3. Koczy, F. F. *The Distribution of Elements in the Sea. Proc. Disposal of Radioactive Wastes.* 1960; IAEA, Vienna, 1960.
4. Menzel, R.G. Soil-plant relationships of radioactive elements. *Health Phys.* 1965; 11:1315–32.
5. Miettinen, J.K. Enrichment of Radioactivity by Arctic Ecosystems in Finnish Lapland. Radioecol. Proc. Nat. Symp. 2nd. 1969. USAEC CONF-670503.
6. Pimental, D. Evolutionary and environmental impact of pesticides. *Bio. Sci.* 1971; 21:109–130.
7. Gustafson, C.G. PCB's- prevalent and persistent. *Env. Sci. Technol.* 1970; 4:814–9.
8. Winters, D.L. et al. Trends in dioxin and PCB concentrations in meat samples from several decades of the 20th century. *Organohalogen Compounds.* 1998; 38:75–87.
9. Borg, K., Wanntrop H., Erne, K., and Hanko, E. Alkyl mercury poisoning in terrestrial Swedish wildlife. *Viltrevy.* 1969; 6:301–379.

5

Fate of Chemicals: Translocation, Transformations, and Sinks

Following their release, chemical contaminants may be converted to different forms and/or transferred within and between the various compartments of the environment (i.e., the atmosphere, the hydrosphere, the lithosphere, and the biosphere). The transformation, degradation, and sequestration of chemicals in the environment may occur via three processes: chemical, e.g., atmospheric oxidation by O_2, and its photochemical reaction products; biological, degradation due to the metabolic action of saprophytic organisms, e.g., bacteria, fungi, and protozoa, which occurs mainly in soil and aquatic sediments; and physical, e.g., solubility and gravitational settlement.

The ultimate retention site for a chemical contaminant, known as a sink, and the removal mechanisms are quite variable. Ideally, the distribution of a contaminant within the environment would be determined by the dynamics among all of its sources, intermediate forms, and sinks. However, many chemicals are not in a steady state, largely because their source strengths vary too rapidly due to economic drivers, seasonal cycles, or climatic and technological changes.

The time between a contaminant release and any transformations or entrapment depends upon the physical and chemical characteristics of the material, characteristics of the environmental compartment into which it is released, and the degree to which it migrates among compartments. There may be several time constants for retention within the various component regions of each compartment. For example, within the atmosphere, the average residence times, that is, the average time a contaminant remains in the atmosphere, may be very different in the lower troposphere (boundary layer), the balance of the troposphere, and the stratosphere. In the biosphere, where contaminants interact among numerous organisms, the complications increase enormously.

AIR CONTAMINANTS

Contaminants emitted into the atmosphere are removed by various natural mechanisms, either while in the primary form in which they were originally released or following their physicochemical conversions into other forms. To a large extent, the fate of an airborne chemical depends upon whether it is in the particulate or gaseous phase.

Particles

Airborne particles are removed primarily via natural processes, often after undergoing alterations in size, number, or chemical composition. The principal processes are gravitational settlement, impaction on and/or interception by earth surface objects, and rainout and washout. Certain of the mechanisms are more effective for particles of one size range than for those of another and for particles within specific regions of the atmosphere.

Gravitational settlement, or sedimentation, increases with the aerodynamic diameter of the particle and is important for those with aerodynamic diameters greater than approximately 20 μm. Below this size, the particle's settling velocity is insignificant compared with its vertical displacement by motions of local convective air currents.

Impaction and interception onto surfaces, for example, buildings, vegetation, and so on, are often effective removal processes for particles suspended near ground level; these particles may be transported to the surface by atmospheric convection. Impaction occurs when the momentum of a particle moves it across flow streamlines, causing it to collide with a surface. Interception refers to contact with a surface because the flow streamline coincident with the center of the particle is less than one particle radius away from the surface. The efficiency of removal by these processes is dependent upon such factors as the rate at which the particles approach the surface and the collection efficiency of the surface. The former is affected by wind velocity and by the general stability of the atmosphere and the latter by the shape, size, texture, and degree of wetness of the deposition surface. Submicrometer particles, which have high diffusivity, may be removed via diffusion to ground-level surfaces.

Sedimentation and surface impaction, which are also known as dry deposition, are the principal removal mechanisms in the lower troposphere. At altitudes above 100 m, the primary removal mechanisms are the precipitation scavenging processes of rainout and washout, collectively, known as wet deposition.

"Rainout" refers to the scavenging and incorporation of contaminants into aqueous droplets within clouds, and their subsequent removal via precipitation, while "washout" refers to the incorporation of the chemical into raindrops or snowflakes below the clouds (i.e., direct removal by collision with falling precipitation). The washout mechanism is effective for particles greater than approximately 2 μm in diameter, while rainout is most effective for those with smaller diameters.

Physical processes that occur in the atmosphere may affect removal efficiency. Hygroscopic particles may increase in size due to the accumulation of atmospheric water vapor, and this growth may enhance their removal by sedimentation and washout. As an aerosol ages, the constituent particles may grow by coagulation with other airborne particles, forming larger particles that may have more appreciable settling velocities. The primary mechanism responsible for coagulation is Brownian motion, and it results in the formation of droplets and solid aggregate particles up to about 0.7 μm in diameter. Particles with diameters greater than 0.3 to 0.5 μm have small diffusivities and act as acceptors of the more rapidly diffusing, smaller particles. Particle growth may also occur due to electrostatic forces. Very small particles, those below 0.1 μm, are referred to as ultrafine particles. They can act as condensation nuclei, which induce the deposition of vapors onto particle surfaces; these particles subsequently coagulate with other particles and are thereby eventually removed from the atmosphere by rainout, washout, or dry deposition.

Because of the large variation in the particle sizes of atmospheric particulate matter (PM), and the dependence of removal mechanisms upon size, no precise values for particle residence times in the troposphere are possible. Furthermore, residence times are highly dependent upon the amount of precipitation in any particular area. Nevertheless, average residence times in the lower troposphere range from approximately five days to two weeks, while particles in the upper troposphere may remain suspended for as long as 30 days. Particle residence times in the lower stratosphere range from four months to one year and for those in the upper stratosphere from one to five years. Stratospheric residence time depends upon latitude, with the shortest times occurring over the poles and the longest over the equator.

Gases and Vapors

The main processes for removal of contaminant gases and vapors from the troposphere are precipitation scavenging, with or without chemical reaction; absorption by or reaction on earth-surface objects; and chemical reactions in the atmosphere and conversion to other gases or particles. In the last case, the level of one contaminant may decline while that of another increases. Another removal mechanism is escape through the tropopause and subsequent destruction in the stratosphere; the stratospheric sink may be significant for gases with long tropospheric residence times. The high-energy, short-wavelength radiation in the stratosphere degrades many molecules by disruption of their chemical bonds.

The precipitant scavenging of gaseous contaminants is not as well understood as is that for particles. Whereas particle scavenging may be thought of as a collisional phenomenon, gaseous scavenging is a diffusional phenomenon, which is dependent upon the concentration gradients of the particular gas across a liquid-air interface. Very often, chemical reactions occur following incorporation of the gas into a water droplet.

Absorption and/or reaction at the surface of the earth provides a sink for some gaseous contaminants. As with surface impaction of particles, the efficiency of surface absorption of gases is dependent upon factors such as meteorologic conditions, which affect the rate at which a gaseous molecule moves toward a surface; properties of the molecule; and the nature of the deposition surface. Certain surfaces are excellent sinks for specific gases, irreversibly absorbing them. Examples include uptake of sulfur dioxide (SO_2) by vegetation and some masonry, ozone (O_3) by rubber, and hydrogen sulfide (H_2S) by lead-(Pb-) based paint.

Atmospheric chemical reactions are an important removal process for many gaseous contaminants in the troposphere by producing stable gaseous end-products or a particulate form that is more easily removed. Some particles may accelerate atmospheric reactions by providing surface catalysis sites for both simple and complex reactions or by sorption of gases from dilute gaseous concentrations, thereby providing discrete sites of higher concentration.

The relative importance of any particular removal mechanism for gases depends upon the specific chemical and environmental conditions. Removal can be described in terms of sink strength, the rate at which the contaminant is removed from the atmosphere via a particular route. There is generally a broad range of sink strengths reported in the literature, largely because their estimation is quite difficult. Rates of removal by various routes are very often extrapolated from limited studies or from models where contaminant levels are unrealistically high. Thus, global sink strength estimates are only rough numbers. The following sections discuss sinks for major gaseous air contaminants.

Nitrogen-containing gases and vapors

The main gaseous compounds of nitrogen in the atmosphere are N_2O (nitrous oxide), NO (nitric oxide), NO_2 (nitrogen dioxide), and NH_3 (ammonia). NO_2 is in equilibrium with its dimer nitrogen tetroxide (N_2O_4). Estimates of tropospheric

TABLE 5–1. Mean Tropospheric Residence Times for Nitrogen Gases

CHEMICAL	RESIDENCE TIME
N_2O	4 years
NO	4–5 days
NO_2	3–5 days
NH_3	7–14 days

residence times for gaseous N compounds are given in Table 5–1. Sink strength estimates are presented in Table 5–2.

The most abundant N compound in the troposphere is N_2O. However, this gas is emitted almost totally by natural sources, is very stable chemically, and thus plays an insignificant role in low-level air pollution chemical reactions. Most of the N_2O is removed via its only known atmospheric reaction, which is photolysis occurring in the upper troposphere and stratosphere and resulting in the production of NO and N_2. Some N_2O is also removed from the lower troposphere by soils, plants, and bodies of water.

NO is removed largely via oxidation to NO_2. Although this may occur by reaction with O_2, O_3 oxidizes NO much more rapidly. For example, at atmospheric concentrations of ~1 ppm of NO, approximately 100 hours are necessary for conversion of 50% of the NO to NO_2; with O_3, the half-life of NO is only 1.8 sec. In

TABLE 5–2. Sink Strengths for Nitrogen Compounds

SINK	SINK STRENGTH (10^{10} KGN/YEAR)
N_2O, NO, NO_2	
Rainout/washout	2–7.5
Dry deposition	1.9–7
Stratosphere	0.03–0.2
Gaseous deposition on surfaces	4.5
NH_3	
Rainout/washout	3–19
Dry deposition	5–7
Gaseous deposition	75
Oxidation in troposphere	7
Stratosphere	0.04

the upper troposphere and stratosphere, photolysis of NO results in the production of atomic nitrogen (N), which may then form the nitrogen dimmer (N_2) following reaction with other molecules of NO.

The primary mechanism for removal of NO_2 is dissolution in cloud and rain droplets. Although numerous schemes have been proposed for subsequent reactions of NO_2, the end result is always production of nitrous acid (HNO_2) and nitric acid (HNO_3), which may then be converted to nitrite and nitrate salt aerosols by reaction with NH_3 or other atmospheric bases or become adsorbed onto other particles. These acid or salt particles are then removed by wet or dry deposition mechanisms. Some NO_2 may react with O_3, also producing particulate nitrites and nitrates. Both HNO_2 and HNO_3 may be photochemically decomposed, resulting in the reformation of NO and NO_2. Minor sinks for NO_2 and NO are vegetation and soils, the latter partly by biological means. However, the NO_2 that is absorbed by soils is eventually oxidized to nitrate, which, upon oxidative decomposition, produces NO_2 once again.

NO_2 is important because of its involvement in atmospheric photochemical reactions. It absorbs solar energy over the entire visible and UV ranges of the spectrum in the troposphere. At wavelengths less than 0.38 μm, NO_2 is photodissociated into NO and O; the latter reacts with O_2 to produce O_3. The photochemical reactions of NO_2 were discussed in detail in chapter 3. In the stratosphere, NO and NO_2 form nitrogen pentoxide (N_2O_5), which combines with water vapor to produce HNO_3 vapor.

Although most sources of NH_3 are natural, it plays a major role in the formation of PM in the atmosphere. In fact, the main sink for NH_3 is aerosol formation. This occurs by reaction with acids or acid-forming oxides in the gaseous phase or after dissolution in rain-water; the latter process also enhances the oxidation of dissolved SO_2 and NO_2. The main aerosol species produced are ammonium bisulfate (NH_4HSO_4), ammonium sulfate ($[NH_4]_2SO_4$), and ammonium nitrate (NH_4NO_3). On the other hand, because of its high water solubility, some NH_3 vapor may be removed from the troposphere by absorption onto wet surfaces such as vegetation, bodies of water, and soils. Some ammonia may also be removed by reaction with hydroxyl radicals to form NO.

Carbon-containing gases

The main carbon- (C) containing gases in the atmosphere are carbon monoxide (CO) and carbon dioxide (CO_2). With the exception of CO_2, more CO is released into the atmosphere than any other contaminant. The major removal pathways of CO are via the soil and gas-phase reactions in the troposphere and stratosphere.

Many soil types have been found to be able to remove CO from the atmosphere, releasing CO_2, likely the result of biological activity. The greatest activity is shown by organically rich tropical soils, while organically poor desert soils have the least.

TABLE 5–3. Sink Strengths for CO

SINK	SINK STRENGTH (10^9 KG/YEAR)
Soil	67–1400
Gas phase oxidation in stratosphere	52–71

There are two atmospheric sinks for CO. One involves the reaction of CO with hydroxyl ($\cdot$OH) and hydroperoxyl ($\cdot$OOH) radicals in the troposphere, producing CO_2 and H atoms. The other involves reaction with hydroxyl radicals in the stratosphere, following CO migration through the tropopause. The importance of the stratosphere as a sink is dependent upon ambient CO levels, the rate of transport into the stratosphere, and hydroxyl levels in the stratosphere. Sink strengths for CO are presented in Table 5–3.

Approximately 30% to 50% of the CO_2 released into the atmosphere, largely due to combustion of fossil fuels, stays within the atmosphere. The remaining 50% to 70% enters sinks in the hydrosphere and biosphere. Green plants may be a temporary sink for CO_2 in the biosphere. They take it up in photosynthesis, but the CO_2 ultimately returns to the atmosphere during respiration and upon oxidative decomposition of dead plants. The largest natural influence upon CO_2 levels in the atmosphere is the exchange with the oceans, which contain approximately 60 times as much CO_2 as does the atmosphere. The average residence time of CO_2 in the atmosphere before it transfers to ocean waters is around 10 years, but the actual approach to equilibrium levels between atmosphere and oceans may be much slower, inasmuch as atmospheric CO_2 levels have been increasing, reaching 400 ppm_v in 2016.

CO_2 is destroyed by photochemical decomposition in the mesosphere, the region above the stratosphere that is approximately 30 to 50 miles above the earth's surface.

Sulfur-containing gases and vapors

The primary gaseous compounds of sulfur (S) present in the atmosphere are hydrogen sulfide (H_2S) and SO_2. Another S species of importance is sulfur trioxide (SO_3).

H_2S is produced primarily from natural sources and is not an important general air contaminant in terms of anthropogenic emissions. Its only significant mode of removal from the atmosphere is oxidation to SO_2. Most of this oxidation of H_2S occurs by reactions with O_3, although reaction with atomic O may be significant in the stratosphere and under conditions of photochemical smog, where significant quantities of atomic O are produced by the photolysis of O_3. The reaction system in this latter case results in the production not only of SO_2 but also of SO_3 and H_2SO_4. Tropospheric residence times for H_2S range from hours to days.

The sinks for SO_2 are precipitant scavenging, diffusion to and absorption onto earth-surface features, and chemical conversion into other S species. Washout and rainout are the major mechanisms for removal of atmospheric SO_2, largely because SO_2 is very soluble in water. Scavenged SO_2 undergoes a series of chemical reactions, ultimately forming sulfuric acid (H_2SO_4) and its ammonium salts.

Under conditions of high relative humidity and in the presence of catalytic surfaces on particles, SO_2 may undergo oxidation. Salts of certain metals, for example, iron (Fe) and manganese (Mn), serve as the catalysts, either by acting as condensation nuclei or by undergoing hydration; both actions result in the production of liquid droplets. Both SO_2 and O_2 readily dissolve in these droplets, producing sulfite (SO_3^{-2}) and eventually strong acid droplets. The H_2SO_4 may react with other atmospheric constituents to produce ammonium (NH_4^+), sodium (Na), and calcium (Ca) salts, and numerous other sulfate species.

The presence of NH_3 in water droplets also enhances the oxidation of SO_2 in solution. As a droplet becomes highly acidic, the rate of SO_2 oxidation decreases, due to the decreased solubility of SO_2 in acid solutions. NH_3 dissolved in the droplet increases SO_2 solubility, and if sufficient levels of NH_3 are present, oxidation is not impeded by accumulation of H_2SO_4, since conversion to $(NH_4)_2SO_4$ occurs. Calcite dust and other airborne alkalis other than NH_3 are also involved in the rapid oxidation of SO_2 to sulfate. The rate of catalytic oxidation of SO_2 decreases if the water concentration in the atmosphere falls below the level necessary to maintain catalyst droplets, namely ~70% relative humidity.

In clean, dry air, SO_2 is only very slowly oxidized via homogeneous (i.e. gas phase) reactions to SO_3 vapor. The most important transformation of SO_2 under low-humidity conditions is photochemical oxidation. Photochemical reactions have been discussed in chapter 3, and only a general description is presented here. In the initial event, absorption of light results in activation of a molecule of SO_2, which then proceeds to react with O_2 or O_3 at a faster rate than would unactivated SO_2 molecules. The resultant SO_3 reacts almost immediately with water vapor to produce H_2SO_4.

The rate of photochemical oxidation of SO_2 is fairly slow; estimates range from 0.006%/min to a theoretical maximum of 0.03%/min.[1] However, the rate may be greatly accelerated by the co-presence in air of certain reactive intermediate species in the photooxidation of hydrocarbons under the influence of NO_x, for example, NO and olefins.

In the stratosphere, SO_2 reacts with atomic O, resulting in the formation of SO_3 vapor, which reacts with water vapor to form H_2SO_4 vapor. The latter combines with more water to produce H_2SO_4 droplets, which are removed by precipitation.

Aside from atmospheric sink reactions, SO_2 may also be removed by other mechanisms. Some plants may absorb SO_2, and soils are also able to absorb significant amounts of SO_2, which may then be oxidized to sulfate. Other sinks are

TABLE 5–4. Sink Strengths for Sulfur Compounds

SINK	SINK STRENGTH (10^9 KG/YEAR)
Gaseous absorption, SO_2	
Vegetation	15–75
Ocean	25–100
Washout/rainout (land)	70–85
Dry deposition (land)	10–20
Wet and dry deposition, SO_4^{-2} (oceans)	70–100

absorption by the hydrosphere and uptake by carbonate stones, for example, in buildings, in moist atmospheres. Sink strengths for SO_2 are presented in Table 5–4.

Hydrocarbon gases and vapors

A broad range of gaseous hydrocarbons are emitted into the atmosphere. Certain of these may be taken up via soil microbial action. Most organic molecules are oxidized only very slowly in clean atmospheres. The mechanism involves reaction chains that are initiated by removal of a hydrogen atom from the molecule, producing a radical that combines with an O_2 molecule to form a peroxide. The latter, being reactive, may go on to remove a hydrogen from another organic molecule.

In the presence of light sensitizers or initiators, hydrocarbons may be degraded by photochemical reactions. These initiators provide the energy necessary for radical production from the organic molecule. Photochemical reactions are the primary degradation processes for reactive hydrocarbons, for example, olefins and polycyclic aromatic hydrocarbons. These molecules rapidly undergo transformations under the influence of sunlight and in the presence of atomic O, O_3, SO_2, and NO_2. The result is conversion to other organic molecules. The reaction rates for various hydrocarbons are quite different, and there are few data on the rates for many species or for the products formed.

Up to about 10% by weight of the reactive hydrocarbons emitted into the atmosphere eventually are converted to, or adsorbed onto, particles that are then removed by wet or dry deposition processes. The remainder of the hydrocarbons eventually undergo chemical degradation, becoming oxidized to other hydrocarbons and, if oxidation is complete, to CO_2 and H_2O. For substituted hydrocarbons, the noncarbon constituents are found in various salts.

Methane (CH_4) is the predominant atmospheric hydrocarbon; however, the paraffin group, of which CH_4 is the first member, is much less reactive than other hydrocarbons. The main sink for CH_4 is oxidation to form CO and CO_2, through the action of hydroxyl radicals in the troposphere.

Except for CH_4, the lack of precise estimates of emissions and ambient levels of hydrocarbons prevents an accurate estimation of atmospheric residence times. The residence time of CH_4 in the troposphere is 1.5 to 2 years; those for higher molecular–weight hydrocarbons are on the order of days to months. In the stratosphere, many hydrocarbons react with atomic O and may eventually be oxidized to CO_2 and H_2O.

Ozone

Ozone is a constituent of photochemical smog atmospheres. A major O_3 sink is the surface of the earth, primarily soils and vegetation; absorption by oceans may also occur to some extent. Ozone is a powerful oxidizing agent. It is involved, especially in contaminated atmospheres, in many reactions in atmospheres containing NO_2, such as photooxidation of hydrocarbons as previously discussed.

WATER CONTAMINANTS

Chemical transformation and physical or biological removal of contaminants from the atmosphere may be thought of as natural purification processes that produce nonreactive, both biologically and chemically, products and/or otherwise result in removal from the atmosphere of the chemical agent. Analogous self-purification processes occur in aquatic environments.

The self-purification processes by which natural waters tend to rid themselves of contaminants involve physical, chemical, and biological mechanisms that occur simultaneously and interact with each other. At one time, these processes were sufficient to prevent permanent effects on many bodies of water, but now this is not always the case. Self-purification processes in aquatic environments are quite complex, especially due to the interaction of biota. Each specific waterway, be it estuary, stream, river, lake, or ocean, has its own capacity for self-purification and recovery due to its unique combination of physical, geochemical, and biological characteristics. Nevertheless, the basic processes in all environments are essentially the same.

The resultant residence times for chemical contaminants in the hydrosphere are extremely difficult to determine because of the various types of aquatic environments and the complex physical, chemical, and biological factors that determine the fate of the contaminant; these differ to some degree in these different environments. As noted previously, the oceans often represent the ultimate sink for contaminants originally contained in flowing terrestrial waterways. Unless they are removed into the atmosphere, the ultimate fate of chemical contaminants entering the oceans is usually removal to the sea floor. Thus, the residence time in the marine environment is generally considered to be the time the contaminant remains in the ocean water between its introduction and its incorporation into the sediments.

Residence time depends upon the degree of reactivity of the specific contaminant, a factor that also determines the region in the marine environment in which the chemical reaches its sink. As a generalization, chemically reactive elements, for example, Fe and Al, and those involved in biological cycles, for example, P and N, have shorter residence times then do those contaminants that tend to be less reactive or are inert in solution, for example, alkali metal ions. The first two groups tend to become associated with PM and/or biota prevalent in the coastal ocean regions and are largely removed there by deposition processes. The latter group tends to reach the open ocean waters, where they accumulate until they eventually undergo downward transport in the waters. Contaminants that do not undergo biological degradation may have residence half-times in the ocean of thousands of years.

Physical Self-Purification

The physical mechanisms involved in self-purification of water are the processes of dilution, mixing, sorption (adsorption and absorption) and sedimentation. Dilution of chemicals occurs due to diffusion, advection (i.e., movement of one parcel of water with respect to another) and turbulence (i.e., eddy diffusion). Mixing is primarily due to turbulence, which results from temperature gradients in large bodies of water, and from bed friction in rivers and streams. In coastal areas and estuaries, the action of tides also affects the mixing of contaminants.

Sorption of chemicals may alter their behavior and often results in a purification of the water. Suspended particles are quite ubiquitous in natural waters, occurring in a wide assortment of sizes, shapes, chemical compositions, and concentrations. In terms of sorption and binding of chemicals, the colloidal particles are the most important, with clay being the most significant class of common minerals that occur as colloidal matter in natural waters. Because of their strong tendency to sorb chemicals from the water, clays may effectively immobilize dissolved chemicals, purifying the water. In addition, some microbiological degradation processes for organic wastes may also occur at the clay-particle surfaces, giving clay a role in biological purification.

A wide variety of dissolved chemical effluents may be sorbed and bound by suspended particles. This includes various organic chemicals, such as herbicides and nonvolatile organics, as well as dissolved inorganic materials, such as trace metals.

The PM, either natural or that discharged into water, along with any sorbed chemicals, may eventually be removed from suspension by sedimentation. This process depends upon the particle size, shape, and density and the mixing and flow characteristics of the waterway. In turbulent streams, suspended particles may be carried for long distances before they settle.

Sedimentation of small colloidal particles may be enhanced by aggregation, which involves the processes of coagulation and flocculation. Coagulation is the aggregation of colloidal particles of the same material. Flocculation is dependent upon the presence of bridging compounds that form chemically bonded links between the colloidal particles and enmesh the particles within large floc networks.

Movement of suspended particles into and within different water bodies may affect their sedimentation. The size of colloidal particles may change due to coagulation when they move from freshwater into saltwater; the salt ions act to reduce the normal electrostatic repulsion between the particles. This process may be accompanied by the uptake and/or loss of some of the sorbed contaminations.

Particles that deposit on the sediments may remain there, undergoing no removal processes. However, very often accumulation and precipitation processes are reversible. Bottom sediments undergo a continuous process of leaching and ion exchange with the overlying water. For example, trace metal ions sorbed on particles deposited in the sediments may be released by ion-exchange mechanisms. Sediments that move from aerobic to anaerobic environments may undergo changes in sorbed chemical species.

Nevertheless, bottom sediments are often the ultimate sinks for many contaminants. The deposition may occur in the waterway into which the contaminant discharge occurred or in downstream areas. In this regard, oceans are enormous sinks, often representing the ultimate "downstream" locale for flowing waterways. Rivers deliver billions of tons of dissolved and suspended PM each year to the oceans and seas; much of this is subsequently deposited in the sediments, often in coastal regions.

Chemical Self-Purification

Specifics of the chemical reactions of many contaminants in aquatic environments are unknown; the nature of chemicals is often influenced by biota. In addition, chemical reactions are also affected by water temperature and pH. It is beyond the scope of this book to discuss the fate of specific chemicals in water. However, certain general types of chemical reactions are often involved in the self-purification of inorganic contaminants, and these are presented here. Purification of many organic contaminants involves biological processes, and these are discussed in a subsequent section. Furthermore, in aquatic chemistry, interactions occurring solely in solution are of less significance than those that occur at solid–water and gas–water interfaces. In this regard, colloidal solids are of great significance in chemical interactions in the aquatic environment.

Acid–base reactions are important in the assimilation of acid and alkaline wastes discharged into water. Acids may react with numerous minerals, such as those in crystalline rocks, clays and other silicates, and limestone and other carbonate rocks. The capacity of a waterway to neutralize acid is primarily a function of its content of bicarbonate, carbonate, and hydroxide ions. Alkaline contaminants are purified mainly via reactions with silica, bicarbonate, and free carbonic acid in the water, resulting in the production of silicate and carbonate salts. The rates of acid–base reactions differ in different aquatic systems and are important determinants of the area of the waterway over which unassimilated acids and alkaline wastes traverse.

Nonbiologically mediated oxidation-reduction reactions are also of great significance in the purification of aquatic environments; many hydrocarbons and heavy metals undergo these types of reactions.

Metal ions may be removed from solution by the formation of insoluble complexes known as coordination compounds. Some naturally occurring organic complexing agents, for example, humic and fulvic acid, are found in certain aquatic environments, and these agents strongly bind heavy metals. Synthetic chelating agents (e.g., nitrilotriacetic acid [NTA] and ethylenediaminetetraacetic acid [EDTA]) are used in detergents and industrial processes and are often discharged into surface waters where they may form complexes with metals. On the other hand, metal ions may also be solubilized by formation of complexes.

Some dissolved chemicals may be precipitated under the influence of other agents in the water (coprecipitation), while some may leave by volatation. Others may be hydrolyzed in the aquatic environment. For example, hydrolysis of esters results in products that may then be degraded by biological processes. Some organic chemicals may undergo light-induced transformation reactions in water.

Biological Self-Purification

Although some chemicals may undergo a limited number of reactions that may serve to remove them from the water others, for example, nitrate (NO_3^-), undergo few reactions without the action of microorganisms. Biological purification processes generally involve the assimilation of putrescible organic wastes entering waterways. The process involves the transformation of the chemicals into innocuous products due to action of aquatic biota, primarily microorganisms; it usually occurs at the expense of dissolved oxygen (DO) and of some species that are normally present in the biota of the uncontaminated waterway.

Because of the multitude of microbial types that exist in aquatic environments, the overall process is quite complex. However, most aquatic environments show, to some extent, common biological reactions, which may vary due to differences in flow and mixing characteristics, and according to the concentration and composition of the specific contaminant.

As an example of biological self-purification, consider the case of a discharge of organic matter into a stream. The discharge initially results in an increase in the biochemical oxygen demand (BOD); the levels of DO in the stream below the point of discharge therefore decrease. Depending upon the initial waste load, a certain time and distance along the flow path downstream are necessary before stream reaeration processes return the DO to near air-saturation levels. By considering factors of BOD loading, reaeration processes and rate of flow, an oxygen sag curve may be developed (Fig. 5–1).

Suspended organic solids eventually settle out to the bottom of the waterway. These may be subject to microbial decomposition, although the processes differ from those occurring in flowing water. Decomposition in bottom sediments varies, with the depth of deposit, from aerobic to anaerobic.

Dissolved solids are also removed or diluted as the stream flows. As various nutrient cycles occur, organic compounds are broken down, and levels of inorganic nutrients, such as ammonia, nitrate, and phosphate, are increased. This results in changes in biota, since species differ in their ability to tolerate increased nutrients; some are favored and others are disadvantaged. Bacteria and fungi predominate in the zone where DO becomes a severely limiting factor. Further downstream, these decline and algae may bloom, benefitting from

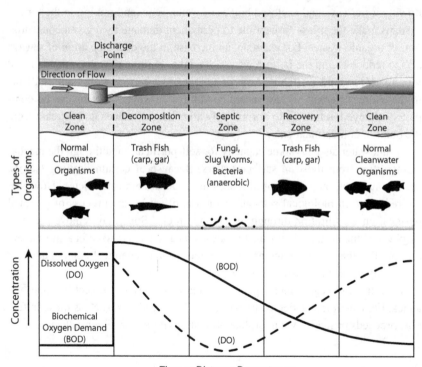

Time or Distance Downstream

FIGURE 5–1. Dissolved oxygen (-_ _-_ _-_) versus oxygen demand curves (——) following pollutant release into a body of water. The rate of recovery depends on flow rates and the pollutant load.

the increase in mineral nutrients. Eventually, these also decline. Low DO levels cause reduction in the number of species of larger invertebrates; highly specialized sewage organisms, for example, tubificid worms, midge larvae, and other undesirable forms that can tolerate these conditions, increase in number due to the large supply of food and little interspecies competition. Increasing dilution occurs with flow downstream and, as the river DO and minerals return to normal, the number of species able to survive increases and the strong numerical predominance of particular species tends to disappear. A gradual return to the community found upstream of the discharge occurs. The water has purified itself and recovered from the discharge. The entire scheme of biological purification is shown in Figure 5–1.

Natural microbial purification is not a fast process, and heavily contaminated streams may have to travel fairly long distances before any significant degree of purification occurs. Exact rates are quite difficult to estimate because of the compositional and mixing complexities of most waterways and of the interactions of biological, chemical, and physical processes.

There are generally significant differences in biological reactions between flowing waterways and enclosed bodies of water. Flow and mixing in the former systems make them less susceptible to permanent damage by a given concentration of organic wastes. For example, an increase in the decomposition of sewage acts to reduce DO in the bottom waters of a lake. In flowing waters, the constant mixing acts to hinder establishment of such distinct surface-to-bottom gradients in the entire waterway. Under severe conditions, however, streams can be completely changed, returning to normal only after long distances downstream from the point of contaminant entry.

Groundwater also has some capacity for self-purification, although the mechanisms differ from those in surface waters. Because of certain limiting factors, such as sunlight, the variety of microbes in groundwaters is restricted. However, the reduction in biological self-purification is offset by the process of physical purification via filtration through the soil and rocks. Soils may remove particles by physical means and chemicals by various means, as discussed in a subsequent section. But there is a limited ability to remove many chemical compounds that commonly occur in industrial wastewaters.

Aside from sewage, other organic contaminants may be subject to microbial attack. For example, the degradation of many hydrocarbons, such as those in fuel oils, proceeds primarily via microbial utilization of the constituent alkanes.

SOIL CONTAMINANTS

Terrestrial soil is a sink for numerous chemicals. The fate and persistence of chemicals in soil is a complex function of physical, chemical, and biological factors; the main ones are the specific chemical and its formulation, soil type and specific physicochemical properties, type of cover vegetation, degree of soil cultivation, and nature of the microbial population. The major routes of removal of contaminants from soil are degradation via microbial action, which is discussed in the next section; chemical degradation, for example, hydrolysis; evaporation and volatilization from the surface; and uptake by vegetation. Most contaminants are not effectively removed by leaching.

Uptake by plants may only be a temporary sink, since the contaminant may return to the soil in plant litter. Temporary removal may also be afforded by microbial assimilation without degradation, which may immobilize the contaminant until the death of the microorganism. Chemicals may also be immobilized in soil via various adsorption processes.

Colloidal soil particles can adsorb various chemicals. Neutral organic molecules, for example, may be adsorbed by physical mechanisms, such as via van der Waals forces. Polar or polarizable organic ions, especially cations, and inorganic metallic ions, such as Pb^{+2}, Hg^{+2}, and $^{90}Sr^{+2}$, may be adsorbed via ion exchange. Other processes implicated in adsorption of contaminants, particularly pesticides, are hydrogen bonding and coordination, the latter being formation of a complex between the chemical and an exchangeable ion on the soil particle. Some metals, such as copper and zinc, may become tightly bound to soil organic matter by chelation. Other soluble inorganic chemicals may precipitate out of solution in soil water as oxides or hydroxides. Thus, a wide variety of immobilization processes may involve all types of chemicals. Immobilization affects the subsequent fate of the contaminant in the soil in any number of ways. It may hinder volatilization and leaching, act to catalyze chemical degradation, prevent biological degradation by sequestering the contaminant from the microbes, and enhance biological decay by concentrating the chemical near microbes.

Specific components of certain soils may affect the fate of contaminants. For example, part of the humic fraction in many agricultural soils is composed of lipids; contaminant-lipid interactions in these soils often serve to immobilize fat-soluble pesticides.

There are few reliable data on the residence times of chemicals in soils, since these depend upon so many physicochemical and biological factors. Type of soil is, of course, a major factor. For example, many pesticides tend to persist longer in soils with higher organic matter than in sandy soils, since the former tend to bind the residues much tighter.

CONTAMINANTS IN THE BIOSPHERE

The discussion in this chapter has so far concentrated on the physicochemical processes by which contaminants are removed from air, water, and soil environments.

In this section, microbially mediated transformations and their relation to the bio-degradability of chemicals is discussed.

Biogeochemical Cycles

The biological cycles for C, N, and S are discussed in the following sections. These cycles are not independent of each other but, as with most environmental processes, they interact.

Carbon cycle

There are two main reservoirs for inorganic C: the atmosphere and the hydro-sphere. In both, C naturally exists in the form of CO_2; interchange occurs via

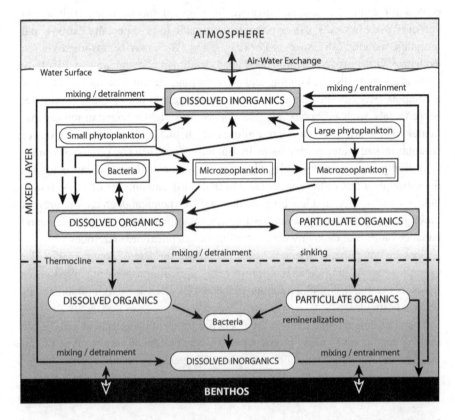

FIGURE 5–2. The carbon cycle in ocean water. Biologically mediated transformations of C begin with small plankton, which primarily recycle CO_2 within the euphotic zone, and the larger plankton, which generate most of the flux of organic carbon (OC) in particulate and dissolved forms from the deep ocean. Some of the carbon that reaches the deep ocean is re-mineralized into dissolved inorganic form, some is consumed at the benthic surface, and a small portion is buried in the sediments of the ocean floor.

diffusion, evaporation, and precipitation. The carbon cycle in oceans is shown in Figure 5–2.

Atmospheric or hydrospheric CO_2 is fixed into organic carbon as biomass via photosynthesis, primarily by green plants on land and by phytoplankton in the sea. The organic carbon then moves through various trophic levels and is eventually released back to the reservoir as CO_2 via the respiratory activity of plants and animals, in the processing of waste products and remains of dead biota by decomposers, and via the process of combustion. Some C is trapped in deposits of the remains of plants and animals and becomes peat, coal, and oil; some in the shells of aquatic organisms; and some in formation of carbonate rocks in the oceans. The weathering of these rocks and the combustion of oil and coal release this formerly trapped C into the cycle once again.

Nitrogen cycle

The N cycle is shown in Figure 5–3. Although the nitrogen dimmer (N_2) is the dominant gas in the atmosphere, very few living organisms are able to use

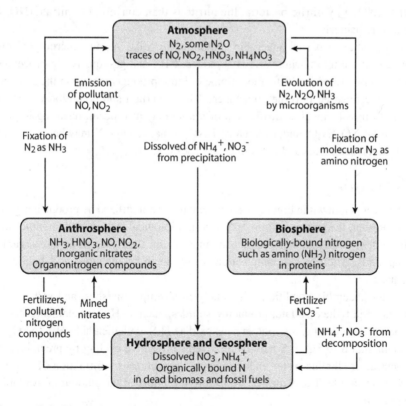

FIGURE 5–3. The environmental nitrogen cycle.

gaseous nitrogen. Rather, N must first be fixed into an inorganic compound, primarily nitrate, for utilization by the producers in biological processes. The primary mechanism for fixation is biological. Various groups of free-living aerobic and anaerobic microorganisms, primarily bacteria, and also certain symbiotic microorganisms, for example, the root-nodule bacteria of legumes, are able to fix atmospheric N_2. Fixation occurs primarily in terrestrial ecosystems, although certain organisms in aquatic ecosystems, for example, blue-green algae, are also capable of N_2 fixation. Physicochemical processes in the atmosphere may also result in fixation; these include electrification (lightning) and cosmic radiation.

The N contained within the bodies of plants and animals exists as ammonium ions, in amino-compounds, such as proteins, and in nucleic acids. Mineralization of this nitrogen occurs via the processes of ammonification and nitrification. In the process of ammonification, numerous aerobic and anaerobic microorganisms metabolize bound nitrogen, releasing NH_3 gas or ammonium (NH_4^+) compounds. The subsequent conversion of the NH_3 and NH_4^+ salts back to nitrate (NO_3^-) is called nitrification. This occurs in two stages and involves two classes of aerobic bacteria. The first step is oxidation of NH_3 to nitrite (NO_2^-) by nitrite bacteria; the nitrite is then converted to nitrate (NO_3^-) by nitrate bacteria.

Under total or partial anaerobic conditions in soils, various microorganisms convert nitrate into nitrite, various N oxides, NH_3, and gaseous N_2, in processes collectively termed denitrification. Some of these products remain in the soil and others are released into the atmosphere. Under certain soil conditions, such as high acidity, chemical denitrification of nitrate may also occur, resulting in production of NO_2, N_2O, and, eventually, HNO_3. Finally, some N may be trapped in deep ocean sediments and in sedimentary rocks.

Sulfur cycle

The S cycle is shown in Figure 5–4. Elemental S is not utilized by producer organisms. Rather, the principal form of S used in biological systems is inorganic sulfate (SO_4^{-2}), although some organisms may obtain their S from certain organic compounds. Microbial decomposition results in mineralization of S bound in biomass.

Under anaerobic conditions, especially in swamps, marshes, and soils, various bacteria reduce sulfate, producing sulfides, such as H_2S, or elemental S. In aerobic environments, microorganisms oxidize H_2S to elemental S, H_2S to SO_4^{-2}, or elemental S to SO_4^{-2}. S may be removed from active cycling by precipitation in neutral or alkaline water under anaerobic conditions by formation of ferrous and ferric sulfide. The S may be trapped to the limits of the amount of available Fe present.

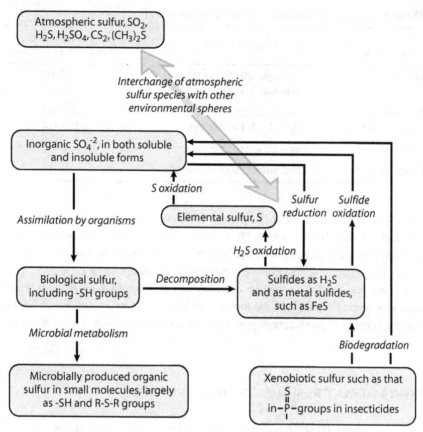

FIGURE 5-4. The environmental sulfur cycle.

Biodegradability of Contaminants

It should be clear from the previous discussion that the ultimate mineralization and degradation of a chemical in any biogeochemical cycle is dependent upon micro-organisms, primarily bacteria. Microbial metabolism is, therefore, responsible for the mobility, or lack of it, of chemicals in the biosphere. These microbes purify streams into which organic contaminants are dumped, degrade pesticides in soil, decompose sewage sludge, and also have a role in the degradation of oil spilled into waterways. Contaminants that may be degraded by microbes are termed bio-degradable. Although these chemicals are primarily organic in nature, they yield inorganic end products. It should, however, be mentioned that microbial conversion may result in products that are more hazardous than was the original material.

On the other hand, certain chemicals released into the environment are resistant to microbial decomposition or else are degraded at such slow rates that they build up in the environment. These types of contaminants are termed nonbiodegradable,

TABLE 5–5. Residence Times of Selected Pesticides in Soil

DESIGNATION	CHEMICAL GROUP	RESIDENCE HALF-LIFE (YEARS)
Degradable	Organophosphate insecticides	0.02–0.2
	Carbamate insecticides	0.02
Moderately degradable	Urea herbicides	0.3–0.8
	2,4-D; 2,4,5-T herbicides	0.1–0.4
Persistent	Organochlorine insecticides	2–5
Permanent	Lead, arsenic, copper pesticides	10–30

refractory, or persistent and include many chemicals. Notable examples are organochlorine insecticides, such as DDT, and numerous synthetic polymers, such as PCB, PVC, teflon, polyethylenes, and mylar. The residence times in soil for selected biodegradable and persistent pesticides are presented in Table 5–5.

Various reasons for resistance of chemicals to microbial degradation have been proposed. They include the inability to be attacked by a microbe having an enzyme capable of producing degradation or the complete absence of an essential microbial enzyme.

CONTAMINANT RESIDUES IN THE ENVIRONMENT

Following their release, contaminants may be dispersed throughout the environment until they reach a sink. The dispersal may result in global distribution, as well as high levels on a local or regional scale. However, the actual concentrations of contaminant residues in the air, water, soil, and biota vary greatly in different parts of the world due to differences in dispersion patterns and source strengths and characteristics. Based upon the discussions in this and the previous chapters, it should be clear that the distribution and therefore resultant residue levels are a complex function of physical, chemical, and biological processes acting on the contaminant and within the components of the environment. Thus, it is difficult to generalize as to global residue levels.

The long-range transport of chemical contaminants often results in their presence in regions far removed from any known anthropogenic source. This is especially true for those refractory chemicals that do not degrade and may, therefore, travel long distances before reaching some ultimate sink. For example, DDT has been found in Antarctic snow and ice, and one estimate is that approximately 2.4 × 10⁶ kg were collected in the snow over a 22-year period.[2] These residues were probably transported from Europe and North America via northeasterly trade winds and were removed from the atmosphere by wet deposition.

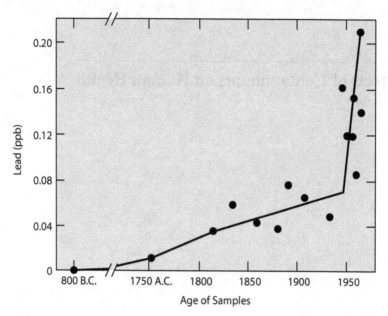

FIGURE 5–5. The concentration of lead found in dated levels of a Greenland glacier. (*Source*: National Academy of Sciences. Lead: Airborne Lead in Perspective. p. 7. Washington, DC, 1972.).

Another classic example is Pb, which has been analyzed in chronologic layers of snow strata in Greenland.[3] Annual ice layers from the interior of northern Greenland have shown a substantial rise in Pb concentration (Fig. 5–5). The layer corresponding to 1750 represents the Pb concentration at the beginning of the Industrial Revolution. The concentration at that date was already 25 times greater than the natural level. By the early 1970s, the Pb concentration in Greenland snow was approximately 400 times the natural level.

Additional evidence reflects increasing Pb concentrations in other regions. For example, the preindustrial Pb content for marine water was estimated at 0.02 to 0.04 μg/kg.[4] However, ocean surface waters in many areas away from anthropogenic sources show Pb concentrations as high as 0.20 to 0.35 μg/kg, and only deep waters (below 1000 m) remain relatively uncontaminated.

REFERENCES

1. Seinfeld, J. H. *Atmospheric Chemistry and Physics of Air Pollution.* 1986; New York: Wiley.
2. Peterle, T. J. DDT in Antarctic snow. *Nature.* 1969; 224:620–623.
3. Nraigu, J. O. Tales told in lead. *Science.* 1998; 281:1622–1623.
4. Chow, T. J. Isotope analysis of seawater by mass spectrometry. *J. Water Poll. Control Fed.* 1968; 40:399–411.

6

Effects of Contaminants on Human Health

The earliest historical evidence that environmental chemicals could affect health was based on associations between certain diseases and occupational exposures. It is now clear that chemicals in environmental media can also influence the health of the general population as well. This chapter presents a broad overview of the health effects of chemical contaminants in the environment.

METHODS FOR ASSESSING HEALTH EFFECTS

Health effects in humans resulting from exposure to environmental chemicals can be assessed using two broad types of studies, namely observational and experimental. These can provide either direct or indirect evidence of a relationship between chemical exposure and health outcomes. Observational, or epidemiological, studies look at a defined population that is exposed to a specific chemical or mixture of chemicals in the environment in relation to development of adverse health effects.

Recognizable health effects in population studies are generally divided into three categories: mortality, morbidity, and functional decrements. Mortality incidence refers to the number of deaths per unit of population per unit time and/or to the ages at death. Data are sometimes available on cause-specific mortality.

Morbidity refers to nonfatal cases of reportable disease or dysfunction. Functional decrements include permanent losses in biological capacities of organ systems.

Chemicals can cause deaths in populations to occur within a short time after exposure. They can also result in residual disease and/or dysfunction. In many cases, the causal relationships are well-defined, and it is often possible to develop quantitative relationships between short-term exposures and subsequent acute responses. In addition, long-term exposures at lower levels can lead to chronic effects, such as irreversible organ damage and functional disability, and/or premature mortality (i.e., lifespan reduction).

The number of people exposed to chemical contaminants at low levels is, of course, much greater than the number exposed at levels high enough to produce overt responses. Furthermore, low-level exposures are often continuous or repetitive over periods of many years. The responses, if any, are likely to be nonspecific (i.e., an increase in the frequency of chronic diseases that are also present in nonexposed populations). As a result, for example, small increases in the population incidence of heart disease or lung cancer attributable to a specific chemical exposure are often difficult to detect, since these diseases are present at high levels in nonexposed populations. In smokers, the increased incidences are likely to be influenced more by cigarette exposure than by the chemical in question.

Some common types of epidemiological study designs are cohort, case-control, cross-sectional, and ecological. A cohort, or panel, study involves selection of a specific study population that is followed over time to detect any increase in the incidence or prevalence of adverse health indications in the fraction of the cohort that is exposed to the chemical of interest for comparison to the incidence or prevalence in the fraction that is unexposed. It is critical in this type of study to assure, to the extent possible, that the exposed and unexposed portions of the cohort are as similar as possible in terms of all other parameters, such as age and gender, as well as prior or pre-existing diseases and smoking and occupational histories, since such parameters may also affect the observed health outcomes of interest.

A case-control study involves initial identification of some health effect of interest that occurs in a specific group of people and identification of a group that does not show the health effect. The two groups are then compared in terms of prior exposure to the chemical or chemicals of interest. The main difference between this type of study and the cohort study is that the latter type follows people over time for development of health effects, while the former type selects populations based upon the presence or absence of the health outcome of interest. In addition, cohort studies generally require much larger numbers of individuals than do case-control studies and are much more costly.

A cross-sectional study involves selection at a specific point in time of a group of people from a defined population and assessment of the health outcome of interest in relation to the exposure of interest. In order to prevent what is termed selection bias, that is, obtaining an unrepresentative sample of the entire population, this type of study generally involves random sampling methods for selection of the individuals to be assessed. This type of study is fairly simple and can provide immediate results, and no follow-up of the individuals is necessary. However, and similar to the case-control type of study, it is not necessarily possible to determine how the adverse health outcome was temporally related to exposure, that is, whether it occurred simultaneous with exposure or developed later. This problem can be alleviated somewhat by obtaining exposure histories for the chemical of concern for the individuals of interest.

An ecological, or aggregated, study examines the frequency of a health outcome in terms of the exposure of interest, either in a number of groups of people at one time or in the same group at different times. This type of study does not allow for linking individual exposures to specific adverse health effects insofar as the entire group of people, and not individuals, define the observational unit.

These study designs have specific advantages and disadvantages, some of which have already been noted. In general, all epidemiologic analyses are beset with inherent problems, often making it difficult to obtain a quantitative link between long-term exposure to contaminants and health effects. One major problem is separation of the effects of a specific contaminant from the possible mediating effects of other factors that also affect health, such as concurrent disease, diet, living conditions, cultural factors, and occupational exposures. The isolation of the effect of only one factor upon health requires data from large populations that ideally differ only with respect to exposure to the contaminant in question. Because of the many unknown factors, assumptions are often made that these factors are identical for all groups or vary randomly with respect to levels of the contaminant. Other problems in epidemiologic studies involve possible interaction between individual contaminants, especially since most communities are exposed to combinations that often make isolation of any one contaminant as the primary culprit of observed effects quite difficult; the accuracy of the classification and reliability of the records of symptoms during the period under study; and reliability and accuracy of the measurements of ambient pollution levels in the area under study and the relationship between these levels and actual personal exposures to the population(s) of concern.

Epidemiological studies are designed to provide some quantitation of the association between exposure and health outcome, and thus they provide statistical (i.e., correlative) conclusions for any such association rather than direct evidence for causation. Therefore, care must be taken in interpretation of epidemiological studies; the finding of an association between specific contaminants and a health effect does not necessarily prove a causal relationship.

Quantitation of the effect of exposure in epidemiological studies are reported using various metrics that aim to provide an indication of the relative likelihood of the exposed group to show the health outcome compared to the unexposed group; the most common of such metrics are risk ratio (i.e., the ratio of risk in the exposed group to that in the unexposed group) or odds ratio (i.e., the odds of the adverse effect in the exposure group to that in the unexposed group). Another approach in epidemiology is to assess excess risk, which is a more absolute metric of effect of exposure. This is an indication of the number of extra cases exhibiting the adverse health outcome that are due to the exposure of interest; however, the underlying assumption here is that there was a causal relationship between exposure and response. In any case, these studies sometimes provide the only available data on the effects of long-term, chronic exposures of large populations to ambient levels of contaminants and, as such, are often used as the basis for policy decisions by governmental agencies.

Specific causality, rather than just correlation, between exposure and response can be better developed with use of experimental approaches for examining exposure-response relationships. Such studies can provide biological mechanistic support for any associations between exposure and health outcomes noted in epidemiological analyses.

There are three common types of experimental studies, often termed toxicological studies: controlled exposures using human subjects, controlled laboratory animal exposure studies, and in vitro studies. Controlled human exposure studies can provide direct evidence of a relationship between a chemical exposure and adverse health outcomes. Such studies are conducted under strictly monitored conditions for exposure concentration, exposure duration, and biological parameters that could affect health outcomes. However, these studies are generally limited to healthy individuals, or at least those with only minimal disease, and thus cannot involve potentially highly susceptible individuals in the study group. Furthermore, ethical considerations require that exposure concentrations and durations be limited to prevent any lasting effect from exposure.

In vivo studies involving laboratory animals can provide direct evidence for adverse health outcomes resulting from exposure to a chemical of interest. These studies can examine more severe outcomes, more invasive endpoints, and longer exposure durations than is possible in human studies. While such studies can provide information on potential health risk to humans, results are subject to some uncertainty due to interspecies differences in exposure parameters (e.g., dose) and biological mechanisms, such as biotransformation, and require some extrapolation to the human exposure situation. In many cases, there can be very large differences between humans and certain animals in the level, or even nature, of response to a specific dose of a specific chemical. Nevertheless, results from these studies can complement controlled human exposure studies in providing evidence for exposure-related adverse health effects and can also provide biological plausibility for health outcomes noted in epidemiological studies.

Another approach in examining response to chemicals is in vitro studies. These involve direct exposure of cells or tissues to the chemical of interest followed by analysis of biological endpoints of interest. They are useful to assess chemical effects on overall cytotoxicity and specific biochemical or functional effects. While the extrapolation from results in such studies to the in vivo situation can often be somewhat uncertain, and they do not allow for evaluation of the toxicokinetics of the chemical being assessed, these can provide support for mechanistic understanding of chemical action on biological systems and do complement the other types of studies already discussed ..

While there are a number of study designs that aim at assessing adverse health outcomes from chemical exposure, the question still remains as to what should be considered as an adverse health outcome. A number of biological parameters may be affected by exposure that are not of any clinical significance. Some examples of what could be considered as adverse are a negative effect on the quality of life, a permanent change in a physiological or biochemical function that affects homeostasis, alteration of growth or development, mortality, development of discrete pathology or illness, and increased susceptibility to effects of other environmental influences. Other exposure-related changes may be considered adverse as biological monitoring becomes more sophisticated and sensitive.

The question often arises as to how one knows if exposure has occurred, especially when there are no overt signs or symptoms. To address this question, markers of exposure or response are often employed, although there is not always a clear distinction between these two categories. In any case, biomarkers can be used in all of the types of studies discussed here. Some examples of biomarkers of exposure are the presence of the chemical or its metabolite(s) in biological samples, for example, in hair, blood, or urine, or production of adducts of DNA. Exposure to volatile organic compounds can sometimes be evaluated by measuring their concentration in exhaled breath. While, as noted, the distinction between markers of exposure and response are not always clear, and some markers are actually both, very often the latter are clear indicators of some change in the function of cells, tissues, or organs. These would ideally be subclinical indicators of potentially permanent damage should exposure continue. Examples of these would be biochemical changes, such as in enzyme activity or induction of stress proteins.

It is not possible to examine all chemicals for potential toxicity. Thus, many regulatory agencies use what is termed structure-activity relationships to estimate relative toxicities of chemicals. This involves assessing the physico-chemical properties of an untested chemical to predict toxicity based upon known toxicity of a similar chemical or chemicals in the same general chemical family. The process is based upon the presumption that the toxicity of a chemical is related to its molecular structure and resultant physico-chemical properties. However, the process can under- or overpredict toxicity, depending on the specific chemical and endpoints being evaluated.

THE DOSE–RESPONSE RELATIONSHIP

The specific parameters of a chemical exposure and the specific effect or effects due to that exposure can be related to each other in a correlative manner termed the dose–response or exposure–response relationship. This fundamental concept in toxicology is essential for understanding potential biological effects from exposure, since it relates the amount, or dose, of the chemical that is necessary to produce an effect or effects of interest, that is, the response. Such relationships are generally derived empirically from the types of experimental studies noted earlier. Evaluation of such a relationship allows for determination of causality, namely whether the chemical of concern results in the response of concern, determination of the lowest dose that results in the response of interest, termed threshold, and the rate at which the response occurs in relation to dose, evidenced by the slope of the dose–response curve.

A point of confusion is the difference between exposure and dose. From a technical standpoint, "exposure" refers to the opportunity to come into contact with some chemical, and "exposure concentration" is the amount in the contact environmental medium. Given this, "dose" refers to the concentration that actually contacts the target site, and different sites can have different doses for the same exposure. Thus, the maximum dose would be some function of the exposure concentration and duration of exposure. In other cases, subcategories of dose are provided, whereby the administered dose is the concentration in the exposure scenario, the internal dose is the amount that enters the organism, and the biologically effective dose is the amount that reaches the target site for that chemical. Typically, however, there is a blurring of the difference between "exposure" and "administered dose" or just merely "dose," and the terms are often used synonymously.

The actual dose, however defined, is dependent upon a number of variables. These include the actual concentration of the chemical in the exposure environment and the duration and frequency of the dose delivery; the former refers to the entire period of time over which the exposure occurs, such as a month or a year, while the latter refers to how often the exposure occurs within this period, such as daily, weekly, and so on. Other factors include characteristics of the exposed individuals, such as age, genetics, health status such as pre-existing disease, route of exposure, and co-exposure to other chemical toxicants (e.g., cigarette smoke). All of these factors need to be considered when examining the effects of chemical exposure on populations or even an individual.

A dose–response curve, in which dose is the independent variable on the x-axis and response the dependent variable on the y-axis, can be generated either for an individual or for a population. The former shows the graded response to various doses of a chemical in one person, while the latter shows the distribution of responses to various doses in a defined population, reflecting variability in individual responses within that population but indicating that increasing dose results in an increased number of individuals in the population showing the response. Figure 6–1 gives examples of dose–response curves for a population. In general, typical dose–response curves are plotted using the log of dose versus the percentage of the population responding at each dose and for the most part generally show a sigmoid shape. What these curves indicate is that, within any population, there can be a wide variability in response in relation to dose, with some segment of the population more sensitive and another more resistant, but that most of the population will respond in a similar manner. It should be noted that different biological effect parameters may show different dose–response relationships even for the same exposure scenario.

As noted, examination of the relationship between dose and response allows for determination of whether or not there is some threshold of dose, defined as the dose below which there would be no observable response for the specific endpoint being examined. For any given chemical, thresholds are generally specific for each specific endpoint assessed. Curve A in Figure 6–1 shows an example of a no-threshold relationship, while Curve B shows the existence of a threshold.

The toxicity of a chemical can be determined by examination of the slope of the dose–response curve. A steep slope, as in Figure 6–1 Curve A, indicates increasing

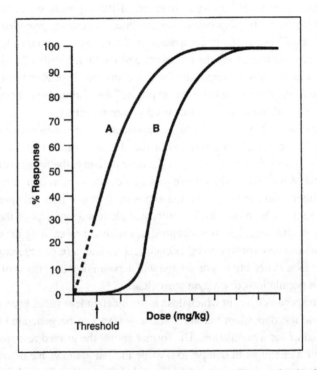

FIGURE 6–1. Examples of population dose response curves. See text for details.

risk of response as the dose increases, while a flatter slope suggests that there is less toxicity as dose increases. Comparison of dose–response curves for chemicals can allow for determination of their relative potency. Thus, in Figure 6–1 chemical A is considered more potent than B since exposure to chemical A results in a greater percentage of the population showing the response per unit increase in dose than exposure to chemical B.

DELIVERY, ABSORPTION, DISTRIBUTION, AND EXCRETION OF CHEMICALS

Exposures to chemicals are often mediated by complex environmental pathways, and uptake may occur via more than one route. People may be exposed to a chemical via the air they breathe, through the water and food they ingest, or by skin contact with air, water, or solid surfaces. Once exposure does occur, depending upon its specific nature, a chemical contaminant may exert its toxic action at various sites in the body. However, the first contact is at a portal of entry: the respiratory tract, gastrointestinal (GI) tract, and skin. At the portal, a chemical may have a topical effect. However, for actions at sites other than the portal, the agent must be absorbed through one or more body membranes and enter the general circulation, from which it may become available to affect internal tissues, including the blood itself. The initial distribution of any chemical contaminant in the body is, therefore, highly dependent upon its ability to traverse biological membranes. There are two main types of processes by which this occurs: passive transport and active transport.

Passive transport is absorption according to purely diffusional processes in which the cell has no active role in movement across the membrane. Since biological membranes contain lipids, they are highly permeable to lipid-soluble, nonpolar, or nonionized agents and less so to lipid insoluble, polar, or ionized materials. Many chemicals may exist in both lipid-soluble and lipid-insoluble forms; the former is the prime determinant of the passive permeability properties for the specific agent.

Active transport involves specialized energy-dependent mechanisms, and in these the cell actively participates in movement across the membrane. These mechanisms include carrier systems within the membrane and active processes of cellular ingestion, namely, phagocytosis and pinocytosis. Phagocytosis is the ingestion of solid particles, while pinocytosis refers to the ingestion of fluid containing no visible solid material. Lipid insoluble materials are often taken up by active-transport processes. Although some of these mechanisms are highly specific, if the chemical structure of a contaminant is similar to that of an endogenous substrate, the former may be transported as well.

In addition to its lipid-solubility characteristics, the distribution of a chemical contaminant is also dependent upon its affinity for specific tissues or tissue components. Internal distribution may vary with time after exposure. For example, immediately following absorption into the blood, inorganic lead is found to localize in the liver, the kidney, and red blood cells. Two hours later, approximately 50% is in the liver. A month later, approximately 90% of the remaining lead is localized in bone.[1]

Once in the general circulation, a contaminant may be translocated throughout the body, during which it may become bound to macromolecules; undergo metabolic transformation (biotransformation) in specific organs; be deposited for storage in depots, which may or may not be the sites of its toxic action; or be excreted. Toxic effects may occur at any of several areas involved in these processes. On the other hand, the biological action of a contaminant may be terminated by storage, metabolic transformation, or excretion.

Figure 6–2 provides a framework connecting pollutant sources, personal exposures, dose delivery to target organs and tissues, and health effects resulting from such exposures. Figure 6–3 shows various routes of absorption, distribution, and excretion of toxicants.

Delivery of Chemicals into the Body

The portal of entry presents the first possible site of action of, and first line of defense against, environmental chemicals. Some environmental agents have multiple portals of entry, with the specific physicochemical properties in the exposure environment determining the significant entry path. The specific portal may be an important factor in determining whether there will be any adverse effect from exposure, in that the nature of any such depends not only upon quantitative and qualitative parameters of dose but also in many cases upon the route of exposure as well.

Respiratory tract

The respiratory tract is the prime portal of entry for airborne chemicals. It may be separated into two functional zones: the conducting airways and respiratory airways. The former extends from the upper respiratory tract (i.e., the nasopharynx, pharynx, and larynx) through the tracheobronchial tree to the terminal bronchioles and serves as a conduit system for the transport of air into and out of the gas-exchange region of the lungs. The latter zone, often called the alveolar or pulmonary region, consists of respiratory bronchioles, alveolar ducts, alveolar sacs, and alveoli and is involved in the mutual gas exchange of oxygen (O_2) and carbon dioxide (CO_2) between air and the pulmonary circulation. The very thin cellular layer between blood and air in the alveoli reduces the effectiveness of this region of the lungs as a barrier to the systemic uptake of many inhaled chemicals.

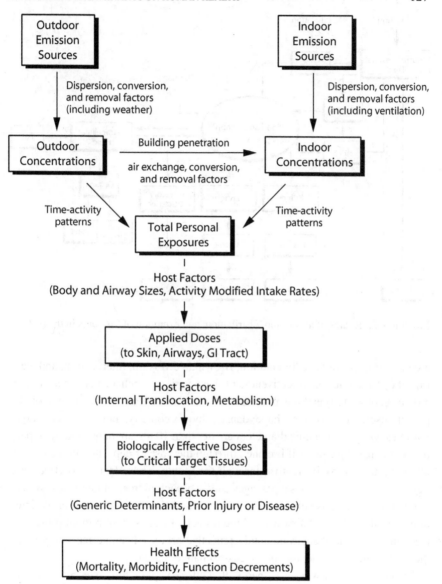

FIGURE 6–2. Framework for personal exposure assessment and exposure-response. (Source: Modified from NAS, 1985.).

The fate of a specific contaminant within the tract is dependent upon the form in which it exists, i.e., gaseous or particulate matter. Deposition is a main factor in determining the fate of inhaled particles that are not subsequently exhaled. A variety of physical mechanisms act to remove suspended particles in various regions of the respiratory tract. The main processes affecting deposition onto the airways

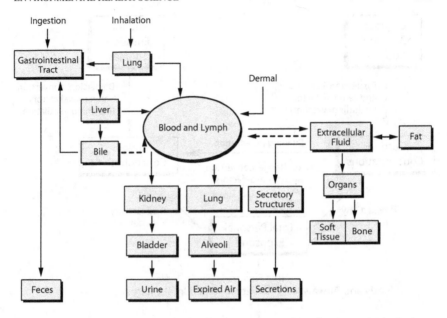

FIGURE 6–3. Routes of absorption, distribution, and excretion of toxicants in the body.

are inertial impaction, sedimentation, Brownian diffusion, interception, and electrostatic attraction. The effectiveness of these various mechanisms depends upon
airway geometry, breathing rate and pattern, and certain characteristics of the
particle itself, such as size, shape, density, hygroscopicity, and electrical charge.
Particles with aerodynamic diameters greater than 10 μm are deposited by impaction in the nasal passages if breathed through the nose or deposited in the mouth,
pharynx, and larynx if breathed through the mouth. Those from 2 to 10 μm are
largely deposited in the nasal passages and/or bronchial tree, while those less than
2 μm penetrate extensively to the bronchioles and alveoli. Deposition of particles
in the nasal and/or oral airways and the bronchial tree is an important protective
mechanism for those environmental agents that produce health effects only when
they penetrate to and deposit in the gas-exchange region.

The conductive airways are lined with a ciliated epithelium, overlaid by a thin
fluid layer composed largely of water and glycoproteins and formed by secretions
of specialized glands and individual cells. This "mucous blanket" is propelled by
the cilia toward the pharynx, where it may be swallowed or expectorated. A similar
blanket in the nasal passages and sinuses also moves toward the pharynx. The rate
of mucous transport varies in different regions of the tracheobronchial tree, and the
composition and/or consistency of the fluid lining also differs in different regions.

Water-insoluble particles that deposit on the nasal passages and tracheobronchial tree may be cleared from the respiratory tract via this mucociliary system,
with residence times ranging from minutes up to approximately 24 to 48 hours.

Frequently, the rapidity and degree of clearance plays a major role in determining the risk from particulate matter exposure. Extended contact time between deposited toxic particles and the epithelium may lead to local damage or an increased degree of absorption and resultant systemic effects.

Water-soluble particles that deposit in conductive airways may dissolve in the mucus, moving with it out of the respiratory tract, or they may be absorbed through the epithelium into the circulatory and/or lymphatic system. The alveolar region of the lung does not contain cilia or mucus. Particles that reach and deposit within this zone are, however, also subject to clearance but by different mechanisms, with half-times varying from days to years. Deposited particles may be taken up, via phagocytosis, by specialized cells known as alveolar macrophages, which are present on the alveolar surface. These cells may then follow a number of possible subsequent clearance routes. The macrophages, and also free particles, may penetrate the alveolar epithelium, entering the interstitial tissue where they may remain (parenchymal sequestration) or become absorbed into the lymphatic system. Macrophages carried in this system may be trapped in lymph nodes, which act as dust stores of the lung, or they may eventually drain with the lymphatic fluid into the systemic blood.

Another possible clearance mechanism is migration of the particle-laden macrophages to the level of the mucociliary blanket and subsequent clearance via the mucociliary system. Free particles and macrophages may also move to the level of the mucociliary system by mechanisms that may involve respiratory motion of the alveolar walls, capillary action, tensile forces between alveolar fluid and mucus, the influence of ciliary beat in bronchioles, and the shearing effect of layers of mucus. Soluble particles depositing in the alveoli may be removed by direct translocation across the alveolar epithelium into the blood.

For inhaled gases and vapors, the site and extent of absorption is, for the most part, determined by their solubility characteristics in water. Highly soluble gases, for example, sulfur dioxide (SO_2) are largely absorbed in the upper respiratory tract, while those that are less water soluble, for example, nitric oxide (NO_2) and ozone (O_3), reach the lower airways. Absorption is also dependent upon airflow rate and the partial pressure of the gas in the inspired air. Gases dissolved in the fluid lining of the tracheobronchial tree may be cleared with the mucus.

Other defense mechanisms of the respiratory system include sneezing, which serves to cleanse the upper respiratory tract, and coughing, which acts to clear the larger bronchial airways. Constriction of bronchi due to the contraction of surrounding smooth muscle following exposure to certain chemicals may act as a defense by decreasing further penetration to the deeper lung areas.

Gastrointestinal tract

Chemical contaminants that are waterborne or transported via food reach people via the GI tract. Ingestion may also contribute to the uptake of chemicals that were initially inhaled, since material deposited on or dissolved in the bronchial mucous blanket is eventually swallowed.

The GI tract may be considered to be a tube running through the body but whose contents are actually external to the body. Unless the ingested material affects the tract itself, any systemic response depends upon absorption through the mucosal cells lining the lumen. Although absorption may occur anywhere along the length of the GI tract, the main region for effective translocation is the small intestine, although some chemicals can be absorbed through the stomach as well. The enormous absorptive capacity of this region is due to the presence in the intestinal mucosa of projections, termed villi, each of which contains a network of capillaries; the villi result in a large effective total surface area for absorption.

Although passive diffusion is the main absorptive process, certain active transport systems allow essential lipid-insoluble nutrients and inorganic ions to cross the intestinal epithelium. These carrier systems may also be responsible for some contaminant uptake, especially if there are chemical similarities between the contaminant and nutrients. For example, lead may be absorbed via the system that normally transports calcium ions.

Materials absorbed from the GI tract enter either the lymphatic system or the portal blood circulation; the latter carries material to the liver, from which it may be actively excreted into the bile or diffuse into the bile from the blood. The bile is subsequently secreted into the intestines. Thus, a cycle of translocation of a chemical from the intestine to the liver to bile and back to the intestines, known as the enterohepatic circulation, may be established. Enterohepatic circulation usually involves contaminants that undergo metabolic degradation in the liver.

Various factors serve to modify absorption from the GI tract, enhancing or depressing its barrier function. A decrease in GI mobility generally favors increased absorption. Specific stomach contents and secretions may react with the contaminant, possibly changing it to a form with different physicochemical properties, for example, solubility, or they may absorb it, altering the available chemical and changing translocation rates. The size of ingested particles also affects absorption. Since the rate of dissolution is inversely proportional to particle size, larger particles are absorbed to a lesser degree, especially if they are of a fairly insoluble material in the first place. Certain chemicals, for example, ethylene diamine tetra-acetic acid, also cause a nonspecific increase in absorption of many materials. Finally, as a defense, spastic contractions in the stomach and intestine may serve to eliminate noxious agents via vomiting or by acceleration of feces through the GI tract, respectively.

Skin

The skin is generally a very effective barrier against the entry of environmental chemicals into the body. In order to be absorbed via this route (percutaneous absorption), an agent must traverse a number of cellular layers before gaining

access to the systemic circulation. Skin consists of two structural regions, the epidermis and the dermis, which rest upon connective tissue. The epidermis consists
of a number of layers of cells and has varying thickness depending upon the region
of the body; the outermost layer is composed of keratinized cells. The dermis contains blood vessels, hair follicles, sebaceous and sweat glands, and nerve endings.
The epidermis represents the primary barrier to percutaneous absorption, with the
dermis freely permeable to many materials. Any passage through the epidermis
would occur by passive diffusion. Some fat-soluble environmental chemicals are
readily absorbed through the skin, for example, phenols, carbon tetrachloride,
tetraethyl lead, and organophosphate pesticides. Other chemicals, for example,
dimethyl sulfoxide and formic acid, can alter the integrity of skin and facilitate
penetration of other materials by increasing skin permeability. Moderate changes
in permeability may also result following topical applications of acetone, methyl
alcohol, and ethyl alcohol. In addition, cutaneous injury may enhance percutaneous absorption. In any case, the main factors affecting percutaneous absorption are degree of lipid solubility of the chemical, site on the body, local blood
flow, and skin temperature. Although the main route of percutaneous absorption
is through the epidermal cells, some chemicals may follow an appendageal route,
for example, entering through hair follicles, sweat glands, or sebaceous glands.

Interspecies differences in percutaneous absorption are responsible for the selective toxicity of many insecticides. For example, the pesticide DDT is about equally
hazardous to both insects and mammals if ingested but is much less hazardous
to mammals when applied to the skin. This is due to the poor absorption through
mammalian skin, compared to its ready passage through the insect exoskeleton.

Fate of Chemicals in the Body

Once absorbed through the portal of entry, chemicals will enter the blood, lymph,
or other body fluids and then may be translocated to storage sites, biotransformation sites, or sites of excretion. The specific distribution of any chemical depends
upon the net result of a number of processes, including rate of uptake, rate of
elimination, distribution of blood flow to various tissues, rate of storage site
uptake, and any metabolic changes that occur.

Metabolism: Biotransformation

Chemical contaminants that enter the body may undergo metabolic conversion, a
process known as biotransformation. At one time it was believed that all contaminants underwent biotransformation into less toxic states. Thus, these reactions
were often referred to as detoxification mechanisms. However, it is now clear that
certain reactions can result in products that are more toxic than the parent (metabolic toxification). These chemical reactions are mediated by enzymes and result
either in an alteration of the parent compound or formation of new product(s) via

the combination of the parent with various endogenous substrates. Some of these enzymes are nonspecific, that is, they act upon various substrates, and are capable of catalyzing a variety of biotransformation reactions. Other chemicals require specific enzymes for metabolic transformation. In addition, the enzymes generally do not act upon lipid insoluble material. Contaminants that have structures similar to certain normal substrates in the body may undergo biotransformation by enzymes that normally catalyze transformation of these substrates. For example, the enzyme monoamine oxidase catalyzes the metabolism of endogenous amines such as epinephrine; it is also involved in the reactions of foreign, short-chain amines, such as benzylamine.

Many biotransformation reactions are specifically catalyzed by enzymes that are members of a special group, namely the cytochrome P450 family. These enzymes are involved in metabolism of a very wide variety of exogenous chemicals as well as some endogenous ones. In some cases, their production is actually induced by the chemical that they are metabolizing

The primary site of biotransformation is the liver, although some transformation reactions may also occur other organs as noted later. Most biotransformation mechanisms fall into two main classes of reactions: Phase I and Phase II. Phase I reactions include hydrolysis, oxidation, and reduction processes and generally result in converting the chemical into a more polar metabolite by introducing or unmasking a functional group. These enzymes generally reside in the microsomes of liver cells.

Phase II reactions, or conjugations, involve the production of a new product from the parent compound or a metabolite from Phase I and an endogenous substrate, generally a polar or ionic moiety. In many cases, the conjugation results in enhanced excretion via the kidneys. Conjugation includes numerous classes of reactions, for example, acetylation, alkylation, acylation, methylation, and esterification. The enzymes catalyzing these reactions are mostly located in the cytosol of liver cells.

A number of biotransformation reactions have been found that do not fall into either of these basic classes. These miscellaneous mechanisms include aromatic dehydroxylation, aliphatic dehydroxylation, nitrile hydrolysis, dehalogenation, cyclization, and ring scission.

More than one reaction type may be involved in the metabolism of a specific chemical. For example, Phase I and II reactions often occur sequentially, whereby the former prepares a functional radical enabling subsequent conjugation with a polar compound. The exact type of biotransformation reactions that an agent undergoes depends upon the particular structure of the chemical and also upon its route of entry. Route is important because liver cells are not the only sites for biotransformation mechanisms. Enzymes in blood plasma, the kidneys, the respiratory tract, the GI tract, and skin are capable of catalyzing metabolic transformation reactions. Certain components of the detoxification system and the extent of biotransformation activity do, however, differ in these different tissues.

Various host and environmental factors serve to modify biotransformation mechanisms. Host factors include age (e.g., infants have incompletely developed enzyme systems), nutritional status, and genetic deficiencies (e.g., lack of certain enzymes or excess enzymatic activity). The environmental factors involve co- or prior exposure to certain chemicals that can result in induction or inhibition of biotransformation enzymes. A variety of chemicals can increase the amount and activity of enzymes. These include drugs, such as phenobarbital; organochlorine insecticides; certain herbicides; the food additive BHT; and various polycyclic aromatic hydrocarbons, for example, benzo(a)pyrene. If detoxification is the result of metabolism, induction would tend to be protective; if, however, metabolic toxification is the result, induction may increase harmful effects in the host. Other chemicals may nonspecifically inhibit microsomal enzymes. These include organophosphate insecticides, carbon tetrachloride, O_3, and carbon monoxide (CO). Enzyme inhibition would result in the persistence of a chemical that would normally be metabolized. Induction and inhibition of biotransformation enzymes are important when one realizes that people are generally exposed to more than one chemical contaminant; thus, exposure to some may increase susceptibility to what alone would be relatively nontoxic levels of other contaminants due to modulation of the amount or activity of biotransformation enzymes.

Some contaminants may be altered by undergoing bacterially mediated and nonenzymatic reactions. For example, GI flora may reduce aromatic nitro groups into potentially carcinogenic aromatic amines. Carcinogenic nitrosamines may be formed in the low pH environment of the stomach from secondary or tertiary amines, which naturally occur in a variety of foods, and nitrite, which is used as an additive in smoked meats. Another example is the development of a disease called methemoglobinemia in infants who ingest water and foods having high nitrate content. The condition, which results in a decrease in the oxygen-carrying capacity of blood, occurs due to differences in the pH and bacterial content of the GI tract of infants compared to adults, resulting in greater conversion of nitrate to nitrite, the actual causative agent, in the former. Bacterially mediated reactions in the GI tract may also result in the prolongation of action of certain chemicals via deconjugation, causing reabsorption of a chemical that had been previously conjugated in the liver and secreted with bile into the intestine.

Storage

Some contaminants tend to concentrate in specific tissues due to physicochemical properties, such as selective solubility, or to selective absorption onto or in combination with macromolecules such as proteins. Storage of a chemical often occurs when the rate of exposure is greater than the rate of metabolism and/or excretion and depends upon properties of the chemical. For example, fat-soluble chemicals tend to be stored in adipose tissue.

Storage or binding sites may not be the sites of toxic action. For example, CO produces its effect by binding with hemoglobin in red blood cells; on the other hand, inorganic lead is stored primarily in bone but acts mainly on the soft tissues of the body. If the storage site is not the site of toxic action, selective sequestration may be a protective mechanism, since only the freely circulating form of the contaminant produces harmful effects. Until the storage sites are saturated, a systemic buildup of free chemical may be prevented. On the other hand, selective storage limits the amount of contaminant that is excreted. Since bound or stored toxicants are in equilibrium with their free form, as the contaminant is excreted or metabolized, it is released from the storage site.

Contaminants that are stored may remain in the body for years without effect, for example, organochlorine pesticides. On the other hand, accumulation may produce illnesses that develop slowly, as occurs in chronic cadmium poisoning. Table 6–1 lists the major storage sites, with examples of chemicals selectively stored there.

Excretion

The major role in elimination of toxicants from the body is played by the kidneys. However, extra-renal routes may be important in elimination of certain specific substances; these routes are the lungs, GI tract, and skin and elimination may also occur via body secretions, such as saliva and milk.

When the blood is filtered by the kidneys, contaminants are subject to the same removal mechanisms as are normal metabolic end products. As the filtrate travels through the renal tubules, an agent may be reabsorbed into the blood or remain in the filtrate to be excreted in the urine. In general, lipid-insoluble substances are poorly reabsorbed, although reabsorption via active transport may occur regardless of solubility characteristics. As noted, biotransformation is a mechanism for the conversion of chemicals into metabolites that may then be more readily excreted by the kidneys than the parent compound. They may occur, for example, by conversion of a highly lipid-soluble agent into a less lipid-soluble metabolite. On the other hand, biotransformation reactions may also produce metabolites with increased lipid solubility.

Contaminants can appear in the feces if they are not absorbed after ingestion, are excreted into bile, or are excreted along the GI tract. The GI tract is the primary

TABLE 6–1. Examples of Storage Sites for Some Contaminants

STORAGE SITE	EXAMPLE OF CHEMICAL SELECTIVELY STORED	MECHANISM OF STORAGE
Fat	DDT, PCB	Dissolution
Bone	Inorganic Pb, ^{90}Sr	Substitutes for calcium in bone matrix
Kidney	Cd	Binds with intracellular protein

route of excretion for many trace metals, for example, chromium (Cr), manganese (Mn), lead (Pb), and certain large molecules, for example many pesticides.

Since the circulation from the GI tract goes to the liver prior to entering the general circulation, the liver may serve to remove materials following GI absorption, preventing distribution to the systemic circulation. The contaminant or its metabolites may then be excreted into the bile, entering the small intestine. Generally, lipid-insoluble substances are more readily excreted into bile; once in the intestines, the agent may be reconverted into a more lipid-soluble form and be reabsorbed, or it may be excreted with the feces.

The lungs can serve as an organ of excretion for substances that exist in the gas phase at body temperature. Ingested volatile organic compounds, such as carbon tetrachloride (CCl_4), may be partially excreted from the lungs as vapors via diffusion. In general, gases with low solubility in blood at body temperature will be more readily excreted through the lungs than will those with high solubility.

Although of minor importance, contaminants may be excreted in sweat, tears, saliva, and milk. Generally, lipid-soluble materials are passively diffused into these secretions. Contaminants in breastmilk may be passed along to the suckling infant. Excretion into saliva may result in reabsorption via the GI tract. Contaminants can also be eliminated by incorporation into hair and nails, which are periodically trimmed.

Excretion rates can be quantitated using various parameters. These include biological half times, the amount of time after exposure needed to reduce the concentration in the body to half of the initial dose, or clearance and elimination rates, the amount removed per unit of time.

BIOLOGICAL RESPONSES

The action of chemical contaminants covers a range of effects, from a mere nuisance to extensive tissue necrosis and death and from generalized systemic effects to highly specific attacks on single tissues and even individual enzyme systems. Multiple sources and multiple insults via different environmental routes may increase the risk of disease from environmental chemicals. Biological effects, which may be reversible or irreversible, can be the result of exposure to an individual agent or be due to interaction among different agents.

A specific chemical may produce more than one effect upon its target, while completely different agents may result in similar, or even identical, pathological manifestations. A chemical may produce a local effect at the initial site of contact and/or it may produce systemic effects distant from the route of initial contact or absorption. Furthermore, a specific chemical may affect different sites when exposure is by different routes, such as inhalation versus ingestion. For example, chemicals entering via the latter route are subject to what is termed the first pass effect in which they are subjected to biotransformation in the liver before entering the general circulation; this can reduce the systemic absorption of the parent compound in favor of absorption of any metabolites, leading to different toxicity from the same chemical entering by another route. Finally, chemicals may produce acute effects after a short exposure of hours to days or chronic effects with prolonged exposure of months to years. In addition, an effect may persist even after a short-term exposure has ended, and this too is considered a chronic response.

A number of host and environmental factors serve to modify the effects of chemicals; the ultimate response is the result of the interaction between these factors. One of the main host factors is age; for example, older age groups tend to be more susceptible to morbidity and mortality during periods of increased air contamination. This is usually due to chronically reduced cardiovascular and respiratory function, resulting in the inability to cope with additional stresses. Newborns and infants are more sensitive to some toxicants than are adults, partly because of the inability to synthesize specific biotransformation enzymes and due to the high cell turnover rates that occur during liver maturation of organ systems. Other factors are state of health (e.g., concurrent disease or dysfunction may result in enhanced toxicity following exposure or may make an organ more susceptible to damage); nutritional status and dietary habits (e.g., an adequate dietary supply of trace metal nutrients and proteins is critical for the synthesis of many enzymes involved in biotransformation); immunological status; gender and other genetic factors (e.g., enzyme-related differences in biotransformation mechanisms, such as deficient metabolic pathways, and inability to synthesize certain detoxification enzymes); psychological state (e.g., stress, anxiety); cultural factors; and co-exposure to other chemicals (e.g. cigarette smoking, which may affect normal defenses or may potentiate the effect of other chemicals).

Environmental factors that affect biological response include the concentration, stability, and physicochemical properties of the agent in the exposure environment and the duration, frequency, and route of exposure. Thus, acute and chronic exposures to a chemical may result in different pathological manifestations.

Chemicals can interact with target sites in a myriad of manners. For example, they can interfere with energy production or with various regulatory mechanisms and mediators; they can also induce apoptosis, modulate immune responses, or alter genetic components of cells. Given this, however, specific organs can manifest gross results of chemical exposure in only a limited number of ways, although there are numerous diagnostic labels for the resultant diseases. Thus, similar responses may occur in different systems. The following sections discuss broad types of biological responses resulting from chemical exposure.

Some chemicals affect one specific target in one organ or organ system, while others may have a broader systemic affect, producing responses in multiple organ systems. Table 6–2 presents the main target sites for selected chemicals that may

TABLE 6–2. Main Target Sites of Selected Systemic Chemical Agents

CHEMICAL	RESPIRATORY TRACT	HEMATOPOIETIC SYSTEM	SKIN	GASTROINTESTINAL TRACT	LIVER	KIDNEY	BONES, TEETH	ENDOCRINE SYSTEM	NERVOUS SYSTEM (CENTRAL AND/OR PERIPHERAL)
Mercury	X	X		X	X	X			X
Lead		X		X	X	X			X
Cadmium	X	X		X	X	X	X	X	X
Arsenic	X	X	X	X	X	X		X	X
Fluoride		X		X			X		
Molybdenum		X			X				
Selenium		X	X	X	X	X	X		
Organophosphate pesticides		X	X						X
Organochlorine pesticides					X				X
Carbamate pesticides									X
Carbon tetrachloride					X	X			
Chlorinated biphenyls			X		X				X

be classified as systemic toxicants. Some of the metals listed are essential to life at low levels but are quite toxic at high concentrations. In addition, the specific form in which a chemical exists in the exposure environment may affect its toxicity. For example, selenium as selenate and methylated forms of mercury (Hg) are much more toxic than are other forms of these elements.

Irritancy

A pattern of generalized, nonspecific tissue inflammation, irritation, or destruction may result at the area of direct chemical contact. This type of reaction is caused by a class of chemicals termed irritants. It is a portal of entry response, which occurs whether or not the agent is absorbed into the systemic circulation. A list of some common chemical irritants is presented in Table 6–3.

Primary irritants are those that produce no systemic effect, generally because the irritant response is much greater than any systemic effect, for example, inhaled mustard gas in the respiratory tract or ingested acids in the GI tract. Other irritants can also have significant systemic effects following absorption, for example, hydrogen sulfide (H_2S) absorbed via the lungs, causing respiratory paralysis.

Exposure to irritants may result in death if critical organs are severely damaged. On the other hand, the damage may be reversible or it may result in permanent loss of some degree of function, such as impaired gas exchange capacity in the lungs or malabsorption of nutrients in the GI tract.

The respiratory tract is a very common site of irritant exposure. Acute exposure results in a common inflammatory response in all affected regions: rhinitis, pharyngitis, and laryngitis in the upper respiratory tract; bronchitis in the bronchial tree; and edema and chemical pneumonia in the alveolar region. Longer duration exposure may result in a chronic inflammation of small bronchi and hyperplasia (an increase in size) and/or metaplasia (transformation of one cell type into another) of mucus-secreting glands and goblet cells. Other responses also include deposition of connective tissue, alteration in structural proteins, and loss of lung elasticity.

TABLE 6–3. Some Chemical Irritants

Sulfur dioxide	Ozone
Sulfuric acid and other strong acids	Peroxyacetylnitrates
Ammonia and other strong bases	Aromatic hydrocarbons, e.g., benzene,
Nitrogen dioxide	ether
Hydrogen fluoride	Particulate sulfates, e.g., zinc sulfate,
Formaldehyde	ammonium sulfate
Acrolein	Organic selenides
Chlorine	Numerous pesticides

Changes in respiratory function without development of a specific disease state have been seen in response to exposure to many inhaled irritants, for example, SO_2, NO_2, and O_3. These changes are often due to constriction of bronchi caused by a reflex response or by the release of endogenous chemicals, such as histamine. Another important effect of exposure to many pulmonary irritants is alteration of the normal functioning of the mucociliary and/or alveolar clearance systems; some chemicals showing this effect include NO_2, SO_2, and H_2SO_4.

The skin is also a portal of entry frequently subject to environmental chemical irritation. However, most environmental skin disorders arise from occupational exposures to irritants. Most reactions to these are eczematous in nature and include dermatitis, ulceration, and granuloma formation.

Fibrosis

A number of environmental agents are etiological factors in the development of a group of chronic lung disorders termed pneumoconioses. This general term encompasses many fibrotic conditions of the lung, that is, diseases characterized by scar formation in the interstitial tissue. Pneumoconioses are due to the inhalation and subsequent selective retention of certain dusts in the alveolar region, from which they are subject to interstitial sequestration.

Pneumoconioses are characterized by specific fibrotic lesions, which differ in type and pattern according to the dust involved. For example, silicosis, due to the toxicity of crystalline free silica, is characterized by a nodular type of fibrosis, while a diffuse fibrosis is found in asbestosis due to exposure to long fibers of asbestos. Certain dusts, such as iron oxide, produce only altered radiology (siderosis) with no functional impairment, while the effects of others range from minimal disability to death. Some important dusts involved in fibrosis of the lung are listed in Table 6–4.

Dusts initially inhaled may be translocated to other parts of the body. For example, fibrotic nodules due to occupational inhalation exposure to asbestos can also be found in the spleen, pleura, and peritoneum, possibly due to translocation by lymph or blood or to swallowing of fibers carried to the pharynx via the mucociliary system. The fate of inhaled fibers, and their potential effects in terms of producing lung fibrosis, lung cancer, and cancers of the plueral and peritoneal mesothelium, are highly dependent on fiber length, fiber diameter, and their biopersistence (i.e., dissolution rate).

Asphyxiation

Some chemicals utilizing the respiratory tract as a portal of entry can cause death by their action in preventing an adequate oxygen supply from reaching body tissues. These chemicals are known as asphyxiants. Some "biologically inert" or "nontoxic" chemicals, known as simple asphyxiants, prevent tissues from receiving enough

TABLE 6–4. Pulmonary Fibrotic Agents

MATERIAL	DISEASE DESIGNATION
Inorganic Fibers and Dusts	
Crystalline silica (quartz, cristobolite)	Silicosis
Asbestos (chrysotile and amphibole fibers)	Asbestosis
Talc	Talcosis
Coal (mine dust)	Coal workers' pneumoconiosis
Kaolin	Kaolinosis
Graphite	Graphite lung
Organic Fibers and Dusts	
Cellulose	Bagassosis
Cotton	Byssinosis
Flax	Byssinosis
Hemp	Byssinosis
Metallic Fumes	
Tin oxide	Stannosis
Iron oxide	Siderosis
Beryllium oxide	Berylliosis

oxygen simply by displacing oxygen in the air. Examples of these are CO_2 and CH_4. Other materials, known as chemical asphyxiants, prevent oxygen from being carried by the blood, for example CO, or from being utilized by the tissues, for example, hydrogen cyanide (HCN). Chronic exposure to asphyxiants may result in systemic disorders due to a persistent impairment of the oxygen supply to vital organs.

Allergenicity and Other Immune Responses

The interaction of chemicals with the immune system can be categorized as either stimulation (i.e., an allergic response) or suppression. Allergic agents, termed allergens, generally affect the respiratory tract and skin, although they may also affect other organs such as the intestines (allergic colitis). Some chemical allergens are listed in Table 6–5.

While the immune system is normally protective, these hypersensitivity reactions can lead to severe illness and even death. In some cases, the initial exposure to an allergen results in the antibody-mediated release of specific mediators that modulate the ultimate response. This is called immediate hypersensitivity. On the other hand, an immune reaction may show a delayed response, termed delayed hypersensitivity, which is mediated by immune cells rather than antibodies.

TABLE 6–5. Some Chemical Allergens

ORGANIC CHEMICALS	METALS
Herbicides	Nickel
Organophosphate insecticides	Cobalt
Carbamate herbicides	Arsenic
Toluene diisocyanate	Chromium
Acrolein	Beryllium salts
Epoxy resins	

The common respiratory allergic reactions are bronchial asthma, reactions in the upper respiratory tract involving the release of histamine or histamine-like mediators following immune reactions in the mucosa, and a type of pneumonitis (lung inflammation) known as extrinsic allergic alveolitis. Skin reactions include urticaria (hives); atopic dermatitis, which is due to ingestion or inhalation of an allergen; and contact dermatitis, due to direct skin contact with an allergen. In addition to these local reactions, a systemic allergic reaction (anaphylactic shock) may follow exposure to some chemical allergens.

Some irritants are also sensitizing agents, in that initial exposure produces no response at the site of contact but subsequent exposure can result in significant irritant response. This type of reaction is most common in skin, producing what is termed allergic contact dermatitis. Examples of chemicals that can be sensitizing agents to skin are epoxy resins, isocyanates, certain metals (e.g., hexavalent chromium), and diphenols (e.g., resorcinol). Some chemicals can result in allergic irritant response after exposure to sunlight; these are termed photosensitizers and include such chemicals as petroleum-based tars.

Chemical-induced immunosuppression can result in an increased susceptibility to pathogens and even cancer formation. Examples of chemicals that can result in immunosuppression are polycyclic aromatic hydrocarbons, nitrosamines, some metals (e.g., cadmium, mercury, and lead), and some pesticides.

Genotoxicity

Genotoxic chemicals act by altering the hereditary material (i.e., DNA) in the chromosomes of the cell. Alteration of genetic material can result in modulation of the expression of specific genes, usually either inactivating them or causing overexpression, which can then be manifested in some disease state. Some chemicals will react directly with DNA, while others must undergo biotransformation or reaction with other cellular components, such as lipids and proteins, to become genotoxic. Chemical-induced alteration of DNA and subsequent genetic activity may be due to mechanisms which include production of adducts or breaks in the

DNA backbone, or production of DNA–DNA or DNA–protein cross-linkages. Some specific types of genotoxic action are discussed next.

Mutagenicity

A mutagen is a chemical that acts directly on DNA, producing a change in the nucleotide sequence, which then can result in errors in replication, resulting in some phenotypic alteration. If these changes occur in the germline, namely the sex cells, they can produce pathology in the offspring. Mutations that occur in other cells are termed somatic mutations, and these can affect the individual but do not affect the offspring. A variety of environmental chemicals have been found to be mutagenic, at least in test systems of cell cultures of mammalian and nonmammalian cells; these may, therefore, pose potential hazards to humans (Table 6–6). While many mutagenic chemicals react directly with the DNA, others are not mutagenic per se but undergo biotransformation into mutagenic compounds.

As noted, some mutagens react directly with DNA bases, whereas others are mutagenic via alternate mechanisms. A base analog can substitute for another normal base in the replicating DNA, resulting in transition mutations and altered base pairing during subsequent DNA replication. These are classified as either purine analog or pyrimidine analogues. The most common bases analogues are 5-bromouracil and 2-aminopurine. A chemical that is termed an intercalating agent can be inserted between the DNA bases, resulting in frameshift mutations during replication; examples are ethidium bromide and proflavine. Exposure to some chemicals is associated with production of reactive oxygen species, such as superoxide, and these in turn can react with the DNA directly or interfere with DNA repair.

Alteration of transcription factors

Transcription factors are proteins that modulate transcription of genetic material from the DNA to the messenger RNA. Alteration of these can affect DNA structure or function, thus affecting gene expression. These factors can be modulated by various endogenous mediators, such as hormones, and if a chemical alters these mediators then transcription factor activity can also be affected. Increased

TABLE 6–6. Some Known or Suspected Chemical Mutagens

DDT	Sodium arsenate
2,4-D	Cadmium sulfate
2,4,5-T	Lead salts (some)
Benzene	Nitrite
PAHs	Bromine
Aromatic amines	Nitrous acid

activity of transcription factors has been implicated in some cancer development. Various heavy metals act by altering transcription factors.

Alteration of epigenetic factors

Epigenetics refers to factors that attach to DNA and that interact with DNA but do not change the basic DNA sequence. They act by interfering with gene expression, turning genes on or off, thus affecting the production of proteins. Examples of epigenetic mechanisms include DNA methylation and histone acetylation or phosphorylation. Such changes can actually result in increased sensitivity of DNA to direct damage (i.e., mutation), effectively increasing susceptibility to further chemical exposure.

Teratogenicity

A teratogen produces a generative change during early embryonic development, resulting in anatomical defects or other functional or biochemical developmental errors manifested in the fetus. These agents can produce their effect by interfering with normal apoptotic processes occurring during development, inhibiting function of certain cells, producing damage to DNA or maternal toxicity. Numerous chemicals, especially drugs, have been shown to be teratogenic in experimental animals (Table 6–7); and these too may be toxic to humans.

TABLE 6–7. Some Known or Suspected
Chemical Teratogens

Dioxisn
Organic mercury
Phthalic acid esters
Methyl Mercury
Lead
Ethyl Alcohol
Carbamates
Cadmium sulphate
Sodium arsenate
Phenylmercuric acetate
Acetaldehyde
Acetonitrile
Benzene

Carcinogenicity

Cancer is a general term for a group of related diseases characterized by the uncontrolled growth of certain tissues. Its development is due to a complex process of interacting multiple factors in the host and the environment. In some cases, specific cancer causing chemicals, termed carcinogens, are involved, especially in occupational environments. In fact, the first documented cases of cancer produced by exposure to environmental chemicals, in this instance coal tar, were scrotal cancers in men and boys employed as chimney sweeps in eighteenth-century Britain.[2] However, there are numerous carcinogens present in the ambient environment.

One of the great difficulties in attempting to relate exposure to a specific chemical to cancer development is the generally long latent period, typically from 15 to 40 years, between onset of exposure and disease manifestation. Research studies of the carcinogenicity of chemicals, therefore, generally involve bioassay techniques in experimental animals. Although the ultimate relation of any positive results to humans is not always clear, evidence for carcinogenicity in such model systems generally results in the classification of the chemical as a potential human carcinogen.

Certain chemicals are carcinogenic as they exist in the environment; these are termed complete carcinogens. On the other hand, other environmental agents, termed procarcinogens, become carcinogenic only after they undergo conversion to some other form. Most of these procarcinogens are converted by biotransformation reactions, generally via the cytochrome P450 enzyme family, although some undergo nonenzymatic hydrolytic reactions to produce carcinogenic intermediates. It is not always clear whether a specific chemical is a complete or procarcinogen.

Some known and suspected human carcinogens are listed in Table 6–8. Many other environmental chemicals have been shown to produce cancer in laboratory animals and thus may be potential human carcinogens. These include N-mustards, epoxides, small-ring lactones, urethane, thioamides, nitrosamine, and azo dyes. The list is quite long.

Current evidence suggests that most organic chemical carcinogens belong to a limited number of chemical classes and, within each class, specific structures appear to be associated with carcinogenicity. However, it is not yet possible to always predict from chemical structure whether a specific chemical is carcinogenic, since very slight differences in molecular structure and orientation result in wide variability of carcinogenic potential.

The site of development of malignant tumors may be at various areas of the body other than the initial portal of entry. Thus, for example, inhaled asbestos produces cancer at its portal of entry, the lung, but also in the peritoneum; azo dyes produce tumors in the liver, their site of biotransformation; radium produces bone cancer at its storage site; aromatic amines produce tumors in the bladder, their site of excretion.

TABLE 6–8. Some Environmental Chemicals Carcinogenic to Humans

Organic Chemicals	Inorganic Chemicals
Benzidine	Arsenic (trivalent)
4-aminobiphenyl	Chromate
_-naphthylanine	Nickel carbonyl
_-naphthylamine	Beryllium
Benzene	Cadmium oxide
Vinyl chloride monomer	Asbestos
Bischloromethylether	
Soot, tar (probably due to polycyclic aromatic hydrocarbons)	**Radionuclides**
	Strontium-90
	Thorium dioxide
	Radium-226
	Radon and daughters

Chemical carcinogenesis is generally a multistage process, which involves initiation, promotion, and then progression. Initiation results from exposure to a carcinogen, which can be either direct acting (i.e., do not required biotransformation) or indirect acting (i.e., procarcinogens that require metabolic conversion into a carcinogen). In either case, the chemical interacts with DNA, interfering with normal cell division and gene transcription. Promoting agents serve to enhance the clonal expansion of the initiated cells and includes epigenetic processes that modulate survival of tumor cells and their further proliferation. Promoter chemicals are not themselves carcinogenic and include such materials as phorbol esters, some phenols, and some hormones. Finally, progression involves development of cells that will continue to produce tumors independent of any further exposure to the initiator or promoter.

Some changes in DNA resulting from carcinogen exposure can be repaired by various cellular proteins, preventing the progression of tumor formation. One example is the tumor suppressor protein p53. Tumor-suppressor proteins act to alleviate the potential for cancer and tumor formation by modulating cell growth, either through negative regulation of the cell cycle or by promoting apoptosis. Other proteins involved in repair of DNA damage include DNA polynucleotide ligases, which fix single-strand breaks by repairing broken phosphate bonds, and enzymes that regulate excision repair (both nucleotid and base excision repair) and double-strand break repair.

Alteration of Cell Signaling Pathways

In order to maintain their homeostasis, cells receive and process extracellular chemical signals that originate from other cells. These signals act to regulate various cellular functions as environmental conditions around the cell may change. These signaling molecules bind to a receptor protein, which are most commonly located within the cell membrane, resulting in receptor activation that initiates the response by transmitting a subsequent signal to the cell interior via internal signaling pathways. These pathways involve a cascade of intracellular signaling proteins, the signal transduction cascade, which can modulate the signal as it is spread, or integrate signals from other pathways.

One of the more toxicologically important of the signaling pathways are the protein kinase cascades. The kinases are enzymes that attach phosphate to certain specific amino acid residues on protein molecules. This phosphorylation serves to modulate the activity of enzymes involved in the signaling pathway, either activating or inactivating them. The ultimate downstream result is modulation of the activity of an "effector" protein, which is generally a transcription factor or enzyme.

Chemicals may affect signaling pathways in a number of manners, namely by activating them, inhibiting them, or producing some other disruption along the cascade pathway. Chemicals can affect the receptor, the effector, or they can produce some disruption at various intermediate sites along the pathway. A chemical may affect more than one pathway and at various sites along each. The resultant overall toxicological response would be due to the net balance of effect along all of the linked cascades. Many of the signaling cascades result in activation of genes that control cell survival, division, and apoptosis, and the resultant net effect is manifested as an imbalance in these processes.

Interaction at the receptor level would result in alteration of downstream signaling. An example of a chemical that disrupts the process at this point is plasticizer diethyl-(2-ethylhexyl) phthalate. Because of the tissue site of the receptors affected by this chemical, exposure has resulted in abnormal apoptosis of germ cells.

Another manner in which environmental chemicals can affect the signaling cascade is via activation of protein kinases themselves. For example, many heavy metals can induce production of reactive oxygen species, which in turn can activate protein kinases, thus triggering a signaling cascade. Such activation can upregulate proteins involved in control of the cell cycle and could be involved in pathogenesis of cancer. Dioxin is an example of another chemical that activates protein kinases. Some heavy metals can inactivate mechanisms that normally act to inhibit certain pathways, thus effectively indirectly activating them. Finally, chemicals may interfere with the end product of the cascade, such as a transcription factor; heavy metals are an example of these.

INTERACTIONS OF CHEMICALS

Most people are generally exposed to a chemical mixture, namely more than one chemical contaminant at any one time; thus, interactions of chemicals, as these may relate to health effects, are quite important. Interaction of chemicals can refer to either a qualitative or quantitative change in toxicity with exposure to a mixture compared to what would be expected with exposure to single chemicals, resulting in responses that can be greater or less than additive. Additivity occurs when exposure to a mixture results in responses that are the sum total of the effects from each of the individual chemicals if given alone. This includes the case where there is exposure to low concentrations of chemicals that would have no effect if given alone yet in combination produces a response.

Sometimes the biological response following simultaneous exposure to a combination of chemicals may be much greater than would be expected from the additivity of the action for each individual agent. This is termed synergism. Examples of synergism are combined effects of many solvents on the nervous system following inhalation or combined effects of alcoholic beverages with some inhaled organic compounds. Another type of interaction, termed potentiation, occurs in a binary mixture when one chemical has an effect and the other does not when administered separately, but the latter enhances the effect of the former when exposure is to both chemicals. An example of potentiation is gas-particle interaction in the lungs, whereby gases adsorbed onto inhaled particulates may reach deeper lung areas than they normally would if inhaled in the gas phase alone.

An important area of concern is the synergistic/potentiation effect of certain environmental chemicals in the pathogenesis of cancer. As previously noted, agents termed promoters enhance the carcinogenic effect of other chemicals, although they are not carcinogenic themselves. In some classic examples, the combination of SO_2 with benzo(a)pyrene resulted in respiratory tract tumors in hamsters and rats, while no tumors were found following inhalation at similar concentrations of either agent alone or to SO_2 at any level.[3] Asbestos workers and uranium miners who smoked cigarettes were found to have a significantly greater risk of developing lung cancer than did their nonsmoking counterparts.[4,5]

In another type of interaction, a chemical may act in an antagonistic manner with respect to another; that is, there is a decrease in toxic effect below that expected by sole exposure to the latter or the combined effect is less than additive compared to the response when each is administered alone. Sometimes the former type of interaction is termed inhibition. As opposed to synergism, exposure to both agents does not have to be simultaneous to produce antagonism. For example, there is some evidence that selenium induces protection from mercury toxicity, while Zn seems to be protective of lead toxicity. Finally, combined exposure to various organic solvents can result in mutual reduction of metabolism.

A related type of interaction is termed masking, and this occurs when the components of the mixture produce opposite or competing responses at the same target, thus diminishing the effects of each other.

A further type of interaction involves environmental chemicals and viable agents. A nonspecific response to certain chemicals, primarily irritants, can be a depression of the immune response, making the host more susceptible to viable pathogenic agents. Increases in the susceptibility of experimental animals to respiratory tract infections, for example, bacterial pneumonia, have been observed following exposures to oxidant gases such as O_3 and NO_2.

A special type of interaction is termed tolerance. Tolerant individuals are not killed by levels of certain chemical exposures that are lethal to their nontolerant counterparts. Exposures of rodents to sublethal levels of O_3 afforded protection against mortality in subsequent exposures to the same levels of O_3. Treatment with low concentrations of O_3 can also confer protection against the acute effects of other chemicals, for example NO_2. This phenomenon, termed cross-tolerance, also occurs between many other oxidants.

Examination of the effects of mixtures, especially if there are more than two components, is made difficult by the various ways in which the toxicological interaction may be manifested due to interaction with various biological systems. Thus, chemical interaction may affect absorption, storage, or biotransformation of one or more components of the mixture, for example by interacting with metabolizing enzymes, or it may inhibit or increase excretion. How these interactions ultimately affect health clearly depends on the relationship between the chemicals involved and the target affected.

The mechanistic bases for toxicological interactions are direct chemical–chemical or a change in toxicokinetics (e.g., modulation of absorption, distribution, metabolism or excretion of one chemical by another) or toxicodynamics (e.g., alteration by one chemical of the target for another or alteration of chemical binding sites, or both, of another). A chemical–chemical interaction mechanism occurs if one chemical interacts directly with another, resulting in a change in the chemistry of one or more chemicals in the mixture. An example of this that would result in a less than additive response would be reaction of chelating agents with metals, while an example of greater than additivity would be reaction of ingested noncarcinogenic nitrites and amines to form carcinogenic nitrosamines.

A toxicodynamic interaction would occur when there is alteration by one chemical of the target site for another. If the interaction occurs at the same target (i.e., if both chemicals have the same target site), then the result is generally antagonistic. An example of this would be blocking of acetylcholine receptor sites by atropine thus preventing toxicity from organophosphate pesticides. On the other hand, interactions that involve different targets or binding sites for the various chemicals in the mixture may result in greater or reduced toxicity from the mixture compared to the components alone.

A toxicokinetic interaction involves modulation of absorption, distribution, metabolism, or excretion of one chemical by another. These changes may involve enhancement or reduction of the specific function noted. For absorption, these changes may occur in respiratory ventilation parameters or GI mobility, for example. Thus, zinc (Zn) in the diet can inhibit some Pb toxicity or that due to inorganic Hg by reducing dietary absorption of these metals.

Alterations in the distribution of a chemical may result from a change in receptor site or the release from storage sites. For example, selenium (Se) can reduce cadmium (Cd) toxicity by changing the sites of Cd distribution, and arsenic (As) affects the distribution of copper (Cu) in the body.

A change in chemical metabolism is a common basis for interaction, and this usually involves the P450 family of enzymes and may involve either an inhibition or induction of enzyme activity. For example, ethanol is a competitive inhibitor or alcohol dehydrogenases, resulting in a decrease in the metabolism of methanol to the toxic intermediates formaldehyde and formic acid.

Finally, alteration in excretion can occur due to effects in the respiratory tract, kidneys, or GI tract, whereby one chemical can alter excretion of another by altering pulmonary clearance mechanisms or urinary output. For example, arsenic enhances the excretion of Se by the liver into the bile, and Pb can increase the urinary excretion of Zn.

REFERENCES

1. Mehaffey, K.R., McKinney, J., and Reigart, J.R. Lead and Compounds, Chapter 14 in *Environmental Toxicants: Human Exposures and Their Health Effects*, 2nd ed. (M. Lippmann, Ed.), New York: Wiley, Interscience, 2000, 481–521.
2. Pott, P. Chirurgical observations relative to the cancer of the scrotum, 1775. Reprinted in *Nat. Cancer Inst. Monogr.* 10:7–13, 1963.
3. Kuschner, M. The causes of lung cancer. *Am. Rev. Respir. Dis.* 98:573–590, 1968.
4. Selikoff, I.J., Hammond, E.C., Chung, J. Asbestos exposure, smoking and neoplasia. *J. Am Med Assoc.* 204: 106–112, 1968.
5. Archer, V.E. and Wagoner, J.K. Lung cancer among uranium miners in the United States. *Health Physics* 25:351–371, 1973.

7

Effects of Contaminants on Environmental Quality

Chemical contaminants may affect the abiotic environment in various ways; the result is often expressed in terms of changes in environmental quality. This term, however, is difficult to define, since environmental quality means different things to different people and is related to cultural and social attitudes. To some, a quality environment is a virgin forest, while to others it may be a thriving metropolis. Yet all would probably agree that no matter what specific environment is ideal, certain factors can reduce its quality. A quality environment may thus be defined as one that does not adversely affect the health or welfare of its human inhabitants or its ecological stability.

AIR CONTAMINANTS

Effects on Animals

Like humans, both domestic and wild animals are subject to the acute and chronic health effects of air contaminants. During various acute air pollution episodes, for example, many animal pets and livestock became ill, and some died from cardiopulmonary disorders. For example, in London during the Killer Fogs of December 1883 and December 1952, prize cattle in the annual Smithfield Cattle Shows died of bronchitis and pneumonia.

Chronic poisonings of animals are more common than are acute intoxications and arise from inhalation exposures or livestock feeding on forage upon which

TABLE 7–1. Effects of Some Air Contaminants on Animals

CHEMICAL	SYMPTOMS
Arsenic	Colic, ulcers, hair loss, scleroderma (increase in thickness of upper layer of skin), bone malformation
Lead	Nervous system disorders, swollen joints, emaciation
Fluoride	Loss of appetite, weight loss, gastrointestinal disturbances, decreased milk production, lameness, worn teeth, bone disorders
Organic mercury	Nervous system disorders, listlessness, vomiting
Selenium	Loss of hair, abnormal hoof growth, systemic effects
Molybdenum	Gastrointestinal disturbances, limb stiffness, hematological disorders

air contaminants have accumulated. Chronic poisoning of livestock due to arsenic (As) from smelters, as well as arsenic poisoning of game animals and bees, were reported as long ago as the early part of the twentieth century.[1,2] Airborne fluorides have caused more damage to domestic animals on a worldwide basis than any other air contaminant. One of the earliest descriptions of fluoride poisoning was provided in 1937, and involved livestock near an aluminum (Al) smelter in Italy.[3] Numerous insects have been victims of fluoride poisoning due to contaminant deposition on plant surfaces.[4]

Another air contaminant of historical importance in terms of damage to animals is lead (Pb). For example, in Germany in 1955, cattle and horses near a foundry developed symptoms of Pb poisoning,[5] while cattle and horses grazing within 5 km of a smelter in Canada were found to have Pb-induced damage.[6] Captive animals in city zoos are also subject to poisoning; as they licked their fur, these animals ingested Pb accumulated primarily from automobile exhaust emissions.[7] Table 7–1 describes the effects of selected air contaminants upon animals.

Acidification of surface waters in some regions by deposition of airborne acid aerosols has led to a decline in fish and amphibian populations. A decrease in water pH is often associated with an increase in availability of certain metals, such as Al and mercury (Hg), which themselves may have adverse effects. For example, Al can affect the diffusion of oxygen (O_2) within the gills of fish.

While air pollutants can affect animals directly, they can also result in indirect effects by producing alterations in ecosystems. For example, an adverse impact on vegetation would result in changes in animal populations, and certain metals that deposit in soils could affect the viability of soil invertebrates upon which other organisms depend.

Effects on Vegetation

In the early history of air pollution, there were well-documented cases of the destruction of vegetation around certain industries. Today, the scope of the problem has changed, with more widespread, but generally less severe, effects predominating. Although industrial sources do account for a certain amount of injury to vegetation, the larger problem is associated with more ubiquitous air contaminants, such as ozone and sulfur dioxide, which are common to urban centers. For example, large natural forests in southern California have been severely depleted by oxidant damage to pine trees, and pollution-related damage to vegetation has also occurred in the corridor extending from Washington, DC, to Boston.

Phytotoxic air contaminants may cause damage to natural plant communities as well as to agricultural crops and ornamental vegetation. Different species of plants, different varieties of one species, and even different parts of one plant can vary widely in their sensitivity to a specific chemical. Sensitivity is affected by a number of factors, such as the age of the plant, amount of moisture in the soil, air temperature, plant nutritional status, and intensity of sunlight. In addition, some contaminants are associated with specific manifestations in particular plants; this often allows identification of the type and range of a specific chemical.

Contaminants may enter plants following wet or dry deposition on aerial structures (leaves, stems) or from the soil via the roots. Most gases enter via leaves through the stomata, small openings between cells on the lower leaf surfaces. Particles are generally not injurious unless they are corrosive or deposit in very heavy layers.

As in animals, injury to vegetation may be acute or chronic. Short-term exposure to relatively high concentrations of phytotoxicants produce necrotic patterns on leaves due to cell collapse and perhaps eventually plant death. More commonly, however, plants are subject to long-term exposure to lower contaminant levels, resulting in numerous types of chronic injury, the most common of which are stunting of growth, destruction of leaf tissue, and chlorosis, a reduction and loss of chlorophyll. Other responses are premature aging, leaf abscission, small fruit, and flowering or fruiting abnormalities. Genetic and biochemical changes, such as alterations in the activity of certain enzymes, may occur without any visible injury. Chronic exposures may also increase the susceptibility of plants to other environmental stresses, such as pests and disease. On the other hand, in some cases a contaminant may be beneficial to vegetation; large increases in atmospheric carbon dioxide (CO_2) may enhance photosynthetic activity in terrestrial forests, resulting in increased biomass. Some plants may accumulate chemicals. For example, during the time in which leaded gas was used, high levels of Pb were found in vegetation near highways.[8]

Table 7–2 lists some of the more important phytotoxic air contaminants and presents typical effects and some examples of the more sensitive plants. The list is not all-inclusive. Other contaminants, such as Hg vapor, ammonia (NH_3), hydrogen sulfide (H_2S), sulfuric acid (H_2SO_4), and herbicides, may affect plant life. Plant activity may also be indirectly affected by specific classes of contaminants, such as the restriction of growth due to alterations in soil chemistry caused by acidic rainwater.

Table 7–2. Effects of Air Contaminants on Vegetation

CHEMICAL	SYMPTOM	SENSITIVE PLANTS[a]	EXAMPLES OF CONCENTRATION FOR SENSITIVITY
Chlorine	Bleaching, leaf-tip and margin necrosis, leaf abscission, spotting chlorosis	Radish, alfalfa, peach, buckwheat, corn, tobacco, oak, white pine	Radish, 1.3 ppmv
Fluorides	Leaf-tip and margin necrosis, chlorosis, dwarfing, leaf abscission, decreased yield	Gladiolus, tulip, apricot, blueberry, corn, grape, blue spruce, white pine	Gladiolus, apricot, 0.1 ppbv
Nitrogen oxides	Brown spots on leaf, suppression of growth	Azalea, sunflower, mustard, tobacco, pinto bean	Pinto beans, 3 ppmv
Sulfur dioxide	Bleached spots on leaf, chlorosis, suppression of growth, early abscission, reduced yield	Barley, pumpkin, alfalfa, cotton, wheat, lettuce, apple, oats, aster, zinnia, birch, elm, white pine, ponderosa pine	Alfalfa, barley, cotton, 0.3 ppmv
Ozone	Reddish brown flecks on upper surface of leaf, bleaching, suppression of growth, early abscission, premature aging	Alfalfa, barley, bean, oat, onion, corn, apple, grape, tobacco, tomato, spinach, aspen, maple, privet, white pine, ponderosa pine	Tomato, tobacco, 0.5 ppmv
Oxidant gases, e.g., peroxyacetyl nitrate	Glazing, silvering or bronzing of lower surface of leaf	Pinto bean, mustard, oat, tomato, lettuce, petunia, blue grass	Petunia, lettuce, 0.2 ppmv
Unsaturated hydrocarbons, e.g., ethylene	Leaf abscission, dropping of flowers, loss of flower buds, epinasty, chlorosis, suppression of growth	Orchid blossom, carnation blossom, azalea, tomato, cotton, cucumber, peach	Orchids, 0.005 ppmv Tomatoes, 0.1 ppmv

[a]Certain varieties of these plants are sensitive.

Effects on Materials

Significant damage to materials by air contaminants has occurred in many areas. Corrosion rates, for example, are significantly higher in contaminated urban and industrial atmospheres than in rural atmospheres. Such damage may be the result of various mechanisms. These include abrasion destruction and the deposition of particulate matter on surfaces, resulting in the soiling of their appearance. The most important mechanism, however, is chemical attack. This may be direct, that is, the contaminant reacts directly with the material, as in the tarnishing of silver (Ag) and the blackening of Pb-based paints by H_2S, or indirect, that is, damage due to a product of chemical conversion following absorption of the contaminant, as in the absorption of sulfur dioxide (SO_2) by leather and conversion to sulfurous acid (H_2SO_3) and H_2SO_4. The basic processes that lead to the deterioration of masonry surfaces are illustrated in Figure 7–1.

A number of factors affect the extent of deterioration. Degree of moisture is one of the most important. Oxides of sulfur (S), carbon (C), and nitrogen (N) in

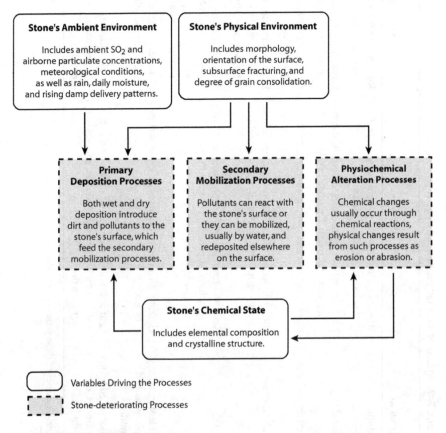

FIGURE 7–1. Stone deterioration processes.

moist air may be converted into acidic forms, leading to corrosion of numerous materials. Acidic rainfall in certain regions of the world has resulted in enhanced corrosion. On the other hand, rain may decrease corrosion rates to the extent that corrosive contaminants are washed away. Another mediating factor in material deterioration is temperature, which affects the rate of the chemical reactions responsible for deterioration and also has an influence on the degree of moisture condensation on surfaces.

Other factors affecting deterioration include sunlight, which may promote photochemical reactions; air movement, which carries contaminants to surfaces; and position of the surface in space with respect to deposition or accessibility to a contaminant. Table 7–3 presents a survey of the types of damage to materials caused by major air contaminants. While the corrosion of steel, zinc (Zn), and limestone is today substantially lower than it was about 25 years ago, the extent of corrosion of Zn and limestone remains essentially unchanged from that time through the turn of the twenty-first century.[9]

Air contaminants have been responsible for damage to art and cultural treasures in many areas of the world. For example, in Athens, the Acropolis and other ancient temples in the Parthenon have been subject to cracking and decay due to the growth of acidic salts penetrating into cracks in the building structure.

Effects on Climate

One of the most potentially devastating manners in which air contaminants may affect the environment is via alteration of climate. This refers to a significant change in climate parameters, such as temperature, precipitation, or wind patterns, that occur over extended periods of time (e.g., several decades or longer). As discussed in chapter 2, the earth–atmosphere energy balance is maintained by minor chemical constituents; this suggests that small changes in their levels due to human activities could have major effects on this balance. The balance may also be subject to perturbation by increasing anthropogenic release of waste heat and by increasing fine particle load with an associated scattering of incident solar radiation. The potential effects of each of these are discussed next.

Waste heat

Waste heat released into the atmosphere can alter local climate, primarily by causing an increase in the average surface temperature in urban areas. The primary sources of waste heat are electric power–generating facilities, space heating, and changes in surface heat absorbing capacity.

On a global scale, the rate of energy used and released as heat is small compared to the solar input to the earth's surface; net solar radiation is over 6,000 times greater than the anthropogenic source strength. However, since energy use is usually concentrated on a local or regional scale in highly populated areas,

TABLE 7–3. Effects of Air Contaminants on Materials

CHEMICAL	PRIMARY MATERIALS ATTACKED	TYPICAL DAMAGE
Carbon dioxide	Building stones, e.g., limestone	Deterioration
Sulfur oxides	Metals	
	Ferrous metals	Corrosion
	Copper	Corrosion to copper sulfate
	Aluminum	Corrosion to aluminum sulfate (white)
	Building materials (limestone, marble, slate, mortar)	Leaching, weakening
	Leather	Embrittlement, disintegration
	Paper	Embrittlement
	Textiles (natural and synthetic fabrics)	Reduced tensile strength, deterioration
Hydrogen sulfide	Metals	
	Silver	Tarnish
	Copper	Tarnish
	Paint	Leaded paint blackened due to formation of lead sulfide
Ozone	Rubber and elastomers	Cracking, weakening
	Textiles (natural and synthetic fabrics)	Weakening
	Dyes	Fading
Nitrogen oxides	Dyes	Fading
Hydrogen fluoride	Glass	Etches, opaques
Solid particulates (soot, tars)	Building materials	Soiling
	Painted surfaces	Soiling
	Textiles	Soiling

natural and anthropogenic source inputs in these regions may be on the same order of magnitude.

Urban heat release has affected the character of the local climate, producing what is known as the "urban heat island." Average urban temperatures in many areas often exceed those in surrounding rural areas by 1° to 2°C, with nighttime differences as great as 5°C. The thermal capacitance for solar input of buildings and streets in built-up areas is also a factor in formation of this heat island.

Local or regional heat islands could conceivably result in changes in atmospheric motions that would be global in scope; however, anthropogenic heat output would have to increase by approximately 50-fold before there would be climate changes comparable to the natural year-to-year variations on a global scale of general atmospheric circulation patterns. It is more likely that local heat islands would disturb the character of natural regional climates. One possibility is that very large concentrations of surface heating could, under appropriate environmental conditions, trigger instabilities leading to convective storms, for example, thunderstorms, hailstorms, and tornadoes. Thermal loading may also contribute to an increase in general cloudiness.

Particle load

Particulate matter (PM) suspended in the troposphere may affect the radiative energy balance and therefore climate, either directly or indirectly. The possible direct climatic effects of tropospheric PM load are quite conflicting. By absorbing and scattering incoming solar radiation, particles tend to reduce the amount of this radiation reaching the ground, resulting in a reduction in heating, both within and below the PM layer. By absorption of the long-wave terrestrial radiation, particles tend to increase the heating in the absorption layer, enhancing the greenhouse effect. The net effect depends upon a number of factors. These include optical characteristics of the aerosol, such as refractive indices of the PM at various wavelengths; particle concentration; particle size distribution; and reflectivity of the underlying earth surface. For example, particle size distribution affects the relative proportion of radiation that is scattered in the forward versus the backward direction (i.e., toward space), as well as the interaction with specific wavelengths of radiation: the greater the particle size, the greater the potential to interact with longer wavelengths.

Particles may also alter climate indirectly by influencing the type, structure, formation, location, or optical properties of clouds. This could affect the earth–atmospheric energy balance, since clouds are a contributory factor both to the amount of solar radiation that is reflected back into space and to the greenhouse effect. Aerosols can act as condensation nuclei for water vapor, increasing cloud droplet number and reflectivity to incoming solar radiation. On the other hand, increased cloud cover would also absorb more outgoing terrestrial radiation. The net effect of increasing cloud cover is therefore not clear. However, increased PM levels and effects on cloud formation may be a factor in the increased precipitation that occurs in, and downwind of, urban areas compared to upwind and remote rural areas.

Carbon dioxide and other greenhouse gases

Even though it is a trace constituent of the atmosphere, CO_2 plays a major role in the earth–atmosphere energy balance. Because it absorbs terrestrial long-wave radiation and reradiates most of it back to earth, increases in atmospheric CO_2 can increase the earth surface temperature. Global background levels of atmospheric CO_2 prior to the Industrial Revolution were estimated at 290 to 300 ppmv, a level believed to have prevailed for about 10,000 years.[10] Currently, the average global level is approximately 400 ppmv,[11] an increase due primarily to the combustion of fossil fuels. The annual rate of increase in CO_2 concentration jumped from 0.7 ppmv/year in the late 1950s, when the first reliable measurements were made, to over 1.5 ppmv/year by the end of the twentieth century. While numerous models of future trends of atmospheric CO_2 levels have been made, actual levels achieved will depend upon whether the current rate of fossil-fuel consumption continues, declines, or levels off. Nevertheless, most models do predict that the trend of increasing atmospheric CO_2 levels will continue.

Given this, numerous models have attempted to ascertain the magnitude of any temperature change due to increased of levels of CO_2. Overall, they suggest that increases in average global temperature will be somewhere in the range of 0.28°C to 4.8°C, with a most likely increase of at least 1.5°C for all scenarios except the one representing the most aggressive mitigation of greenhouse gas emissions.[12]

Other gaseous contaminants, for example, fluorocarbons that were formerly used as refrigerants, methane (CH_4), nitrous oxide (N_2O), and ammonia (NH_3), have infrared absorption bands and absorb in the spectral regions where water (H_2O) vapor and CO_2 are poor absorbers, thus partially interfering with the atmospheric window that transmits most of the reradiated infrared radiation from the earth's surface and lower atmosphere to space. Thus, anthropogenic emissions of these gases may also affect climate by increasing the surface temperature of the earth. The concentration of CH_4 has more than doubled since preindustrial times, reaching approximately 1,800 ppb in 2014, while the concentration of N_2O rarely exceeded 280 ppb over the past 10,000 years but started to rise in the 1920s, reaching 327 ppb in 2014.[13]

Over the course of the earth's history, the climate has changed numerous times, and this has been attributed to various natural processes such as alterations in incident solar radiation or changes in the earth's orbit. However, since the Industrial Revolution, anthropogenic activity has the potential to produce climatic changes that are global in extent. Since 1901, the average global surface temperature has risen an average of 0.08°C. Worldwide, 2016 was the warmest year on record, and the period 2006 to 2016 was the warmest decade on record since observations began. In fact, 2015 was the first time that the average global temperature was 1°C or more above the 1880–1899 average. However, temperature changes are not consistent in all parts of the world. For example, between approximately 1880 and 1940, the mean earth-surface temperature of the Northern Hemisphere increased by approximately 0.6°C, while from about 1940 to 1970, a cooling of 0.2° to 0.3°C occurred in this same region. Furthermore, the mid-century cooling at high northern latitudes was accompanied by a warming at high southern latitudes. It is not entirely clear to what extent these warming and cooling trends were due to anthropogenic activity.

Evidence for a secular increase in mean atmospheric temperature has become much stronger since the mid-1980s, as illustrated in Figure 7–2. An indication of the warming trend at the end of the twentieth century was the observation that there was ice-free ocean at the North Pole during the winter of 1999–2000 for the first time in recorded history. Furthermore, global sea levels have risen approximately 17 cm in the past century, while the rate in the past decade was nearly double that of the past century. Despite these unusual observations, any theory of climatic change is difficult to prove, especially when fluctuations are within the range of normal variability. In addition, many factors control the energy systems that determine climate, and much uncertainty exists in our knowledge of climate cause-and-effect links.

The effects upon global climate of any single contaminant is hard to predict, since possible synergism and numerous feedback mechanisms in the climatic

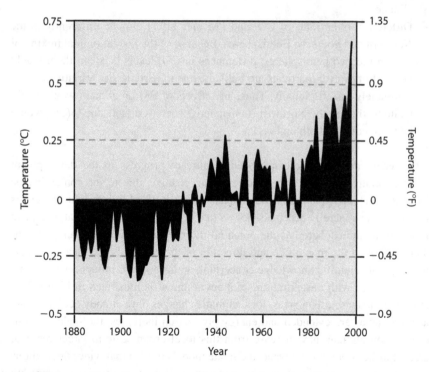

FIGURE 7–2. Global atmospheric temperature anomalies (deviations from average); 1880–1998. "Zero degrees" represents the overall average during each time period. Seven of the ten warmest years have occurred in the 1990's. (*Source*: National Climate Data Center/NESDIS/NOAA.).

system are either unknown or hard to model. Some of the possible mitigating feedback mechanisms are as follows.

- An increase in earth-surface temperature due to high CO_2 levels would likely be accompanied by an increase in cloudiness. This increased cloudiness, by acting to decrease incoming solar radiation, could counteract any warming trend. On the other hand, CO_2 is less soluble in water at higher temperatures, so that increases in sea-surface temperature due to increasing atmospheric CO_2 levels could result in a decrease in ocean uptake, enhancing the warming trend by upsetting the ocean–atmospheric CO_2 balance.
- If earth-surface temperatures increase, snow and ice cover may melt to some extent. This would decrease surface albedo, increase absorption of solar radiation, and thus enhance any warming effect. On the other hand, cooling of the surface could increase ice cover, increasing the albedo, decreasing absorption, and enhancing the cooling effect. Differential heating of the earth sets up heat gradients, which drive wind and ocean currents. Altered gradients due to temperature changes could feed back on winds to change many other aspects of climate.
- Different mixing rates of CO_2 and PM may affect climate differently in the Northern and Southern Hemispheres. Because of the fast latitudinal mixing of CO_2 in the two hemispheres, differences in CO_2 levels between them would be small; however, particles are unlikely, due to particle size variations, to be completely mixed globally. Thus, any effects of PM on climate would likely be intensified in the Northern Hemisphere, whereas effects due to CO_2 would be more equal in both regions.

Over recent decades, there has been significant progress in the development of mathematical models to predict climatic changes due to air contaminants. Nevertheless, inadequacies of the models still often result in inconclusive results However, the generally accepted view is that continued release of heat, CO_2, and PM into the atmosphere has the potential to produce significant changes in climate, although the magnitude and direction of the change is not clear. The current state of scientific knowledge concerning anthropogenic influences on global heat exchange with our extraterrestrial environment is illustrated in Figure 7–3. However, in dense urban areas, local climatic changes have already occurred, and as urban areas are extended and energy production increases, the scale of influences upon climate may increase from this local urban scale to larger regional areas. Furthermore, most advanced climate models predict that wider fluctuations in weather, for example, more severe storms, will precede dramatic changes in average temperature (see Table 7–4).

Temperature changes may also affect the evaporation–condensation pathway of the hydrologic cycle, with possible significant alterations in the amount and

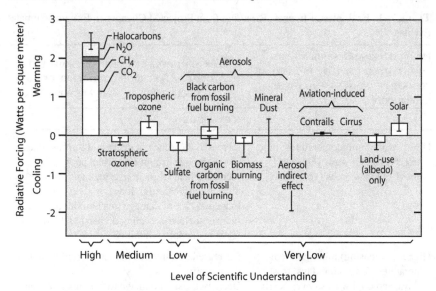

FIGURE 7–3. Estimated global mean radiative forcing exerted by gases and aerosols for the year 2000, relative to 1750.

global distribution of precipitation. The widespread effects of the El Niño and La Niña patterns of water flow and temperature in the eastern Pacific Ocean at the end of the twentieth century produced such changes.

A slight but persistent change in temperature could have serious ecological consequences. It could tip the balance in favor of certain organisms that may have been held in check by previous climatic conditions. Changes in patterns of winds and currents may also disrupt previously adapted organisms. Finally, any increase in the instability of global weather patterns is a definite threat to world food production.

While most of the potential effects of air contaminants upon climate involve the troposphere, certain alterations of the stratosphere may also affect climate. For example, jet-plane flight in the stratosphere, on a large-scale basis, increases the cloudiness of the upper atmosphere, via formation of contrails (i.e., trails of water vapor). Increasing stratospheric moisture content could intensify the atmospheric greenhouse effect. Solid particles injected into the stratosphere, because of their longer residence time than those in the troposphere, may also have an effect upon climate. To date, however, the main stratospheric aerosols are volcanic dust. Although these dusts warm the stratosphere as a result of absorption, surface air may be cooled due to a reduction in radiation reaching the earth's surface. As an example of the complexity in attempting to model future trends in climate change, the increase since 2000 in the abundance of PM in the stratosphere seems to have offset some of the current warming influence of increasing tropospheric CO_2.[14]

TABLE 7–4. Examples of Impacts Resulting from Projected Changes in Extreme Climate Events

PROJECTED CHANGES DURING THE TWENTY-FIRST CENTURY IN EXTREME CLIMATE PHENOMENA AND THEIR LIKELIHOOD[a]	REPRESENTATIVE EXAMPLES OF PROJECTED IMPACTS[b] (ALL HIGH CONFIDENCE OF OCCURRENCE IN SOME AREAS)[c]
SIMPLEX EXTREMES	
Higher maximum temperatures; more hot days and heat waves[d] over nearly all land areas (very likely)	Increased incidence of death and serious illness in older age groups and urban poor
	Increased heat stress in livestock and wildlife
	Shift in tourist destinations
	Increased risk of damage to a number of crops
	Increased electric cooling demand and reduced energy supply reliability
Higher (increasing) minimum temperatures; fewer cold days, frost days, and cold wavesd over nearly all land areas (very likely)[a]	Decreased cold-related human morbidity and mortality
	Decreased risk of damage to a number of crops, and increased risk to others
	Extended range and activity of some pest and disease vectors
	Reduced heating energy demand
More intense precipitation events (very likely[a] over many years)	Increased flood, landslide, avalanche, and mudslide damage
	Increased soil erosion
	Increased flood runoff could increase recharge of some floodplain aquifers
	Increased pressure on government and private flood insurance systems and disaster relief
COMPLEX EXTREMES	
Increased summer drying over most mid-latitude continental interiors and associated risk of drought (likely)[a]	Decreased crop yields
	Increased damage to building foundations caused by ground shrinkage
	Decreased water resource quantity and quality
	Increased risk of forest fire
Increase in tropical cyclone peak wind intensities, mean and peak precipitation intensities (likely[a] over some areas)[c]	Increased risk to human life, risk of infections, disease epidemics, and many other risks
	Increased coastal erosion and damage to coastal buildings and infrastructure
	Increased damage to coastal ecosystems such as coral reefs and mangroves
Intensified droughts and floods associated with El Niño events in many different regions (likely)[a] (see also under droughts and intense precipitation events)	Decreased agricultural and rangeland productivity in drought- and flood-prone regions
	Decreased hydropower potential in drought-prone regions

(continued)

TABLE 7–4. *(continued)*

PROJECTED CHANGES DURING THE TWENTY-FIRST CENTURY IN EXTREME CLIMATE PHENOMENA AND THEIR LIKELIHOOD[a]	REPRESENTATIVE EXAMPLES OF PROJECTED IMPACTS[b] (ALL HIGH CONFIDENCE OF OCCURRENCE IN SOME AREAS)[c]
COMPLEX EXTREMES	
Increased Asian summer monsoon precipitation variability (likely)[a]	Increased flood and drought magnitude and damages in temperate and tropical Asia
Increased intensity of mid-latitude storms (little agreement between current models)[d]	Increased risks to human life and health Increased property and infrastructure losses Increased damage to coastal ecosystems

[a]Likelihood refers to judgmental estimates of confidence used by TAR WGI: very likely (90%–99% chance); likely (66%–90% chance). Unless otherwise stated, information on climate phenomena is taken from the Summary for Policymakers, TAR WGI. TAR WGI = Third Assessment Report of Working Group 1.[21]

[b]These impacts can be lessened by appropriate response measures.

[c]High confidence refers to probabilities between 67% and 95%.

[d]Information from TAR WGI, Technical Summary.

Source: Intergovernmental Panel on Climate Change (IPCC). Climate change 2001: Impacts, adaptation, and vulnerability. Contribution of working group II to the third assessment report of the IPCC. Cambridge: Cambridge University Press, 2001.

Ozone layer

A great amount of interest and international action has involved concern related to the depletion of the stratospheric ozone (O_3) layer due to releases of certain air contaminants. About 90% of the O_3 present in the earth's atmosphere can be found in the stratosphere. Much of this O_3 is created over the equator, where the sun's rays are most direct, and is transported by global air currents toward the poles. This stratospheric O_3 has quite different health and environmental implications than does tropospheric O_3. In the boundary layer of the troposphere, which we share with vegetation and other animals, O_3 is an important component of community air pollution.

Stratospheric O_3 is beneficial because of its ability to screen out potentially harmful solar radiation. The ultraviolet (UV) region of the electromagnetic spectrum is composed of UV-C, UV-B, and UV-A, which in total comprises the wavelengths between 280 and 400 nm. Ozone differentially removes wavelengths of UV-B between 295 and 320 nm; UV-A in wavelengths above 350 nm is not removed, nor is visible light (400–900 nm). Ozone removes all UV-C. Wavelengths between 295 and 300 nm are generally more biologically damaging than are other wavelengths in UV-B and even more so than UV-A radiation.

Ozone concentration is a balance of processes that produce it and those processes and reactions that remove it. Natural formation of stratospheric O_3 involves two steps, whereby oxygen molecules (O_2) are split into single oxygen atoms (O), which, in turn, react readily with nearby atoms or molecules. When O atoms combine with available O_2 molecules, O_3 is formed.

Stratospheric O_3 levels fluctuate periodically. Under some circumstances, the fluctuations are extreme. For instance, atmospheric conditions can at times isolate a portion of the stratosphere, effectively keeping it from mixing with surrounding O_3-rich and warmer pockets of air. Should there be a high concentration of O_3-depleting air contaminants in the isolated pocket, O_3 destruction can outpace O_3 replacement and result in a sharp decrease in O_3 concentration and, consequently, less screening out of UV-B rays. This extreme O_3 depletion occurs during the southern spring (September through November) creating the Antarctic "ozone hole," which is actually not a hole but a thinning of the O_3 layer in the stratosphere that was first detected in 1976. Atmospheric conditions there clearly favor the annual phenomenon, in that extremely cold winter temperatures support the formation of a "polar vortex," an impenetrable core in the atmosphere where O_3 depleting substances accumulated throughout the winter can be activated by springtime solar radiation to break down O_3. Additionally, ice crystals present in this region of the stratosphere can act as reaction "platforms" for enhanced O_3 breakdown. This process is illustrated in Figure 7–4. Eventually, the O_3-depleted air moves over populated areas.

It was determined that the observed depletion of the O_3 layer was the result of the presence of chlorine derived mainly from a class of compounds known as chlorofluorocarbons (CFCs). These chemicals, because of their low toxicity and

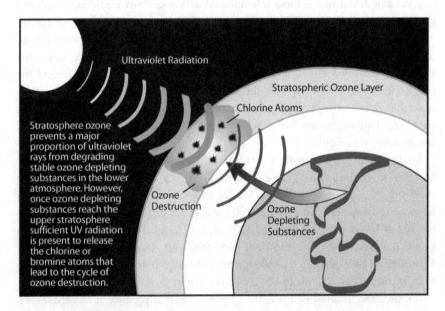

FIGURE 7–4. Basic processes affecting stratosphere ozone depletion.

chemical stability in the troposphere, were widely used since the 1960s as refrigerants and propellants in spray cans. Once released into the troposphere, the CFCs drifted into the stratosphere, where the ultraviolet radiation dissociated the compound, releasing chlorine. The O_3-depleting substances do not just destroy one O_3 molecule; rather, a series of reactions similar to a chain reaction occurs, with chlorine acting as a catalyst, resulting in destruction of O_3 and the formation of diatomic oxygen O_2. The chlorine from the CFCs is then made available to begin the O_3 destruction cycle once again. The bases for concern about stratospheric O_3 depletion are illustrated in Figure 7–5.

The adverse effects of increased UV-B radiation at the earth's surface range from a greater frequency of eye and skin disease in humans and animals, including skin cancer and cataracts, to affecting food chains in natural environments, especially in the polar environments where the O_3 layer was most severely depleted. The first countries to be affected by O_3 depletion over Antarctica were Argentina, Chile, South Africa, and New Zealand. By September and October of 1991, the Antarctic O_3 hole was, for the first time, large enough to allow a large human population in the southern tip of South America to be exposed to increased levels of UV-B. The affected location with the greatest population density was the city of Punta Arenas in southern Chile (~10,000 inhabitants).

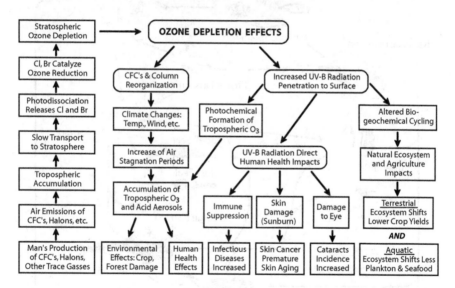

FIGURE 7–5. Ozone depletion processes and effects. Processes involved in stratospheric ozone depletion due to anthropogenic production of CFCs, halons, and other trace gases are shown in the left most pathway. The types of effects caused by stratospheric ozone depletion and consequent increased UV-B penetration to the Earth's surface are shown in the rest of the figure, and are hypothesized to include both direct effects on human health, e.g., increased cancer rates, immune suppression, etc., and other terrestrial and aquatic ecological effects resulting from alterations of biogeochemical cycles.

In terms of sensitive ecosystems in the Antarctic Ocean, plants and animals have adapted to the harsh environmental conditions, and these species are greatly dependent upon one another for survival. Adverse effects to one species in a food web may cause harmful effects to other species and the ecosystem as a whole. A major concern is that UV-B may cause a major decline in plankton that serve as a primary source of nutrients for the Antarctic marine food chain, as illustrated in Figure 7–6. Although injurious UV-B effects have been documented on some individual species within marine ecosystems, the nature and extent of ecosystem responses to UV stress are not well understood. For example, differences in UV-B tolerances between species may be great enough to create a competitive advantage for one particular species over another.

While the connection between stratospheric O_3 depletion and global climate change remains speculative, we do know that CO_2 is a key heat-absorbing compound in the atmosphere and is used up by plants during photosynthesis.

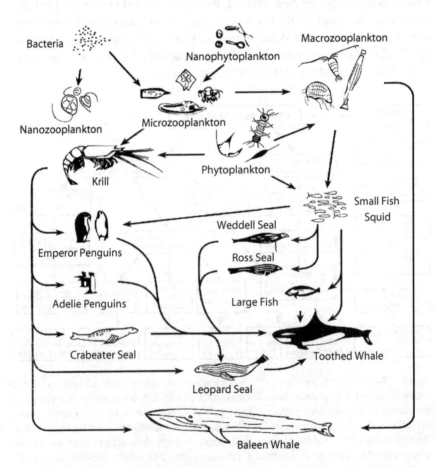

FIGURE 7–6. Antarctic marine food chain. (*Source*: Adapted from EPA/640/K-95/004.)

Significant reductions in plant populations due to increased levels of UV-B radiation may exacerbate the global warming phenomenon.

In view of the long lead times for ground-based emissions of stratospheric O_3-depleting chemicals to reach the stratosphere and the expectation that there will be serious consequences of stratospheric O_3 depletion over a very long period of time, major nations adopted, in September 1987, the Montreal Protocol on Substances that Deplete the Ozone Layer, a landmark international treaty dealing with a worldwide problem in a coordinated way. While recent measurements have shown that stratospheric levels of CFCs are beginning to decline somewhat, those chemicals already in circulation will continue to affect the O_3 layer for decades to come due to the nature of the chain reaction process noted previously. For example, global average ozone levels showed a 3.5% decrease during the period 2006 to 2009 compared with pre-1980 levels, with the decrease in the mid-latitudes of the Southern Hemisphere about twice that observed in the mid-latitudes of the Northern Hemisphere.[15]

Acidification of Surface Waters and Terrestrial Soils

Ordinary rain should have an average pH of approximately 5.7, based upon atmospheric CO_2 in equilibrium with the precipitation. However, large regions of eastern North America (i.e., northeastern United States and southeastern Canada) and southern Scandinavia have received highly acidic precipitation, with a weighted average pH of 4.0 to 4.4 and a range extending as low as 2.8. The acidity is due primarily to atmospheric formation of H_2SO_4 and, to some extent, HNO_3 from SO_2 and NOx, respectively. Acid precipitation occurs especially in areas downwind, up to thousands of kilometers, of dense urban and industrial complexes.

The pH of an aquatic environment reflects both the acidity of the precipitation and the ability to neutralize incoming acid via chemical weathering and ion exchange processes in the watershed. Thus, the susceptibility of any body of water to acidification depends on the ability of the waters and watershed soils to neutralize acid deposition. One measure of this ability is the acid neutralizing capacity (ANC), which characterizes the amount of dissolved compounds that will counteract acidity. Every body of water has a measureable ANC, which depends largely on the surrounding watershed's physical characteristics, such as geology, soil types, and size. Numerous lakes and rivers, especially those with naturally poorly buffered waters (i.e., those with low mineral content) have shown the most dramatic increase in acidity, often having a pH less than 5. With continued acid precipitation, however, even those waters with better buffering capacity may show changes. It does appear, however, that between the early 1990s and 2012, the ANC in lakes in the Adirondack Mountains and New England and in streams in the mid-Atlantic Appalachians in southern New York, west-central Pennsylvania, Virginia, and eastern West Virginia, increased to an extent that many water bodies that were considered acidic in the early 1990s were no longer classified as such in 2012. This trend of increasing ANC in these areas corresponds with a decrease in acid deposition in these regions.[16]

Acidification of waters can affect all aspects of the aquatic environment. Acidified water can impair the ability of fish gills to extract dissolved O_2 from the water and, as noted, can change mobility of certain trace metals, such as Al, cadmium (Cd), manganese (Mn), iron (Fe), and Hg, which, in turn, can place fish and other species at risk. Extensive depletion of fish populations in acidified lakes and streams took place in many of the areas receiving acid rain. The causes were usually direct toxic action of acid waters on eggs and/or larval fish stages, reduction in the number of eggs due to interference with reproductive physiology, and changes in food supply due to the effects of acidity upon other aquatic biota. If the waters are very acidic, a direct toxic action to adult fishes results via interference with ion exchange across the gill membranes.

Like waters, soils having high buffering capacity are not as susceptible to increased acidification. However, any acidification will result in the alteration of some properties of the soil, with an increase in the leaching of certain ions. The ecological effects include a decrease in the growth rate of vegetation and changes in nutrient cycling.

Aesthetics

It is often changes in aesthetics caused by the clearly visible effects of air contaminants that spur the public to demand control of sources of contamination. Some aesthetic effects of air contaminants have already been mentioned, such as the soiling of buildings and damage to ornamental plants. Two others of importance, decreased visibility and odors, are discussed next.

Visibility

One of the most obvious effects of air contamination is the deterioration of visibility. Visibility may be defined, in meteorologic terms, as the greatest distance in a given direction at which it is just possible to see and identify, with the unaided eye, in the daytime, a prominent dark object against the sky at the horizon and, at night, a known, preferably unfocused, moderately intense light source. A reduction of visibility is due both to the absorption and scattering of visible light by gas molecules and particles in the atmosphere, although scattering is the primary mechanism responsible. Scattering is the deflection of the direction of travel of light by airborne matter. Particles in the size range of 0.4 to 0.7 μm are most effective in this regard, because their size range is within the wavelength range of visible light. In addition to size, the degree of scattering is also influenced by particle shape, surface roughness, and refractive index. The coloration of some contaminated atmospheres is due to the selective absorption of specific wavelengths of light by NO_2 and various other atmospheric contaminants. Reduced visibility is not only an aesthetic problem but also a safety hazard. Airline pilots have reported increases since the 1950s and 1960s in ground haze and air pollution domes over some cities; those previously visible from aerial distance of about 30 to 60 km are now often not visible for more than 16 km, or even less during pollution episodes.

Odors

Malodors due to air contaminants are mainly a problem of aesthetics, although property values may be adversely affected in areas where odors are common. Some odorants are merely a nuisance, while others also produce systemic effects, such as nausea and appetite loss. In addition, odors may interact synergistically or antagonistically.

The olfactory system is quite sensitive. Even when chemicals are present in the ppbv range, odors may still be so annoying as to affect personal comfort. However, people do differ widely in their sensitivity to odorants. In addition, the sense of smell is often rapidly fatigued; that is, continued exposure results in loss of smell of the odorant. Many natural sources of odor exist; these are all primarily due to S compounds formed during microbial decomposition of plants and animals, largely in stagnant and insufficiently aerated waters, such as swamps and sewers. Anthropogenic odorants are the result primarily of industrial operations, although mobile combustion sources, such as gasoline and diesel engines, are also contributors in certain areas. The principal industries responsible for odorant release are petroleum refineries, natural gas plants, chemical plants, food processing plants, smelters, paper mills, and tanneries. Table 7–5 lists some common chemical odorants.

TABLE 7–5. Selected Odorants

CHEMICAL	CHARACTER OF ODOR	ODOR THRESHOLD[a]
Acetic acid	Sour	1.0
Ammonia	Sharp, pungent	47
Chlorine	Pungent, irritating, bleach	0.3
Ethyl mercaptan	Decayed cabbage, sulfidelike	0.001
Hydrogen sulfide	Rotten eggs, nauseating	0.0005
Isocyanates	Sweet, repulsive	[b]
Ozone	Slightly pungent, electric sparks	0.05
Selenium compounds	Putrid, garliclike	[b]
Sulfur dioxide	Pungent, mustiness	0.5
Tars	Rancid, skunklike	[b]

[a]This represents odor recognition thresholds for a panel of four trained observers. The concentrations are in ppmv and are those at which all panelists recognized the odor.
[b]Not tested.

WATER CONTAMINANTS

The hydrosphere has long been considered an infinite sink for the disposal of all kinds of wastes. It was assumed that running waters were always able to "purify" themselves and the vastness of the oceans was such that they were able to assimilate any and all of the chemicals and floatable wastes dumped into them. Although water does have certain self-purification capabilities, very often these are simply overwhelmed by contaminant discharge. Today, many aquatic environments are threatened, some to a greater degree than others. These include coastal regions near cities, as well as remote lakes and ocean waters thousands of miles from land areas.

Water quality is a value judgment, since water that is unfit for one use may be fine for another, or water unfit for one species may be quite suitable for another. Even natural waters are never 100% pure. Therefore, an aquatic environment may be considered polluted if the water becomes unsuitable for its intended use or changes occur due to human activities that disrupt the natural ecological balance.

Contaminants may alter aquatic environments in various ways. The type and degree of effect depends upon the type and amount of contaminant and characteristics of the receiving waters, such as temperature, pH, flow rate, degree of dilution and mixing, and presence of other contaminants that may interact with each other.

The effects of water contaminants upon aquatic life may be either direct or indirect. An agent is said to exert direct action if it affects characteristics of the organism itself, for example, growth, reproduction, and physiology. Indirect effects involve alterations of the organism's environment, either abiotic or biotic, so as to make the habitat unsuitable for continued existence or reproduction, for example, by affecting food resources, changing turbidity, or dissolved oxygen (DO) content. Indirect effects may be just as critical to species survival as direct effects.

Contaminants may be quite selective in their action, or they may affect many types of organisms. Often, only certain tolerant species survive, resulting in a change in the population balance. As food sources are destroyed, animals on higher trophic levels die or move on. In addition, stressed individuals are often more susceptible to disease and parasites, further reducing the reproductive potential of the population. Certain chemicals act as repellents to aquatic life, so that, for example, fish are driven out of an area, making the entire region biologically unproductive. Other types of contaminants may result in extreme overproduction and overenrichment of a body of water. The end result of any disturbance is an alteration, often slow and subtle, of the character of the waterway over a period of time.

All contaminants do not adversely affect aquatic life. Some may affect the aesthetics of the water, such as color or the ability to use it for specific purposes, such as drinking or recreation, without affecting natural biotic communities.

As with air contaminants, the aesthetic effects of water contaminants contain components of the entire range of environmental problems. People enjoy high-quality waterways for recreational purposes, or such waters may add a degree of natural beauty to an area. No one likes a beach blackened by floating oil or a lake that smells of rotting sewage. In the following sections, water contaminants are discussed by broad classes; effects upon the physicochemical properties of the aquatic system are presented together with discussions of the effects upon aquatic life.

Waste Heat

Waste heat is released into receiving waters by electric power–generating stations and by industrial processes that require cooling waters. Although cooling towers are often used in attempts to dissipate heat before the water is discharged into the environment, the water is still warmer when released than when it was taken up.

Certain other human activities may affect water temperatures. Removal of forest canopies, removal of brush and shade trees along streams, impounding of river waters, reduction of stream flows, and return of irrigation waters often lead to temperature increases in nearby waterways. The temperature of water naturally varies from season to season, year to year, and even between night and day. Human activities have often served to extend this normal range of variability, sometimes beyond the limits to which native biota are adapted.

Except for birds and mammals, other aquatic organisms are poikilotherms (i.e., "cold blooded"); their body temperature is at or near the temperature of their environment and is subject to change as the environmental temperature changes. Within a certain range of tolerance, however, aquatic organisms can adjust and survive. Problems may arise when the temperature exceeds this limited range. Furthermore, metabolic reactions, like all chemical reactions, are very sensitive to the temperature of the environment in which they occur. As environmental temperature changes, so does the rate of metabolic reactions, which increase with rising temperature. As the metabolic rate increases, so does the demand for O_2. However, with rising water temperature, the solubility of O_2 in the water decreases; thus, just as increased demand occurs, less O_2 is available, and levels of DO may fall below a critical value. Of course, if temperatures are high enough, complete enzyme inactivation and cellular disruption may occur.

Aside from affecting metabolic reactions, temperature influences all facets of life, such as hormonal and nervous control, digestion, respiration, osmoregulation, and behavior. Behavioral changes may be just as detrimental as physiologic changes, for example, attempts to spawn too early in the season or premature migration triggered by changes in temperature.

In addition, many nuisance species of plants and animals thrive at higher temperatures than do more desirable species. For example, blue-green algae survive quite well at around 24° to 34°C, while certain more desirable plankton have their optimal range at about 14° to 24°C. As water temperature increases, these algae will have a selective competitive advantage and they can foul the stream, further reducing DO upon their death and decomposition.

The effects of waste heat also depend, to some extent, upon the season and weather conditions, tending to be more harmful in hot climates than in cold areas. Rapid changes in temperature and intermittent or nondependable fluctuations, such as those due to intermittent discharges, are often more harmful to biota that could otherwise have been able to adapt to slower, more constant changes of temperature greater than their optimal level. Waste heat is also a greater problem in smaller streams and rivers than in larger waterways with better circulation and mixing.

In addition to direct biological effects, waste heat may affect the action of other contaminants in the water. Temperature affects the susceptibility to some chemicals, resulting in increased hazard from these agents. Some examples are the synergistic effects of heat and methyl mercury, PCB, and a number of insecticides.

Suspended Solids

Solids may produce effects on aquatic environments while carried in suspension or after settling to the bottom of the waterway. Suspended solids increase the turbidity of the water, thus decreasing the penetration of light by absorption and scattering of the sun's rays. This directly affects photosynthetic activity in the waterway, which ultimately appears as a reduction in biological productivity; extreme turbidity can result in almost complete cessation of photosynthesis. Increased turbidity may also interfere with the vision of animals, possibly preventing them from sighting their prey. If the concentration of suspended PM is high enough, the feeding of filter-feeding organisms may be affected; in addition, physical injury to delicate eye and gill membranes may occur by abrasion. Once solids settle to the bottom, the benthic environment may be disturbed, for example, by actually burying bottom habitats. Destruction of bottom dwellers and fish larvae, essential parts of certain food chains, could disrupt entire biological communities.

Plant Nutrients: Organic

As described in chapter 5, organic material introduced into receiving waters undergoes a normal process of decomposition by microbial action, which depends primarily on the DO in the water. If sufficient amounts of organic matter are introduced, the rate of oxidative decomposition could exceed the rate of O_2 replenishment, and the DO concentration in the receiving waters will decline. The exact rate and extent of this decline depends upon the differential between the rates of oxidation and replenishment.

Most healthy streams and rivers have DO levels of 5 to 7 mg/L. Values consistently less than 4 mg/L tend to indicate organic overloading. Except for the purest of natural waters, all waters have a measurable biochemical oxygen demand (BOD), which may be as high as 5 mg/L. Many domestic and industrial wastes have BODs of several hundred mg/L which, if inadequately diluted in receiving waters, will produce severe O_2 depletion.

Organic wastes released into waters are, of course, sources of nutrients and food for decomposer organisms. These wastes may encourage their growth, and the resultant increased production may result in other problems, such as health risks and malodors. Suspended microorganisms may increase turbidity and result in discoloration of the water. Organic waste discharges often lead to the growth of bacteria and other microorganisms in the stream bottom, producing what is known as sewage fungus. Not only is this an aesthetic problem, but it may clog water supply inlets or the nets of fishermen, and it may change the benthic environment to make it unsuitable for its natural inhabitants.

Organic waste discharge and resultant O_2 depletion may have both direct and indirect effects upon all aspects of aquatic life. It can result in fish kills due to O_2 depletion or make an area unsuitable for desirable species. In lakes, as DO levels decrease in deep waters, certain species may disappear, while those better able to tolerate reduced O_2 may increase in number. In rivers and streams, the typical example of effects on aquatic life is the succession of communities downstream from a sewage outfall, a process discussed in chapter 5.

The classic scheme for describing the biological effects of organic waste discharge was introduced by Kolkwitz and Marsson.[17,18] This is known as the saprobic system of zones of organic enrichment, which classifies aquatic biota according to areas in which these species are found. Later systems have been patterned after this one, with differences in nomenclature and in delineation of specific zones. Although there are difficulties in attempting to present a rigid and arbitrary classification of an environment as dynamic as a river, the zonation system is of value in ordering, to some extent, interrelationships in rivers receiving organic wastes. The zonal regions proposed by Kolkwitz and Marsson are presented in Table 7–6 in simplified form. It should be kept in mind that in nature there are no clear boundaries between classes, and both the extent and location of any zone within river reaches may vary with time, flow conditions, and seasons of the year.

Plant Nutrients: Inorganic

The primary inorganic nutrients required by plants are nitrates and phosphates, one or both of which are generally the limiting factors to growth. The main effect of an excess of either of these is the "fertilization" of water, resulting in what is

TABLE 7–6. The Zonation of Contaminated Streams

ZONE	LEVEL OF CONTAMINATION	DO	PREDOMINANT BIOTA	BACTERIAL COUNT (PER ML H_2O)
I (polysaprobic)	Heavy	Zero to very low	Bacteria, sludge worms, maggots, some algae	$>10^6$
IIa (mesosaprobic)	Strong, but diminishing	Low to fully saturated	Bacteria, protozoa, worms, rotifers, midge larvae, diatoms, algae, carp	10^4–10^5
III (oligosaprobic)	Low or none	Fully saturated	Diverse plants and animals including game fish	$<10^3$

^aThis zone is actually separated into a heavily contaminated subdivision which adjoins Zone I, and a less contaminated region which adjoins Zone III.
DO: dissolved oxygen.

termed accelerated eutrophication. "Eutrophication" is the term generally applied to the process of natural evolution of a lake. It involves the slow change with time from a biologically unproductive, nutrient-deficient lake with a sparse biotic community (oligotrophic lake) to one that is highly productive, is nutrient-rich, and supports diverse biota (eutrophic lake). The gradual addition and buildup of nutrients in the lake from natural sources in the surrounding watershed and internal mineral fixation leads to this enrichment and biological development. The rate of normal eutrophication in a lake depends upon such factors as soil type, type of vegetation in the drainage basin, and local geology.

At some point in this process, a stable condition results where the rate of biological productivity approximately equals the rate of decomposition. This results in a fairly constant, homeostatic chemical and biologic aquatic environment where small disturbances may cause rates of productivity and decomposition to vary but in balance. Introduction of excessive amounts of inorganic plant nutrients results in accelerated eutrophication in many aquatic environments as these homeostatic mechanisms are overwhelmed.

There are predilective sites for accelerated eutrophication. The prime ones are impounded, either naturally or artificially, bodies of water, such as small lakes and dammed reservoirs. In these, added nutrients may recycle for many years, since there is little removal of any introduced contaminants via water exchange. Semi-enclosed bodies of water such as estuaries and coastal regions of large lakes, especially in those areas downstream to waste discharge sites or runoff from non-point agricultural sources, are also prone to accelerated eutrophication. The rate of accelerated eutrophication is dependent upon the size of the body of water, the rate of nutrient input, and the homeostatic condition that existed prior to any disturbance.

Excess nutrient input initially results in the increased growth of some plankton species, primarily blue-green algae, and certain rooted aquatic weeds. The previously dominant phytoplankton disappear, decreasing the variety of food for herbivores and reducing total plant species diversity. An overview of the major sources of nutrient overload is illustrated in Figure 7–7. It should be noted that eutrophication may also trigger toxic algal blooms, such as red and brown tides.

Upon algal death, oxidative decomposition in the sediments increases the BOD of the water, decreasing the level of DO. Deep stratified lakes that are eutrophic have little or no DO in the hypolimnion. As the deeper waters are robbed of O_2, those fish species that require deep, well-oxygenated waters, for example trout and herring, disappear and are replaced by other species, for example sunfish and carp, which require less O_2 and are supported on shorter food chains. In addition, upon algal death and decomposition, the nitrogen dimmer (N_2) and phosphorus bound in the algae are released back into the water. The spring overturn carries these nutrients to surface layers, where they may promote new blooms. Thus, as mentioned, added nutrients may continually recycle in lakes.

With increasing algal growth, the water becomes more turbid, a condition that restricts the penetration of light to deeper water layers, further depressing photosynthesis below the surface layer. At its worst, the lake becomes stagnant,

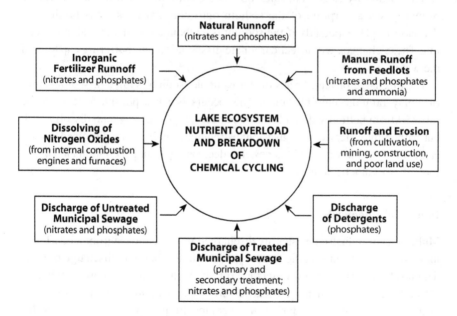

FIGURE 7–7. Major sources of nutrient overload in slow moving lakes.

supporting only anaerobic organisms involved in decay food chains; the decay results in the gradual filling of what is now an odorous (due to NH_3, H_2S, amines), discolored, and generally unsightly body of water. Thus, the biotic and chemical composition of the lake changes, making it unsuitable for domestic, industrial, or recreational uses or even for many species of wildlife.

Although the term "eutrophication" is generally not directly applied to rapidly running waters such as rivers and streams, a response similar to those in lakes does occur following enrichment, with increased algal and bottom growth (chapter 5). In fact, nutrient pollution is the single largest pollution problem affecting coastal rivers in the United States, but this response usually abates if effluent discharge is ended.

Dissolved Solids and Minerals

The total content of dissolved minerals is referred to as salinity. Excessive mineral content may affect the taste of drinking water and be harmful to people who have specific illnesses. "Hard water," that is, water that has excess minerals, primarily calcium and magnesium ions, affects use of the water for domestic, irrigation, and many industrial purposes.

Fresh-water fauna are adapted to low salt levels, and discharges that produce excess salt in fresh-water environments may be lethal. Excess minerals may also affect certain aspects of water chemistry. The addition of chloride and sulfate, primarily as sodium salts, affects the solubility equilibrium of waters, primarily via absorption of the sodium onto clay particles. This produces a decrease in pH, especially if the sodium is exchanged with a less alkaline cation. These changes will affect carbonate dissociation and buffering capacity of the water.

The discharge of inert brines of, for example, sodium sulfate or sodium chloride may promote eutrophication. This occurs because phosphate tends to be released from bottom sediments formed under a region of low total dissolved salt but bathed by waters with higher ionic strength. Finally, some dissolved chemicals may color the water, with a possible decrease in light penetration and, therefore, photosynthetic activity.

Industrial Chemicals

Multitudes of industrial wastes are discharged into aquatic environments. These may be derived from specific factory point sources via direct discharge or accidental release or be general watershed runoff. Many waters contain certain levels of natural toxic chemicals leached from surrounding rocks and soils, for example Hg, or produced during the decomposition of plants and animals, for example, H_2S. Depending upon the specific mechanism that has evolved to handle it, an aquatic organism may be able to tolerate a certain amount of these natural toxicants. However, human activities have often resulted in increasing levels of many natural toxic chemicals to above tolerable limits and have also resulted in the introduction of other chemicals for which the organisms have no defense.

The effects due to discharge of large amounts of toxic chemicals are quite different from those due to sewage discharge described previously. While certain species may become severely depressed by sewage release, others become quite abundant. On the other hand, most toxic chemicals do not benefit aquatic organisms and are probably deleterious to most, except for a few types of microorganisms that may be able to use specific chemicals as a source of energy. Thus, large increases in populations of particular aquatic species are usually not found after toxic chemical release. Rather, the most dramatic effects are fish kills due primarily to acute effects of toxic chemicals. More often, levels of chemicals that are released are not high enough to produce fish kills but rather result in other effects that may or may not be well defined. These may be, for example, expressed in terms of tumors, fungal diseases, systemic pathology, abnormal larval growth, and decreased reproductive potential. Behavioral changes involving migratory, territorial, feeding, and social behavior may also be induced. All of these effects may ultimately result in decreased population size and biologic productivity of the waterway.

Certain chemicals may accumulate in bottom sediments and in food chains. In 1971, for example, the US Food and Drug Administration temporarily banned the sale of swordfish because a large number of samples tested had excessive levels of Hg. High levels of other chemicals often result in the temporary banning of commercial fishing in many areas of the United States. Numerous chemicals, when present in low, sublethal levels, may impart unpleasant tastes to fish and shellfish used as food by humans. Only trace amounts are sometimes enough to produce a noticeable effect without changing the distribution or abundance of biota. For example, unpleasant tastes in fish will occur when chlorophenol is present at levels of 0.0001 mg/L. Other chemicals that impart tastes at low levels are benzene, oil, and 2,4-D. This also affects recreational and commercial fishing and may also make the water undrinkable.

No aquatic environment has an unlimited ability to accommodate contaminants. Even open ocean areas thousands of miles from any population center show evidence of chemical contaminants. For example, waste oil and associated debris have been found in the central Atlantic Ocean, and a decline in plankton primary production since the 1950s has occurred in the North Atlantic. Table 7–7 presents some important industrial chemical contaminants found in aquatic environments and describes their biological effects. The list is by no means inclusive; it would take a whole volume to present the industrial chemicals found in aquatic systems. Bear in mind that

TABLE 7–7. Effects of Selected Industrial Chemical Contaminants in Aquatic Environments

CONTAMINANT	EFFECT UPON AQUATIC ENVIRONMENT
Petroleum	Contains many water-soluble compounds that are toxic to plant and animal species; coats benthic environment when it sinks; oil films decrease light penetration and oxygen absorption; coats body surfaces of aquatic birds and destroys waterproofing; some products may impart unpleasant flavors to fish and shellfish; may act to concentrate other fat-soluble toxic chemicals in the water; interferes with recreational use and is aesthetically unpleasing; oil-laden sediments may move with bottom currents to other areas
Organochlorine pesticides (e.g., DDT)	Reduction in thickness of eggshells in birds which feed upon fish containing residues; decreased survival of fish fry; inhibits phytoplankton photosynthesis; acutely toxic to some aquatic fauna via food chain accumulation
Polychlorinated biphenyls	Similar to DDT; eggshell thinning in aquatic birds
Heavy metals	Toxic to many forms of aquatic life; some may be carcinogenic to fish and shellfish

environmental conditions, such as water pH and temperature, and the presence of other chemicals may modify any effect from each of these contaminants.

Acidity

Aside from acid rainfall, a major source of acidity, especially in streams, is acid mine drainage, primarily from abandoned coal mines. When exposed to air during mining operations, iron sulfide ores (pyrite) become oxidized, resulting in numerous products, including sulfates. Water draining through the mines dissolves these products, and the resulting acidic solution (H_2SO_4) runs off into surface waters and may percolate to groundwater. These acidic waters often contain several kinds of metals, whose solubility in water is increased at the lower pH. The effects of acidity upon aquatic life have been described previously.

DISRUPTION OF ECOSYSTEMS

Natural ecosystems, upon which humans ultimately depend, are complete functional units that create and maintain patterns of energy flow, nutrient cycles, growth, sanitation, and, within limits, self-regulation and self-restoration. If these limits are exceeded, however, disruption and degradation may occur. It may require imbalance in only a few components to upset an entire ecosystem.

Alterations in Biogeochemical Cycles

Biogeochemical cycles are integral components of the ecosystem, serving to maintain steady-state levels of various chemical elements via equilibrium in a number of continuing processes. Disruption of biogeochemical cycles is one of the greatest problems caused by human activity. Many contaminants enter natural cycles, becoming redistributed and/or concentrated at some stage as the cycle strives to regain an equilibrium. What results is too little here and too much there; this redistribution of cycle intermediates may result in an effect on human health and welfare.

The carbon cycle can be influenced in different manners and shows evidence of a global disruption. Exchange of CO_2 between the hydrosphere and atmosphere had resulted in a balanced, relatively constant CO_2 concentration in the latter until about the time of the Industrial Revolution. Since then, the ever-increasing combustion of fossil fuels is returning to circulation carbon that had been fixed into biota, and then subsequently trapped, millions of years ago as coal and petroleum. This has resulted in an upsetting of the dynamic equilibrium between the atmosphere and the hydrosphere. Although the total amount of CO_2 injected into the atmosphere is small when compared to the total amount of CO_2 in global circulation, the atmospheric reservoir is also small. CO_2 cannot be absorbed into the much larger hydrospheric reservoir as fast as it is being generated. The result of this, as discussed previously, is a gradual increase in atmospheric levels of CO_2. In addition, deforestation results in reduction of plants that absorb CO_2.

The nitrogen cycle shows evidence of imbalances on a more local scale. For example, runoff of nitrates from fertilized land areas into lakes and streams may result in overenrichment of these waters. Balanced nutrient cycles would have maintained optimal levels of plant nutrients in these waterways. Organic N from farm wastes and untreated or inadequately treated domestic sewage may ultimately be oxidized to nitrate, again resulting in accelerated eutrophication. Plants that are grown in soils containing high levels of nitrogen fertilizers may assimilate nitrate faster than it is fixed into organic molecules, resulting in an accumulation of nitrate within the plant. These high nitrate levels may adversely affect livestock and people feeding on these plants. Imbalances, especially excesses, at certain stages of the cycle may have other implications in terms of human health. For example, high levels of nitrate in potable water may cause methemoglobinemia in infants. Another manner in which the cycle is affected involves fossil fuel consumption, which releases nitrogen oxides into the atmosphere that can be converted into N_2O, a greenhouse gas.

Prior to the Industrial Revolution, the amount of N_2 removed from the atmosphere via biological fixation was essentially balanced by the amounts returned by denitrification processes. However, the rate of N_2 fixation in the biosphere is now increasing very rapidly due to the large amount of fixation by various industrial processes, such as fertilizer manufacture and various combustion processes, and the increased agricultural cultivation of legumes that fix N through symbiosis with certain bacteria. This accelerated rate of fixation may overwhelm the ability of natural denitrification processes to keep pace or result in increases in denitrification products, producing an imbalance in the N cycle. If sustained for continued periods of time, this imbalance could result in overall global consequences of which we are not yet aware.

Similar to N, steady-state levels of certain S-cycle intermediates normally maintained in undisturbed ecosystems may be upset by human activities. For example, bacterial reduction of sulfate to H_2S under anaerobic conditions is part of the normal S cycle. However, release of sulfate-rich effluents from, for example, paper mills may be responsible for fish kills in waterways, as secondarily derived H_2S reaches lethal levels. Some S compounds released can lead to excess S in soil and waterways, with potential acidification as discussed previously.

Various naturally occurring toxic chemicals have, due to ecological processes controlling their synthesis and degradation, tended to maintain a global distribution that remains fairly constant. As humans produce many of these materials and release them into the environment, an imbalance between degradation and synthesis may occur. A classic example is Hg, which occurs naturally in the environment but generally only in trace amounts. The biogeochemical Hg cycle is shown in Figure 7–8. Under partial to total aerobic conditions in aquatic sediments, microorganisms convert inorganic and phenyl compounds to methyl and, to some extent, dimethyl Hg. These alkyl compounds are the most toxic forms. They are quite stable and, in addition, if taken up by biota, are only very slowly excreted; thus, they tend to become concentrated in biological organisms and may be long-term environmental hazards.

A number of other elements besides Hg may be methylated by microbes in aquatic sediments and terrestrial soils. These include gold, palladium, platinum, tin, thallium, selenium, and tellurium. Like methyl mercury, most of these alkyl metals are quite stable in the aquatic environment. The use of palladium and platinum catalysts in automobile emission control systems and heavy metal alkyls in gasoline additives instead of lead alkyls may thus pose a potential environmental problem.

Disruption of Ecosystem Structure

The introduction of chemical contaminants into the environment often leads to complex changes in overall ecosystem structure and function. However, details of the pattern of alteration have been demonstrated clearly in only a few instances. Although the effects of some contaminants may be more conspicuous than others,

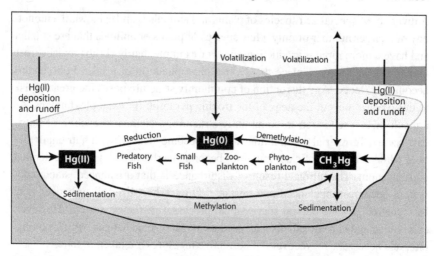

FIGURE 7–8. Fate of mercury in lakes. Mercury entering by direct surface deposition or terrestrial runoff typically includes elemental mercury, Hg(O), oxidized mercury, Hg(II), and extremely small amounts of methylmercury, CH₃Hg. Most lake methylmercury, however, is produced when anaerobic bacteria in the water and sediments methylate oxidized mercury. Methylmercury is bioaccumulated by plankton and other organisms and then passed up the food chain to fish, where it also concentrates. While this diagram represents major processes in the aquatic mercury cycle, the overall cycle is quite complex and cannot be generalized for all lakes.

resultant changes follow similar and broadly predictable patterns for both terrestrial and aquatic ecosystems.

A mature, undisturbed ecosystem is a dynamic unit that maintains some degree of stability. An ecosystem reaches this stage via an orderly development with time, if not disturbed by humans or by natural catastrophes; this evolutionary process is termed ecological succession. Succession involves the interplay between the biotic and abiotic components of the environment, leading toward development of stable communities, with maximum biomass and diversity of organisms in relation to the physical constraints of the environment. Succession proceeds through a sequence of biological community types until a climax community is reached; this is the final and most stable stage. The climax is self-perpetuating if not disturbed, and its complexity results in resistance to perturbations of the physical environment. The biogeochemical cycles are "conservative," that is, only very small amounts of nutrients are lost, as, for example, via drainage water from terrestrial ecosystems.

A classic broad outline of the pattern of ecosystem alteration due to environmental contaminants has been described by Woodwell.[19] The disturbance of any ecosystem is accompanied by a reduction in biotic community structure, with a resultant trend toward simplification and monotony of total ecosystem structure.

A diversity of specialized species of plants and animals is shifted toward a monotony of a larger number of only a few species of plants or animals that are smaller and have a more rapid reproduction rate. For example, hardy shrubs replace trees in a forest, and algae replace the diverse phytoplankton of a lake. Food chains become shortened. Any disruption of community structure poses the greatest risk to those consumers at the apex of the trophic pyramid, for example, humans. As a result of this reduction in community structure, a change in cycling of nutrients occurs. Terrestrial communities tend to become "leaky," with nutrient depletion, while nearby aquatic communities become overburdened due to accelerated nutrient input. The ultimate result of disturbance is that a balanced, stable ecosystem becomes unbalanced and unstable. Often, short-lived disturbances may be "repaired" by the natural successional processes of the ecosystem. However, cumulative and chronic effects may produce more severe changes, with long time intervals necessary for recovery, if corrective measures are taken in time.

While much is still unknown about physicochemical and biological processes in the environment, a summary of the linkage among ecosystem goods and services can now be made in broad terms, as illustrated in Figure 7–9. However, accurate predictions of effects are not always possible, and long-term changes are difficult to assess. This is especially true since individual ecosystems are not discrete entities but are all interconnected with each other in some fashion, especially via food webs.

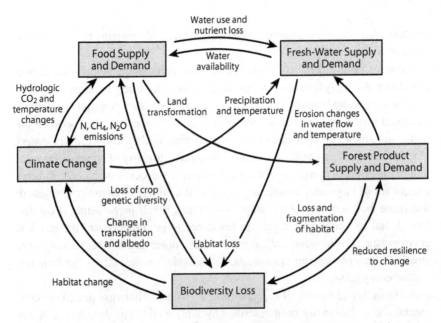

FIGURE 7–9. Linkages among various ecosystem goods and services (food, water, biodiversity, forest products) and other driving forces (climate change).

EFFECTS ON EVOLUTION

The diverse forms of life in the biosphere are the products of evolution, the mechanism of which is natural selection, the power of the environment to select characteristics of the next generation of a population leading to organisms that are better adapted to their surroundings.

Chemical contamination is an environmental change that produces new selection pressures. Given the presence of heritable variation in the population, these new pressures may cause a change in gene frequency and, therefore, a change in the genetic makeup of the population involved. Thus, environmental contamination may be expected to cause evolutionary changes in exposed populations of plants and animals. The evolution of populations resistant to the destructive effects of specific chemicals, or otherwise adapted to life in contaminated areas, enables continued existence of the species under conditions that might otherwise have resulted in their extinction. For example, there is evidence of populations of plants that are specifically adapted to many known chemical contaminants.[20] Thus, changes in population characteristics may be one of the effects of exposure to chemical contaminants, and some of these changes have direct implications for human health and welfare, such as pesticide-resistant insects.

REFERENCES

1. Harkins, W.D. and Swain, R.E. The chronic arsenical poisoning of herbivorous animals. *J. Am. Chem. Soc.* 30:928–946, 1908.
2. Prell, H. Die Schadigung der Tierwelt durch die Fernwirkungen von Industrieabgasen. *Arch. Gewerbepathol.–Gewerbehyg.* 7:656–670, 1937.
3. Bardelli, P. and Menzani, C. La fluorosi (fluorosis), part 2. *Atti Ist. Veneto Sci.* 97:623–674, 1937.
4. Lillie, R.J. *Air Pollutants Affecting the Performance of Domestic Animals: A Literature Review.* Agriculture Handbook No. 380, Washington, DC: US Department of Agriculture, 1970.
5. Hupka, E. *Uber Flugstaubvergiftungen in der Umgebung von Metallhutten.* Wiener. Tierarztl. 42:763, 1955.
6. Schmitt, N., Brown, G., Devlin, E.L., Larsen, A.A., McCausland, E.D., and Savillo, J.M. Lead poisoning in horses. *Arch. Environ. Health.* 23:185–197, 1971.
7. Bazell, R.J. Lead poisoning: Zoo animals may be the first victims. *Science.* 173:130–131, 1971.
8. Chow, T.J. Lead accumulation in roadside soil and grass. *Nature.* 225: 295–296, 1970.
9. Tidblad, J., Kucera, V., Ferm, M. et al. 2012. Effects of air pollution on materials and cultural heritage. *International J. of Corrosion.* Article 496321. http://dx.doi.org/10.1155/2012/496321
10. Barrie, L.A., Whelpdale, D.M., and Munn, R.E. Effects of anthropogenic emissions on climate: A review of selected topics. *Ambio.* 5:209–212, 1976.

11. Royal Society. Is the Current Level of Atmospheric CO2 Concentration Unprecedented in Earth's History. Royal Society.org. 2017.

12. IPCC (2013). Climate Change 2013: The Physical Science Basis EXIT. Contribution of Working Group I to the Fifth Assessment Report of the Intergovernmental Panel on Climate Change [Stocker, T.F., D. Qin, G.-K. Plattner, M. Tignor, S.K. Allen, J. Boschung, A. Nauels, Y. Xia, V. Bex and P.M. Midgley (eds.)]. Cambridge University Press, Cambridge, United Kingdom and New York, NY.

13. USEPA Climate Change Indicators: Atmospheric Concentrations of Greenhouse Gases. 2016. www.epa.gov. https://www.epa.gov/.../climate-change-indicators-atmospheric-concentrations-greenh

14, Solomon, S., Daniel, JS, Neely III, RR, Vernier J-P, Dutton, EG, Thomason, LW The persistently variable "background" stratospheric aerosol layer and global climate change. *Science* 333: 866–870, 2011.

15. WMO. World Meteorological Organization. 2010 *Scientific assessment of ozone depletion*: Geneva, Switzerland. http.://www.esrl.noaa.gov/csd/assessments/ozone/2010/report.html

16. USEPA. Clean Air Interstate Rule, Acid Rain Program, and Former NOx Budget Trading Program: 2012 Progress Report. Environmental and Health Results. 2014. https://www.epa.gov/sites/production/files/2015-08/documents/arpcair12_o2.pdf.

17. Kolkwitz, R. and Marsson, M. Ökologie der pflanzlichen Saprobien. Berichte der Deutschen Botanischen Gesellschaft 26a:505–519, 1908. Translated: Ecology of plant saprobia. In *Biology of Water Pollution*, L.E. Keup, W.M. Ingram, and K.M. Mackenthun, (eds.), Federal Water Pollution Control Administration, US Department of the Interior, 1967, pp. 47–52.

18. Kolkwitz, R. and Marsson, M. Ökologie der tierischen Saprobien. Beitrage zur Lehre von der biologischen Gewasserbeurteilung. *Inter. Revue der Gesamten Hydrobiologie und Hydrogeographie*. 2:126–152, 1909. Translated: Ecology of animal saprobia. In *Biology of Water Pollution Control*, L.E., Keup, W.M. Ingram, and K.M. Mackenthun (eds.), Federal Water Pollution Control Administration, US Department of the Interior, 1967, pp. 85–95.

19. Woodwell, G.M. Effects of pollution on the structure and physiology of ecosystems. *Science*. 168:429–433, 1970.

20. Bradshaw, A.D. Pollution and evolution. In *Effects of Air Pollutants on Plants*, T.A. Mansfield (ed.), Cambridge: Cambridge University Press, 1976, pp. 135–159.

8

Risk Assessment

The human race has always faced dangers, since risk is an inescapable fact of life. At the same time, enlightened societies have traditionally sought to minimize avoidable risks. The greatly increased life expectancy now enjoyed by populations in most parts of the world attests to the success with which modern civilization has been able to reduce some environmental risks to human health and safety.

By the early 1980s, it had become clear that there were myriad risks to public health and welfare from the various chemical agents being released as a result of anthropogenic activities but that aggregate resources for mitigating risks were limited. In response to the need for a set of risk assessment methods and tools to allow estimation of environmental health risks, a strategic approach was developed by the National Research Council (NRC).[1]

While significant uncertainties remained, the NRC framework for risk assessment has been used by various governmental agencies and many industrial organizations, and it continues to expand and evolve as scientific knowledge advances; its major structure is summarized in Figure 8–1. The paradigm basically involves a distinct separation between the overall process described in Table 8–1, that is, risk assessment, which is largely based on the acquisition and interpretation of scientific principles and data, from the process it identified as risk management. The latter is based on executive decisions that

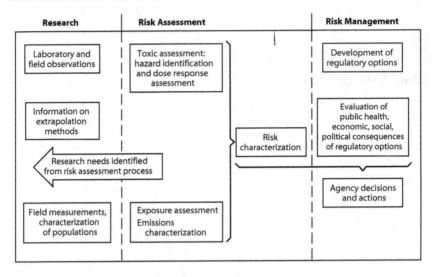

FIGURE 8–1. Paradigm for risk assessment. (*Source*: Science and Judgment in Risk Assessment [NRC, 1994a]. Reprinted with permission.)

recognize the nature and limitations of the scientific risk assessment but also is influenced by the constraints imposed by legislative mandates, the capabilities and costs of implementation of available control options, and the anticipated acceptability of these options in a sociopolitical and legal context. The implication is that scientists with appropriate expertise are best equipped to perform risk assessments but should have no special or dominant role in selecting the means or schedule for the implementation of risk mitigation options. The latter are reserved for responsible officials who can be held accountable by the public and ultimately the court system.

As indicated in Table 8–1, environmental risk assessment is based on hazard identification; dose–response assessment; exposure assessment to determine population distributions of exposures to an environmental stress; and risk characterization, which establishes estimates of the risks for overall populations and for subsegments of the overall population that may be especially sensitive when exposed, each with associated uncertainties.

This chapter reviews environmental risks to human health and welfare from two standpoints: the risk to the individual and the risk to the community. Considered in this context are scientific bases for assessing such risks, the relative magnitudes of different environmental risks as evaluated by knowledgeable experts, the contrasting perspectives in which risks may be perceived by different members of the public, and difficulties in risk communication that complicate societal efforts to protect health and the environment. Options for reducing risks at the individual and community levels are discussed in chapter 10.

TABLE 8–1. The Four Steps of Risk Assessment

STEP	DEFINITION
1. Hazard identification	A review of the relevant biologic and chemical information bearing on whether an agent may pose a carcinogenic hazard and whether toxic effects in one setting will occur in other settings.
2. Dose–response assessment	The process of quantifying a dosage and evaluating its relation to the incidence of adverse health effects response.
3. Exposure assessment	The determination or estimation (qualitative or quantitative) of the magnitude, duration, and route of exposure.
4. Risk characterization	An integration and summary of hazard identification, dose-response assessment, and exposure assessment presented with assumptions and uncertainties. This final step provides an estimate of the risk to public health and a framework to define the significance of the risk.

Source: National Research Council, Committee on the Institutional Means for Assessment of Risks to Public Health. Risk Assessment in the Federal Government: Managing the Process. Washington, DC: National Academy Press, 1983.

The technical means of assessing human exposures to chemical toxicants are discussed in chapter 9, while the means for establishing exposure–response relationships for chemical agents and human health effects were discussed in broad terms in chapter 6. Thus, this chapter is largely confined to a discussion of how risk assessments are performed and how technical risk assessments may differ from risk perceptions of the general public.

THE NATURE OF RISK

Risk is defined as the *probability* of adverse effects on human health resulting from some defined condition of exposure to a particular environmental agent or combination of agents. Risk is distinguished from hazard, which is the *potential* of a chemical to cause some adverse effect under particular exposure circumstances. Thus, while a toxic chemical would be considered to be a potential hazard, if there is no exposure there would be no risk.

Risk may be expressed in various ways, depending on the context in which it is considered. For example, acute responses to short-term peaks of exposure, such as increases in the rates of daily mortality and morbidity, can be expressed in terms of risks to individuals within a population or of overall increments of risk and their distributions within the population. Responses to long-term exposures can be characterized in terms of average annual risk per individual, average lifetime risk per individual, average number of individuals affected annually in a given population, or average loss of life expectancy in affected individuals or in the population as a whole.

For certain types of chronic exposures and their associated health effects, such as pollutant-induced cancers, the risk may also be expressed either as an absolute risk (e.g., absolute increase in the number of cases or the probability of cancer incidence) or as a relative risk (e.g., a relative increase in the background frequency of cancer). Depending on the baseline frequency, or background rate, a small increase in relative risk may be equivalent to a large increase in the number of individuals affected. Conversely, merely a few additional cases of an otherwise rare disorder may result in a large increase in the relative risk.

The importance attached to a given risk depends on the severity, as well as the frequency, of the effect in question. Determinants of severity include such factors as the extent to which the effect is, or is not, symptomatic, painful, disfiguring, incapacitating, reversible, progressive, lethal, and so on; these are the properties that determine its impact on the affected individual, family members, descendants, coworkers, neighbors, and community. In the broadest context, therefore, the measures of severity have many ramifications, including aesthetic, psychosocial, ethical, and economic impacts, as well as impacts on health per se.

Apart from objective measures of the frequency and severity of environmental risks, other qualitative characteristics, such as those listed in Table 8–2, can be important in determining how risks are perceived. Many people not only fail to understand the technical basis for evaluating a given risk, but they may actually distrust and reject it. It has been suggested, therefore, that the definition of an

TABLE 8–2. Psychosocial and Cultural Characteristics Affecting the Perception of Risk

CHARACTERISTICS OF A RISK THAT TEND TO INCREASE ITS ACCEPTABILITY	CHARACTERISTICS OF A RISK THAT TEND TO DECREASE ITS ACCEPTABILITY
Voluntary	Involuntary
Familiar	Unfamiliar
Immediate impact	Remote impact
Detectable by individual	Nondetectable by individual
Controllable by individual	Uncontrollable by individual
Fair	Unfair
Noncatastrophic	Catastrophic
Well understood	Poorly understood
Natural	Artificial
Trusted source	Untrusted source
Visible benefits	No visible benefits

environmental risk should include nontechnical aspects that may be of concern to the public, that is, its "outrage" factors. The importance of such nontechnical factors is illustrated by the marked degree to which public perceptions of a given risk may differ from those of informed experts (Table 8–3).

RISK IDENTIFICATION AND QUANTIFICATION

As outlined in Table 8–1, assessment of an environmental risk to human health involves a sequence of interrelated steps, beginning with identification of the causative agent and exposure situation and culminating in an evaluation of the numbers of persons who are ultimately affected and the severity of the effects.

Hazard Identification

The first step, hazard identification, consists of identifying potentially harmful agents to which persons may be exposed, regardless of the level of exposure. For this purpose, reliance has traditionally been placed primarily on peer-reviewed clinical and epidemiological evidence. For most environmental agents of interest, however, direct toxicity to humans cannot be adequately evaluated from the limited data that are available.[2,3] Instead, the evaluation needs to depend on toxicological approaches, including systematic analysis of pertinent molecular structure-activity relationships, results of in vitro assays, and biological activity in short- or long-term animal bioassays. Principles and procedures for utilizing such methods in predicting toxicity to humans are well developed and are discussed further in subsequent sections of this chapter. At this time, the diversity of potential biological reactions caused by different agents is so large, and the variations in reactivity among different species so great, that the reliability of this approach is often limited.[3] Moreover, for many of the chemicals in commercial production, the available toxicological data do not suffice to enable adequate evaluation (Fig. 8–2).

Dose–Response Analysis

The second step of the process is the dose–response analysis, discussed in Chapter 6, which establishes the mathematical relationship between the dose, or an index thereof, for the agent of interest and any health effects that it may cause. This allows estimation of the nature and magnitude of risks attributable to the agent at the levels of exposure encountered in practice. Since ambient exposure levels are typically many times lower than levels at which any toxic effects may have been documented in laboratory or clinical studies, formulation of the desired risk estimate often requires extrapolation over a broad range of doses and/or animal species, necessitating the use of a dose–response model that may be of uncertain validity.

TABLE 8–3. Perceived as Compared with Real Risks Associated with 30 Widespread Activities and Technologies for two Different Population Groups

ACTIVITY OR TECHNOLOGY	TECHNICAL ESTIMATES (DEATHS/ YEAR)[a]	GEOMETRIC MEAN FATALITY ESTIMATES, AVERAGE YEAR		GEOMETRIC MEAN MULTIPLIER, DISASTROUS YEAR	
		LOWV[b]	**Students**[c]	**LOWV**[b]	**Students**[c]
Smoking	150,000	6900	2400	1.9	2.0
Alcoholic beverages	100,000	12,000	2600	1.9	1.4
Motor vehicles	50,000	28,000	10,500	1.6	1.8
Handguns	17,000	3000	1900	2.6	2.0
Electric power	14,000	660	500	1.9	2.4
Motorcycles	3000	1600	1600	1.8	1.6
Swimming	3000	930	370	1.6	1.7
Surgery	2800	2500	900	1.5	1.6
X-rays	2300	90	40	2.7	1.6
Railroads	1950	190	210	3.2	1.6
General (private) aviation	1300	550	650	2.8	2.0
Large construction	1000	400	370	2.1	1.4
Bicycles	1000	910	420	1.8	1.4
Hunting	800	380	410	1.8	1.7
Home appliances	200	200	240	1.6	1.3
Fire fighting	195	220	390	2.3	2.2
Police work	160	460	390	2.1	1.9
Contraceptives	150	180	120	2.1	1.4
Commercial aviation	130	280	650	3.0	1.8
Nuclear power	100[d]	20	27	107.1	87.6
Mountain climbing	30	50	70	1.9	1.4
Power mowers	24	40	33	1.6	1.3
School and college football	23	39	40	1.9	1.4
Skiing	18	55	72	1.9	1.6
Vaccinations	10	65	52	2.1	1.6
Food coloring	e	38	33	3.5	1.4
Food preservatives	e	61	63	3.9	1.7

(continued)

TABLE 8–3. (*continued*)

ACTIVITY OR TECHNOLOGY	TECHNICAL ESTIMATES (DEATHS/ YEAR)[a]	GEOMETRIC MEAN FATALITY ESTIMATES, AVERAGE YEAR		GEOMETRIC MEAN MULTIPLIER, DISASTROUS YEAR	
Pesticides	e	140	84	9.3	2.4
Prescription	e	160	290	2.3	1.6
Antibiotics Spray cans	e	56	38	3.7	2.4

[a]Based on assessments by technical experts.
[b]League of Women Voters.
[c]College students.
[d]Geometric mean of estimates, which ranged from 16 to 600 per year.
[e]Estimates were unavailable.

Although thresholds are known to exist for many types of acute toxic reactions, no threshold is known or presumed to exist for the mutagenic and carcinogenic effects of cumulative exposures to certain toxicants. For such agents, therefore, an appropriate dose–response model, the selection of which is often

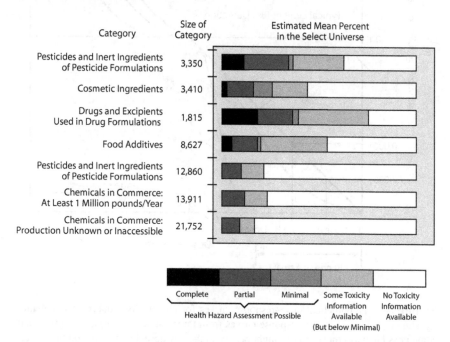

FIGURE 8–2. For most chemicals in commercial production, toxicological data are not sufficiently available for adequate health hazard evaluation.

fraught with uncertainty, must be employed for estimating the magnitude of any risks they may pose. Again, the problem is complicated by a paucity of relevant exposure–response data in human populations and/or dose–response data in toxicological studies. Even in the relatively few instances where human data are available to provide anchor points from which to extrapolate, the data often do not suffice to define the exposure–response relationship in the low-dose domain. In the use of dose–response data from laboratory animals, moreover, there is uncertainty both about the choice of the model for extrapolation to low doses and of the validity of the model for extrapolation of animal responses to humans.[4] Extrapolation options are depicted in Figure 8–3.

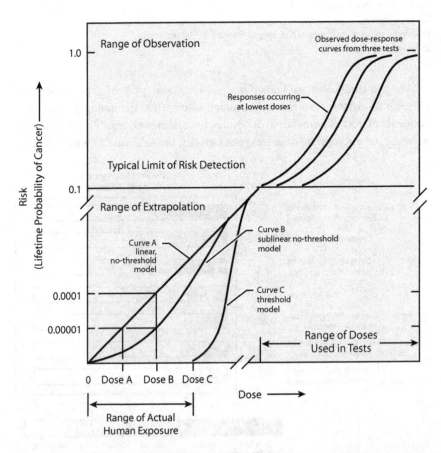

FIGURE 8–3. Hypothetical dose–response curves for chemically-induced carcinogenicity, showing measured dose–response curves from three studies (*top right quadrant*) and some possible ways those curves might behave in the low dose–low risk region (*lower left quadrant*, in The Range of Extrapolation). Note: Not to scale, *lower left quadrant* greatly expanded to illustrate extrapolation options.

Another major source of uncertainty stems from the fact that, in the human environment, chemical agents are usually, if not always, encountered in complex mixtures rather than in the pure form in which they may have been evaluated in most clinical or toxicological studies. Because synergistic or other complex interactions among agents may occur, as discussed in chapter 6, the effects of mixtures of agents can seldom be confidently predicted from what is known about the toxicological effects of any given agent acting alone.

Exposure Assessment

The third step in the risk assessment process, exposure assessment, consists of evaluating the extent to which persons are, or are likely to be, exposed to a particular environmental agent or combination of agents. For the most part, assessments of exposure have relied largely on data from exposure models or from the monitoring of some relevant exposure media (i.e., air, water, soil, food, etc.). Quantitative monitoring of direct exposures to individuals and groups has been limited, in part because of the lack of suitably sensitive, reliable, and practicable methods, as well as validated measures of exposure. Recent advances in analytical techniques, however, and in the development of biomolecular markers will continue to provide future improvements in this area.

Risk Characterization

In the fourth step of the process, risk characterization, the information generated in the first three steps is integrated to derive an estimate of the numbers of persons who may be affected, as well as the types and severities of the effects. To the extent that the utility of the information obtained in each of the preceding steps is constrained by uncertainty, the final characterization of risk derived in the fourth step will, of course, be correspondingly constrained.

Because of the complexity, data requirements, and cost of each step in the overall risk assessment process, as well as the uncertainties inherent therein, detailed and comprehensive attempts at risk characterization have been made for relatively few environmental problems. Examples illustrating some of the uncertainties involved in such assessments are shown in Table 8–4.

RISK COMMUNICATION

People respond to risks as they perceive them. Hence, since experts trained and experienced in evaluating risks can fail to adequately communicate their assessments to the public, efforts to protect health and the environment are sometimes misdirected.[5]

TABLE 8–4. Major Sources of Uncertainty in Risk Assessment

HAZARD IDENTIFICATION	DOSE–RESPONSE ASSESSMENT	EXPOSURE ASSESSMENT	RISK CHARACTERIZATION
Different study types: Prospective, case-control, bioassay, in vivo screen, in vitro screen	Model selection for low-dose risk extrapolation	Contamination scenario characterization (production, distribution, domestic and industrial storage and use, disposal, environmental transport, transformation and decay, geographic bounds, temporal bounds)	Component uncertainties Hazard identification Dose–response assessment Exposure assessment
Test species, strain, sex, system	Low-dose functional behavior of dose–response relationship (threshold, sublinear, supralinear, flexible)	Environmental fate model selection (structural error)	
Exposure route, duration	Role of time (dose frequency, rate, duration, age at exposure, fraction of lifetime exposed)	Parameter estimation error	
	Pharmacokinetic model of effective dose as a function of applied dose	Field measurement error	
	Impact of competing risks		
Definition of incidence of an outcome in a given study (positive-negative association of incidence with exposure)	Definition of "positive responses" in a given study	Exposure scenario characterization	
	Independent vs. joint events	Exposure route identification (dermal, respiratory, dietary)	
	Continuous vs. dichotomous input response data	Exposure dynamics model (absorption, intake processes)	
Different study results Different study qualities Conduct	Parameter estimation Different dose–response sets Results	Integrated exposure profile Target population identification	
Definition of control populations Physical–chemical similarity of chemical studied to that of concern	Qualities Types	Potentially exposed populations Population stability over time	
Unidentified hazards			
Extrapolation of available evidence to target human population	Extrapolation of tested doses to human doses		

TABLE 8–5. Comparison of Factors Relevant to the Cultural Rationality, as Opposed to the Technical Rationality, of Risk

TECHNICAL RATIONALITY	CULTURAL RATIONALITY
Trust in scientific methods, explanations, democratic process	Trust in political culture and evidence
Appeal to authority and expertise	Appeal to folk wisdom, peer groups, and traditions
Boundaries of analysis are narrow and	Boundaries of analysis are broad and include the use of analogy and historical precedent
Risks are depersonalized	Risks are personalized
Emphasis on statistical variation and probability	Emphasis on the impacts of risk on the family and community
Appeal to consistency and universality	Focus on particularity; less concerned about consistency of approach
Where there is controversy in science, resolution follows expertise, status	Popular responses to science; the differences do not follow the prestige principle
Those impacts that cannot be measured are less relevant	Unanticipated or unarticulated risks are relevant

As noted, perceptions of risk involve nontechnical considerations ("cultural rationality") as well as technical considerations ("technical rationality"). Hence, in order to communicate effectively about the nature and magnitude of a given risk, both types of considerations (Table 8–5) must be taken into account. Because the nontechnical considerations are rooted in cultural, anthropological, and ethical traditions, they vary among different groups in society. For communicating a given risk to all audiences, therefore, no single message is likely to be adequate.

Furthermore, for optimal effectiveness in risk communication, the process should involve an iterative exchange of information between the technical risk assessor and any stakeholders who may be directly or indirectly affected. Ideally, such an exchange should begin as early as possible in the process of risk assessment, so that those who may bear the risks can participate adequately in the derivation of the assessment itself. Trust between all who are involved is also critical to the success of risk communication, and it is best fostered through the mutual exchange of information in an open, participatory, consensual process.

In contrast to other risks in daily life, such as various types of accidents the frequencies of which are well documented in recorded statistics, many environmental risks to health are not known precisely, and they can be estimated only on the basis of unproven assumptions and extrapolations. As noted, such assessments are complicated at virtually every step by large uncertainties in the numerical values of measurements or other quantities affecting the risks; the modeling of exposure and/or toxic responses; temporal, spatial, and interindividual differences in exposure and/or susceptibility; and the comparison of societal and personal measures of risk. To the extent that these sources of uncertainty limit the reliability of a risk assessment, each must be made explicit if the comprehensibility, credibility, and utility of the assessment are not to be jeopardized.

Because, as noted, the perception of risk is a complex process, risk is difficult to communicate in a way that places it in proper perspective. Comparisons of quantitative risk estimates, such as have often been presented for the purpose (e.g., Tables 8–6 and 8–7), or attempts to weight risks solely on the basis of their

TABLE 8–6. Situations or Activities Involving a One-in-a-Million Risk of Death

EXPOSURE OR ACTIVITY	CAUSE OF DEATH
Smoking 1.4 cigarettes	Cancer, heart disease
Drinking 1/2 L of wine	Cirrhosis of the liver
Spending 1 hour in a coal mine	Black lung disease
Spending 3 hours in a coal mine	Accident
Living 2 days in New York or Boston	Air pollution
Traveling 6 min by canoe	Accident
Traveling 10 miles by bicycle	Accident
Traveling 300 miles by car	Accident
Flying 1,000 miles by jet	Accident
Flying 6,000 miles by jet	Cancer caused by cosmic radiation
Living 2 months in Denver	Cancer caused by cosmic radiation
Living 2 months in an average masonry building	Cancer caused by natural radioactivity
One chest X-ray	Cancer caused by radiation
Living 2 months with a cigarette smoker	Cancer, heart disease
Eating 40 tablespoons of peanut butter	Liver cancer caused by aflatoxin B_1
Drinking Miami, Florida drinking water for 1 yr	Cancer caused by chloroform
Drinking 30 12-oz. cans of diet soda	Cancer caused by saccharin
Living 5 years at the boundary of a nuclear plant	Cancer caused by radiation
Eating 100 charcoal-broiled steaks	Cancer caused by benzopyrene

Source: Wilson, R. Analyzing the risks of daily life. *Technol. Rev.* 81:41–46, 1979.

TABLE 8–7. Ranking the Risks from Possible Carcinogens on the Basis of Their Estimated Potencies

HAZARD INDEX: HERP (%)[a]	DAILY HUMAN EXPOSURE	CARCINOGEN DOSE PER 70-KG PERSON
	ENVIRONMENTAL POLLUTION	
0.001	Tap water, 1 L	Chloroform, 83 μg (US Avg)
0.004	Well water, 1 L contaminated (worst well in Silicon Valley)	Trichloroethylene, 280 μg
0.0004	Well water, 1 L contaminated, Woburn	Trichloroethylene, 267 μg Chloroform, 12 μg Tetrachloroethylene, 21 μg
0.0002		
0.0003		
0.008	Swimming pool, 1 hr (for child)	Chloroform, 250 μg (Avg. pool)
0.6	Conventional home air (14 hr/day)	Formaldehyde, 2.2 mg Benzene, 155 μg
0.004		
2.1	Mobile home air (14 hr/day)	Formaldehyde, 2.2 mg
	PESTICIDE AND OTHER RESIDUES	
0.0002	PCBs: daily dietary intake	PCBs, 0.2 μg (U.S. Avg.)
0.0003	DDE/DT: daily dietary intake	DDE, 2.2 μg (U.S. Avg.)
0.0004	EDB: daily dietary intake (from grains and grain products)	Ethylene dibromide, 0.42 μg (U.S. Avg.)
	NATURAL PESTICIDES AND DIETARY TOXINS	
0.003	Bacon, cooked (100 g)	Dimethylnitrosamine, 0.3 μg
0.006		Diethylnitrosamine, 0.1 μg
0.003	Sake (250 ml)	Urethane, 43 μg
0.03	Comfrey herb tea, 1 cup	Symphytine, 38 μg (750 μg of pyrrolizidine alkaloids)
0.03	Peanut butter (32 g; one sandwich)	Aflatoxin, 64 ng (U.S. Avg., 2ppb)
0.06	Dried squid, broiled in gas oven (54 g)	Dimethylnitrosamine, 7.9 μg
0.07	Brown mustard (5 g)	Allyl isothiocyanate, 4.6 mg
0.1	Basil (1 g of dried leaf)	Estragole, 3.8 mg

(*continued*)

TABLE 8–7. (*continued*)

HAZARD INDEX: HERP (%)[a]	DAILY HUMAN EXPOSURE	CARCINOGEN DOSE PER 70-KG PERSON
2.8	Beer (12 oz, 354 ml)	Ethyl alcohol, 18 ml
4.7	Wine (250 ml)	Ethyl alcohol, 30 ml
	FOOD ADDITIVES	
0.06	Diet cola	Saccharin, 95 mg
	DRUGS	
[0.3]	Phenacetin pill (Average dose)	Phenacetin, 300 mg
16	Phenobarbital, one sleeping pill	Phenobarbital, 60 mg
17	Clofibrate (Average daily dose)	Clofibrate, 2000 mg
	OCCUPATIONAL EXPOSURE	
5.8	Formaldehyde: Average daily intake	Formaldehyde, 6.1 mg
140	EDB: Average daily intake (high exposure)	Ethylene dibromide, 150 mg

[a]Possible hazard: the amount of rodent carcinogen indicated under carcinogen dose is divided by 70 kg to give milligram per kilogram equivalent of human exposure, and this human dose is given as the percentage of TD_{50} in the rodent (in milligrams per kilogram) to calculate the human-exposure/rodent potency index (HERP).

Source: Ames B.N., Magaw, R., and Bold, L.S. Ranking possible carcinogenic hazards. *Science*. 236:271–280, 1987.

impacts on life expectancy, on the quality of life as in Figure 8–4, or on their economic costs, are likely to be inadequate by themselves.[6] Instead, the strategy for risk communication must take into account the known dynamics of risk perception, which involve certain principles: unfamiliar risks tend to be less acceptable than familiar risks; involuntary risks are less acceptable than voluntary risks; risks controlled by others are less acceptable than risks that are under one's own control; unapparent and undetectable risks are less acceptable than risks that are apparent and detectable; risks that are perceived to be unfair are less acceptable than risks that are perceived to be fair; risks that do not permit individual protective action are less acceptable than risks that do; dramatic and dreadful risks are less acceptable than undramatic and commonplace risks; unpredictable risks are less acceptable than predictable risks; cross-hazard comparisons tend to be unacceptable; and risk estimation is inherently of less interest to people than risk reduction, and neither is likely to be of interest in the absence of real concern about the risk, or risks, in question.

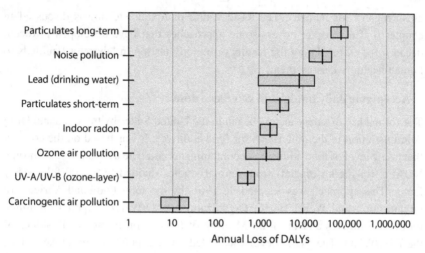

FIGURE 8–4. Annual health loss in disability-adjusted life years (DALYs) for selected environmental exposures in the Netherlands.

APPROACHES FOR HAZARD IDENTIFICATION AND DOSE-RESPONSE ASSESSMENT

Historically, many hazards were identified and quantitated by use of observational studies in human populations, especially in groups of workers with high levels of exposure. With sufficient characterization and quantitation of the exposures and sufficient data on the frequency of health-related responses as a function of exposure level and duration and, based on professional judgement of what effects or degrees of responses were deemed to be adverse health responses, one could establish a no observable adverse effect level (NOAEL) or a lowest observable adverse effect level (LOAEL). There are, presumably, threshold levels for adverse responses somewhere between the two. Threshold levels of exposures that were selected generally included some safety factor, or margin of safety, in recognition of uncertainties in the data, as well as the likelihood that other populations could include individuals with greater susceptibility to adverse responses than the population(s) initially studied.

Toxicological Approach

As numerous new chemicals entered commerce during and after World War II, it was recognized that their potential for producing adverse health effects in production workers, or in the users and consumers of products containing these chemicals, needed to be evaluated prospectively using laboratory animals as surrogates for humans. The basic nature of toxicity testing is discussed in chapter 6. This chapter covers some approaches that have developed in recent decades for extrapolating the results of toxicity testing in laboratory animals to human health risk assessment.

Acceptable daily intake and reference dose

The formal use of safety factors began in the United States by the Food and Drug Administration in the mid-1950s for food additives. It was based on the concept that a safe level of food additives or contaminants could be derived from a chronic NOAEL (in mg/kg of diet) from animal studies divided by a 100-fold safety factor. This approach was adopted, in 1961, by the Joint Food and Agriculture Organization and World Health Organization (FAO/WHO) Expert Committee on Food Additives and the Joint Meeting of Experts on Pesticides Residues of the FAO/WHO. The "safe" level was called the acceptable daily intake (ADI), expressed in mg/kg body weight per day. The procedure involved collecting all relevant data, ascertaining the completeness of the available data set, determining the NOAEL using the most sensitive known indicator of toxicity, and applying an appropriate safety factor to derive the ADI for humans. The ADI approach, as well as the comparable tolerable daily intake approach, are now widely used for some chemical contaminants.

The ADI is defined as the daily intake during an entire lifetime that appears to be without appreciable risk on the basis of all known facts. When setting an ADI, various test data and judgmental factors need to be taken into account, for example, the nature of the effects, the adequacy of the database, age-related effects, metabolic and pharmacokinetic data, available human data, and development of values based on toxic responses of a small number of rather homogeneous laboratory animals in establishing safe doses for what is a heterogeneous human population. Because of these considerations, the ADI generally includes a 100-fold safety factor to account for uncertainties in the database, although exceptions to this value are permitted in accordance with the nature and extent of available data for certain populations (e.g., children) of potential special concern.

In 1988, the US Environmental Protection Agency (EPA) formally adopted, subject to a number of modifications, the ADI approach in its regulatory measures for reducing environmental pollution. However, instead of the terms ADI and safety factor, the terms reference dose (RfD) and uncertainty factor (UF), respectively, have been used for noncancer endpoints. Daily exposure at or below the RfD, or RfC, is considered as acceptable in terms of potential health risk. For inhaled chemicals, the RfD is replaced by the RfC, or reference concentration. The RfD is derived from the NOAEL by the consistent application of generally one order-of-magnitude UFs that reflect the various types of datasets used to estimate RfDs. UFs generally consist of up to a 10 fold factor to account for certain variables, including human variability in sensitivity, uncertainties in interspecies extrapolation; use of the NOAEL obtained from a subchronic animal study rather than a chronic study, and use of the LOAEL in the absence of the NOAEL. In addition, a modifying factor, ranging from less than 1 to 10, is applied when the database includes, for example, a very small number of animals per dose level.

Benchmark dose

The NOAEL is defined as the highest dose level at which no statistically significant effects occur for a biologically relevant endpoint of interest. It is important to keep in mind that the NOAEL is not the same as the true no effect level, $NAEL_{true}$, and that, although the NOAEL could be considered an estimate of the true threshold dose, the quality of the estimate cannot be assessed.

Objections against the use of the NOAEL have been raised. In general, there is a need for consideration of the dose–response relationship as a whole. One of the alternatives proposed has been the benchmark dose (BMD) approach. The BMD is defined as the dose level associated with some predetermined, small change in response compared to controls, called the benchmark response. The only caveat is that while BMD modeling does not make any assumptions about the biological basis for an observed dose–response relationship, it does require that the endpoint shows some increase in response with increasing dose. With sufficient available data, the BMD approach can be employed to derived RfDs.

Basically, the BMD approach involves a regression function fitted to response data to estimate the dose at which adverse effects can be detected. Using regression models for describing the dose–effect relation obviates the need to extrapolate a LOAEL to a NOAEL. This may be considered as a major advantage,[7] because there is no scientific justification for the use of an assessment factor for LOAEL–NOAEL extrapolation.

In the benchmark concept, one needs to postulate a critical effect size (CES). The CES for a critical endpoint is defined as the value of effect-size below which there is no reason for concern, and the associated critical effect dose (CED) is defined as the dose at which the average animal shows the postulated critical effect size defined for a particular endpoint.

A drawback of using dose–effect curves for the evaluation of toxicity is that current toxicological and biological knowledge does not provide a sufficient basis to unequivocally establish the breakpoint between nonadverse and adverse effect size for most health-effect endpoints. Because a single universal CES does not seem a realistic option, a value must be chosen for each separate endpoint. A widespread implementation and acceptance of the value of CES for each of the (most relevant) endpoints would require a broad consensus.

The CED is a true, unknown value, which can only be estimated, to some definable degree of precision, when suitable data for the endpoint of concern are available. The $NAEL_{true}$ may then be defined as the lowest CED of all endpoints, that is, $NAEL_{true} \equiv$ minimum of all CEDs.

The definition of the NAEL refers not only to an unknown but also to a rather theoretical value, because it is unknown to what endpoint it is associated. In practice, one can never be sure whether information on all relevant endpoints for the compound studied is present. Furthermore, the lowest CED in two situations (e.g., animal vs. human) may not refer to the same endpoints. For example, rats may be most sensitive for endpoint A but humans for endpoint B. A drawback to the use of dose–response modeling from a practical point of view is that many toxicity data are not suitable for curve-fitting procedures. A typical study design for some toxicity testing guidelines consider three dose groups and a control. Ideally, more dose groups should be used with each dose group comprising fewer animals.

When data are available for a particular endpoint that does allow for fitting a regression function, the CED may be estimated. Depending on the quality of the data, this estimate has an uncertain degree of imprecision. The complete uncertainty distribution can be estimated by bootstrapping; once a regression model has been fitted, Monte Carlo sampling is used to generate a large number of new data sets from this regression model, each time with the same number of data points per dose group as observed animals in the real experiment. For each generated data set, the CED is reestimated. Taking all these CEDs together results in the required distribution.

Human limit values

Because there will be a certain distribution over all endpoints and substances for each effect, it is possible to extrapolate a CED from one situation to another. Thus, instead of choosing a single, perhaps most sensitive, endpoint from the animal data, each CED distribution that is associated with a relevant endpoint can be extrapolated to the distribution of the associated CED in the sensitive human ($CED_{sens.human}$) by probabilistic combination with the distributions of each effect. This results in a series of distributions for $CED_{sens.human}$, each related to another endpoint. Then this complete set of distributions can be considered as a basis for deriving a human limit value (HLV), for example, by choosing the lowest of each distribution's critical percentile. It has been assumed that there is a complete independence of the various distributions of effects, but this worst-case assumption may not have been valid. When correlations can be demonstrated and quantified, the method allows for these by introducing correlation coefficients.

This approach differs from the benchmark approach in various ways. Crump[8] introduced the benchmark dose level, defined as the lower 95% confidence limit of the CED, as a starting point for extrapolation to the (sensitive) human. By dividing the benchmark dose level by assessment factors for interspecies and intraspecies variation (default values of 10), an HLV can be derived. Instead of this, Vermeire et al.[7] proposed using the entire distribution of the CED rather than the lower 95% confidence limit of the CED. Second, they proposed combining this CED distribution with distributions of extrapolation factors in a probabilistic way. The result of the probabilistic combination of distributions is in the form of an assessment distribution, so that the degree of conservatism is quantifiable in any particular assessment. Their approach allows for deriving a HLV as a function of an a priori chosen degree of conservatism. In addition, it allows for estimating the lower and upper bounds for possible health effects in the sensitive population at a given exposure level.

Epidemiological Approach

Epidemiologic data plays a prominent role in hazard identification and dose–response assessment and may also be used in exposure assessment and risk characterization. Hazard identification is inherently integrative, drawing on all relevant lines of evidence, as do the criteria for causality that are widely applied for interpreting epidemiologic findings. In fact, there are no guidelines for interpretation of epidemiologic data in risk assessments that go beyond the conventionally applied criteria for causality. Because epidemiologic data come from humans, they are usually given strong weighting if they are positive in indicating an adverse effect. On the other hand, it sometimes may be best to give preference to toxicologic data collection over epidemiologic data collection[3] given the cost of epidemiologic research and the ambiguity of findings among some observational studies. However, there are numerous examples of evidence from epidemiologic research that provide definitive identification of hazards, including such well-known causal associations as radon and lung cancer, asbestos and mesothelioma of the lung, vinyl chloride and angiosarcoma of the liver, and active and passive cigarette smoking and lung cancer.

Once a hazard is identified, the second component, that is, the dose–response assessment, is initiated to establish the quantitative relationship between dose or exposure and response. While the dose–response relationship is of prime interest in risk assessment, epidemiologic studies more typically address the exposure–response relationship.

As noted in chapter 6, exposure refers to contact with some material, while dose is the material actually entering the body or the dose that reaches the target tissue. An exposure measure may be considered as a surrogate for dose or may be used to estimate dose. For the purpose of risk assessment, characterization of the exposure–response relationship in the range of relevant human exposures is needed. For some agents, such as environmental tobacco smoke, data on risks are available in the exposure range of interest; for many others, data for estimating dose–response relationships often come from studies of workers, and such data may only be available for levels of dose or exposure above usual environmental levels. For such exposures, exposure–response relationships estimated at the higher exposures are extended downward along an assumed model for the relationship between exposure and dose. When biologic understanding of the mechanism of action is sufficiently certain, a biologic model of the relationship between exposure or dose and response may be assumed for analysis of either toxicologic or epidemiologic data.

For the purpose of risk assessment, epidemiologic data may be used to quantify the dose–response relationship or to guide the selection from among alternative models for the relationship between dose and response. The ability to estimate precisely the quantitative relationship between dose and response may be limited by the extent of the data available and by exposure or dose misclassification. Exposure and dose are often estimated from incomplete information or with the use of surrogates, and comprehensive data in all biologically relevant time windows may not be available. Consequently, misclassification of exposure may bias the description of the exposure–response relationship and confidence intervals may be wide, perhaps with substantially different policy implications from risks at the lower and upper bounds.

In analyzing epidemiologic data on the dose–response relationship, researchers typically have an a priori interest in determining if a dose–response relationship can be shown to be statistically significant and in characterizing the shape of the relationship. Initially, the shape of the dose–response relationship may be explored descriptively, but inevitably statistical models are used to quantify the change in response with increasing exposure. There has been a general reliance on linear, nonthreshold models, particularly for carcinogens. For some agents, for example, alpha radiation, a relatively strong basis exists for assuming a linear model, whereas for other agents, the selection of a linear nonthreshold relationship reflects a policy default that is viewed as "conservative" in protecting public health because of the model assumption that any level of exposure conveys some risk.

Epidemiologic data may also be fit with alternative models of the dose–response relationship when there is uncertainty about the most appropriate shape of this relationship (see Fig. 8–3). The extent of actual data fit may be used to guide the identification of the "best" model. However, epidemiologic data are rarely sufficiently abundant to provide powerful discrimination among alternative models, and sample size requirements for comparing fits of alternative models with different public health implications may be very high. For policymaking purposes, the shape of the dose–response relationship at the lower exposure levels experienced by the population is critical, but epidemiologic data are often very limited or lacking at these levels.[9]

For the purpose of population-based risk assessment, information is desired on the full distribution of exposures in the population. Measures of central tendency may be appropriate for estimating overall risk to the population, but use of central measures alone may hide the existence of more highly and unacceptably exposed individuals. Thus, the upper end of the exposure distribution may also be of interest.[10]

Driven largely by the needs of risk assessment, exposure assessment has matured and has become a field that is broader than its applications in epidemiology. Modern exposure assessment is based on a conceptual framework that relates pollutant sources to effects through multimedia paths of exposure, which ultimately determine the dose. The concept of total personal exposure is central; that is, for health risk assessment, exposures received by individuals from all sources and media need to be considered. To date, epidemiologic data have not played prominent roles in exposure assessments for risk assessments. However, epidemiologic data may contribute to, and also serve to validate, exposure models.

USES OF HEALTH RISK ASSESSMENT

Risk assessment methodology is needed to estimate environmental health risks. The use of risk assessment has become central in many activities of regulatory agencies and of the private sector to set priorities for the regulatory agenda; develop regulatory exposure limits, emission standards, or cleanup standards; and set priorities for research. In the nonregulatory setting, the information generated by risk assessment is used by a wide variety of groups including industry, labor unions, and consumers.

Federal agencies that use risk assessment prominently are the EPA, the Occupational Safety and Health Administration (OSHA), and the Consumer Product Safety Commission. Statutes requiring the assessment of health risks include the Clean Air Act (CAA); the Federal Insecticide, Fungicide, and Rodenticide Act; the Toxic Substances Control Act; the Occupational Safety and Health Act; the Federal Food, Drug, and Cosmetic Act; the Safe Drinking Water Act; and the Comprehensive Environmental Response, Compensation, and Liability Act.

Within governmental agencies, the use of risk assessment may be dictated by statute and may vary even by sections within a statute. For example, Section 108 of the CAA requires the EPA to establish and periodically review National Ambient Air Quality Standards (NAAQS) for pollutants labeled "criteria" pollutants through the preparation of a comprehensive scientific document. These standards are required to protect the public health with an adequate margin of safety without regard to cost. By contrast, Section 112 of the CAA, as applied to point-source pollutants, is technology-driven in its first phase, requiring maximum available control technology (MACT) originally for 189 substances, now reduced to 187, without regard to health risk. Approximately eight to nine years after the application of a control technology, the EPA was directed to determine whether more stringent standards are required to protect public health "with an ample margin of safety." This risk-driven phase requires that the EPA establish such standards for carcinogens that present an excess cancer risk greater than one in a million, even with MACT controls installed. In a third example of the use of risk assessment within the CAA, the EPA is required to identify no fewer than 30 of the 187 toxic substances that present the greatest threat to public health in the largest number of urban areas. The agency is then required to schedule actions that substantially reduce the public health risks from hazardous air pollutants and by these actions achieve a reduction in the incidence of cancer attributable to exposure to hazardous air pollutants emitted by stationary sources by at least 75%. Other statutes are similarly specific in the use of risk assessment.

In addition to governmental agencies, private industry also use risk assessment techniques. For example, some companies use risk assessment principles to set occupational exposure limits for substances that may not be covered by regulations. The American Conference of Governmental Industrial Hygienist[11] publishes threshold limit values, which, while nonbinding, serve as guidelines for safe occupational exposures. These guidelines usually assume a dose–response relationship, with a threshold separating safe and unsafe exposures. Short-term tolerable exposure limits, usually based on acute effects, have also been developed for emergency responses. These include the Acute Exposure Guideline Limits developed by the NRC and the Emergency Response Planning Guidelines developed by the American Industrial Hygiene Association. These exposure guidelines or limits are frequently based on results of occupational epidemiology studies and are discussed further in chapter 11. Risk assessment techniques are also applied increasingly to the environmental arena, where the effects of effluent exposure on nonhuman receptors (e.g., fish, wildlife, and flora) are of concern.

INFLUENCE OF JUDICIAL REVIEWS ON REGULATORY RISK ASSESSMENT

The adoption of new regulatory mandates by a federal agency in the United States often leads to challenges in the courts by affected or interested parties. Some notable examples, discussed next, affected not only the particular regulatory action at issue but also the ways and rates of future decisions on related issues.

Benzene Decision (1980)

In a case involving the OSHA exposure standard for benzene, the US Supreme Court held that OSHA must provide an estimate of the actual risk associated with a toxic substance. Although only OSHA was involved in this case, the decision provided a *de facto* mandate for quantitative risk assessment at all regulatory agencies. The court recognized that OSHA may use assumptions (i.e., science policy) in risk assessment but only to the extent that those assumptions have some basis in reputable scientific evidence.

Vinyl Chloride Decision (1987)

In this litigation involving the EPA's air emissions standard for vinyl chloride, the US Supreme Court interpreted the CAA to require the EPA to first determine a "safe" level of exposure for air pollutants before considering the economic or technological feasibility of achieving reduced emissions. The rule was remanded to the EPA and, in the subsequent rulemaking, the EPA decided to emphasize quantitative risk assessment in the establishment of CAA standards.

OSHA Air Contaminants Decision (1992)

In this case, the US Supreme Court struck down permissible exposure limits (PELs) developed by OSHA for 428 toxic substances on the basis that the assumptions used in the risk assessments supporting the PELs were not substantiated by the available scientific evidence. This case reiterated the lesson of the benzene decision that, although science policy is clearly permissible in risk assessment, science policy decisions must have some basis in fact. Through this rulemaking, OSHA was attempting to update standards that had been set more than 20 years earlier. By requiring a better-substantiated scientific basis for assumptions, a requirement that may not be practical or even possible, this case effectively blocked OSHA's ability to update many of the earlier standards, particularly *en masse*.

Particulate Matter Standard Decision (2001)

In this case, the US Supreme Court dismissed a claim that Congress's delegation of authority to the EPA to set NAAQS was unconstitutional and affirmed that the EPA had made appropriate use of the available scientific information to establish a new NAAQS for $PM_{2.5}$. However, it referred back to the EPA the question of whether a PM_{10} limit was a suitable air quality index for the control of exposures to the coarse fraction of thoracic particulate matter ($PM_{10-2.5}$).

INTERNATIONAL ASPECTS OF RISK ASSESSMENT

Formal risk assessment procedures are increasingly being used by international regulatory bodies. Some of these procedures use different methods for the dose–response step. For example, Maynard et al.[12] suggested a risk assessment procedure based on the use of NOAELs and "uncertainty factors" for carcinogens, along with subsequent risk management steps. Similarly, Hunter et al.[13] described the European Union's process for setting occupational exposure limits, mentioning the NOAEL/uncertainty factor approach for noncarcinogens but not recommending a prescriptive method for carcinogens. Other countries, such as the Netherlands and Canada, rely on a mathematical modeling approach for the dose–response assessment step. While the exact procedures used in the dose–response step vary between countries and for different health outcomes, a common theme is the preference for basing risk assessments on high-quality epidemiologic data, if available. Despite this preference, relatively few risk assessments use epidemiologic data in the hazard-identification and dose–response steps.

REFERENCES

1. National Research Council. *Risk Assessment in the Federal Government: Managing the Process*. Washington, DC: National Academy Press, 1983.
2. National Research Council. *Toxicity Testing: Strategies to Determine Needs and Priorities*. Washington, DC: National Academy Press, 1984.
3. National Research Council. *Science and Judgement in Risk Assessment*. Washington, DC: National Academy Press, 1994.
4. Rodricks, J.V. *Calculated Risks*. Cambridge: Cambridge University Press, 1992.
5. National Research Council. *Understanding Risk. Informing Decisions in a Democratic Society*. Washington, DC: National Academy Press, 1996.
6. National Research Council. *Improving Risk Communication*. Washington, DC: National Academy Press, 1989.
7. Vermeire, T., Stevenson, H., Pieters, M.N., Rennen, M., Slob, W., and Hakkert, B.C. Assessment factors for human health risk assessment: A discussion paper. *Crit. Rev. Toxicol.* 29(5):439–90, 1999.
8. Crump, K.S. A new method for determining allowable daily intakes. *Fundam. Appl. Toxicol.* 4:854–71, 1984.
9. National Research Council. *Health Effects of Exposure to Radon (BEIR VI). Committee on Health Risks of Exposure to Radon, et al.* Washington, DC: National Academy Press, 1998.
10. US Environmental Protection Agency. *Guidelines for Exposure Assessment*. Washington, DC: Office of Health and Environmental Assessment (Publication No. EPA/6002-92/001 FR57:2288–22938), 1997.
11. ACGIH. Threshold Limit Values for Chemical Substances and Physical Agents, and Biological Exposure Indices, Cincinnati American Conference of Governmental Industrial Hygienists, 1999.
12. Maynard, R.L., Cameron, K.M., Fielder, R., et al. Setting air quality standards for quantitative carcinogens: An alternative to mathematical qualitative risk assessment. *Hum. Toxicol.* 14:175–86, 1995.
13. Hunter, W.J., Aresini, G., Haigh, R., et al. Occupational exposure limits for chemicals in the European Union. *Occup. Environ. Med.* 54:217–22, 1997.

9

Environmental Sampling and Exposure Assessment

Exposure assessment is a key element in the overall environmental or health risk assessment. Exposure characterizations are needed to establish baseline exposure levels for receptors and to observe trends in those levels; to identify and evaluate sources of exposure; to provide measures of the success or failure of risk management efforts (e.g., establishment of regulations and success in meeting them); to perform surveillance functions (i.e., using data to identify high-exposure/high-risk groups where interventions might be appropriate and effective in preventing harmful effects); to investigate exposure–response relationships for environmental hazards (e.g., epidemiologic research in human populations); and to inform regulatory decisions by supporting population exposure assessments, cost estimates, and other aspects of regulatory analyses.

There are a number of approaches for making quantitative exposure assessments (Fig. 9–1). Some, such as personal monitoring of airborne toxicant concentrations in the breathing zones of selected individuals, can provide relatively precise information on those individuals but only point estimates for a population at risk. Measurements of concentrations in environmental media can also be precise for the samples that are collected and analyzed but be of somewhat limited relevance for the exposures of individuals in the population. Population distributions of exposure are often estimated using mathematical models with varying

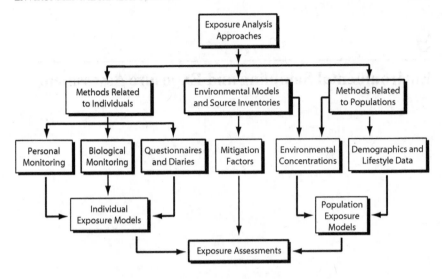

FIGURE 9–1. Possible approaches for analysis of contaminant exposures.

degrees of validation. This chapter discusses methodologies for measurement of contaminant concentrations in environmental media and assessment of exposure.

TYPES OF SAMPLES

There are many kinds of samples that can provide information useful to contaminant evaluations, and it is important not to confuse or misinterpret the contaminant levels found in specific samples. The basic types of samples are source; ambient; personal; and process.

Source samples include samples collected in smokestacks or other vents for airborne effluents and discharge-pipe or channel samples for liquid effluents. They can provide essential information on source strengths but are of very limited value in evaluating exposures of human populations or ecological receptors.

Ambient samples provide information on environmental contaminant levels in representative locations within airsheds and waterways. They can be used to indicate whether established standards are being exceeded and to monitor trends in contaminant levels. The samples are of limited value in identifying or locating specific sources or for estimating exposure levels for individuals within the population.

Personal samples are used to determine air contaminant exposure levels within the immediate environment of specific human individuals. The samplers need to be light in weight and self-contained, so as not to limit the mobility or normal activity of the wearer. They are most widely used to evaluate inhalation exposures of workers and are also used in some community epidemiological studies where indoor exposures are of special interest.

Process samples are samples collected within process equipment and transfer lines, which are taken to characterize mass flowrates or stream composition. Samples up- and downstream of a pollution-control device belong in this category. The concentrations measured may be very high but may bear little relation to quantities discharged to the environment, and even less to ambient environmental levels.

SAMPLING APPROACHES

Environmental samples are generally categorized as air samples, water samples, food samples, soil samples, sediment samples, biota samples (plants and animals), or biological samples of biomarker exposure receptors (e.g., tissues, blood, excreta). Most air and some water sampling is done by extractive techniques, whereby the sampled fluid is passed through a collector that extracts the contaminant onto a suitable substrate and passes the carrier fluid. The resulting contaminant samples are much less bulky and, therefore, are generally easier to package and transport than are samples that include the fluid medium as well as the contaminant. Filters are widely used for extractive sampling of suspended particles in both air and water.

Another technique used in both air and water sampling is adsorptive extraction of organics on granular beds of activated carbon or another solid sorbent. Extractions of ions from water samples can also be performed with ion exchange resins, which exchange and bind ions from the aquatic medium. Other extractive techniques that are primarily limited to air sampling include thermal and electrostatic precipitation and inertial impaction for airborne particles, as well as scrubbing with liquids for both particles and soluble gas contaminants.

In occupational health hazard evaluations, most of the samples are collected on sampling substrates. Biomarker samples can also be used to evaluate the degree of chemical exposures or their effects and are also frequently collected. They may include, as noted previously, blood, urine, exhaled air, and occasionally other materials, such as hair, fingernails, and feces.

In ambient air evaluations, most of the samples are extracted from the sampled air. Other samples that may occasionally be useful are sediment, either as dust fall in a collector or as constituents of surface soil, and biota. Plants can be analyzed for their uptake of contaminant or for pathological features attributable to exposure.

In water contaminant evaluations, bulk samples of water are widely collected, and the samples for analysis are generally extracted in the laboratory rather than in the field. Sediment samples and biota, including rooted plants, shellfish, and finned fish, are also frequently collected. The biota may be analyzed for contaminant burdens, for pathology, or for changes in population density associated with level of contaminant.

When performing measurements to assess exposures, it is necessary to consider the difference between accuracy and precision. Accuracy refers to how close a measurement is to the real or true value, and assurance of accuracy requires proper calibration of sampling and analytical equipment. Calibration, which is defined as the determination of the true values of any instrumental readings, involves comparison of the response of the measuring instrument to that of a reference instrument or to a standardized sample of known concentration. On the other hand, precision addresses how consistent results are when measurements are repeated (i.e., measurement reproducibility); repeated measurements can differ from each other because of random errors. A measurement system can be accurate but not precise, precise but not accurate, neither, or both. For example, if an experiment contains a systematic error, then increasing the sample size generally increases precision but does not improve accuracy. On the other hand, eliminating a systematic error improves accuracy but does not change precision.

In practice, many of the problems in evaluating environmental contaminant levels are not related to the accuracy of laboratory evaluations but rather to biases resulting from the selection of nonrepresentative locations and times for sampling environments and/or to errors introduced in the process of sample collection and field evaluations. Considerations of sampling location, sampling frequency, duration and volume, and the performance characteristics of sampling instruments and collection substrates all influence the results obtained, and poor choices can bias the results.

GENERAL CONSIDERATIONS FOR SAMPLING AND ANALYSIS

Concentrations of contaminants in the environment can vary continuously, and considerably, for many reasons. The rates of anthropogenic contaminant release depend upon the level of human activity and the degree of effort expended in limiting discharges. The source strengths of natural contaminants will vary with time, since they depend upon variable biological, geochemical, and meteorological factors. Ambient concentrations will also depend on the degree of dilution that takes place between the source and the sampling site which, in turn, depends upon the volume of dilution fluid and the mixing characteristics in the fluid stream. Thus, the selection of sampling schedule and location are important considerations in developing a sampling program so as to avoid bias due to temporal and spatial variability in concentration in the environmental medium being sampled.

Temporal variations in environmental contaminant concentration can have several different time constants. There may be rapid fluctuations due to fluid turbulence, which may be superimposed on diurnal cycles associated with variations in sources, strengths, and dilution rates. For example, as discussed in chapter 4, atmospheric dilution depends upon wind pattern and lapse rate which, in the boundary layer, varies with solar flux and radiation from the ground surface. Dilution in aquatic environments depends upon surface-water flow, which varies greatly with rainfall.

Seasonal variations in contaminant concentrations depend on changes in source strength and mixing characteristics. Because of annual growth cycles, source strength is the dominant factor for natural contaminants, such as terpenes from coniferous forests or pollens from plants. The contaminant releases associated with space heating have an annual surge associated with cold weather. On the other hand, the release of chemical raw materials involved in photochemical smog formation has little seasonal variability, and the large observed seasonal patterns in oxidant levels are attributable to differences in solar radiation and atmospheric ventilation that affect photochemical reaction rates in different seasons.

Since concentrations of contaminants can vary spatially as well as temporally, it is often necessary to analyze samples drawn from many different locations in order to adequately characterize the average level in, for example, a given airshed, work environment, stream, or food source. In practice, however, the choice of sampling sites may be limited by such factors as isolation from local sources and flow disturbances; accessibility for sample changing and sampler maintenance; assurance of sample integrity in relation to tampering, theft, or vandalism; and the need for utilities and/or environmental controls for proper sampler performance. These considerations, and the cost of multiple sampling sites, may lead to dependence on a sampling location not adequately representative of the whole. Further evaluations may therefore be needed to establish the correspondence, if any, between contaminant levels measured at a fixed site and those in the larger environment it is supposed to represent.

Sample Handling and Preservation

Certain precautions are necessary in the handling of samples in order to ensure that the characteristics of the sample to be analyzed are not altered between their collection and laboratory analyses. Some important considerations in this regard are to minimize the time interval between sample collection and analysis; to tag or label the sample properly; and to collect supporting data to indicate the purpose of the sampling, the exact identification or location(s) of the sampling point(s), the date(s) and time(s) of sampling, the results of any measurements of ambient conditions at the sampling site, the identity of the individual who collected the sample, and the nature and extent of human activities going on in the vicinity of the sampler that could influence the concentration being measured.

If the contaminant of interest is produced or affected by biological action, or if the material sampled is chemically or thermally unstable or can be adsorbed on the container or substrate surface, the samples will require special handling to prevent, or at least to characterize, changes in composition between sample collection and analysis.

Since available preservation techniques may not be sufficient in all cases, some analyses are usually performed only at the sampling site, for example, pH and DO measurements in aquatic environments.

Sampling Efficiency

It is generally desirable to select the sample collector and its operating conditions so that the fraction of the contaminant of interest that penetrates the collector will be negligible (i.e., high collection efficiency). However, analyses can be just as reliable when the collection efficiency is much lower, provided that it is known and constant. In this case, a correction factor can be applied. The collection of gases and vapors in air, and dissolved chemicals in water, can generally meet this criterion, provided that the ambient temperature is constant, since each molecule of a particular contaminant is essentially equivalent to every other in terms of capture probability. However, the same consideration does not apply to contaminants present as particles, since an additional variable affecting collection efficiency, namely particle size, is involved. Particle samplers should have essentially complete collection to avoid any sampling bias due to this variable.

Sample Size and Analytical Sensitivity

Every analytical procedure has a lower detection limit at which the accuracy of measurement falls below an acceptable level. If the results of the analysis are to be useful, the amount of the specific contaminant in the sample must be as large as, and preferably many times greater than, this lower detection limit. Therefore, for a contaminant of interest, the quantity that will need to be sampled in order to generate useful data on concentration depends on two factors, namely its estimated environmental concentration and the size of the volume sampled.

In extractive sampling, the amount sampled depends on the rate of sample flow and the duration of the sampling interval. Sample size can be increased by either increasing the time interval or the sampling rate. However, freedom to change the rate may be limited in some cases, since sample collection efficiency may be rate-dependent. Sampling rates may also be limited by the size, power requirements, or suction capacity of the sampling equipment.

The selection of an appropriate sample size in a given situation must also involve a consideration of the basic reason for collecting the sample. If the purpose is to compare an ambient level to an established standard or specified action level, the sample should be large enough to permit the determination of a specified fraction of that standard or level. The fraction used is generally 1/10, although equipment limitations may sometimes dictate acceptance of a fraction as high as 1/2.

Duration of Sampling

In the measurement of contaminant levels, it is important to characterize both the total time and specific time intervals of sampling. In deciding when and how long to sample, the kind and degree of temporal variability likely to occur must be considered, as well as whether it is more important to determine peak levels or average levels. In other words, selection of the sampling frequency and duration should take into account the uses to be made of the resulting data.

Samples that are collected within a brief period are called grab samples. Samples collected at a constant rate over a longer period of time are known as integrated samples. While there is no precise demarcation between grab and integrated samples, the former are generally collected in less than a few minutes.

Sampling Substrate

In extractive sampling, the contaminant of interest is deposited onto or into materials that will retain it effectively throughout the balance of the sample collection process and through any subsequent transport and storage prior to laboratory analysis. The material that retains the extracted substance is known as the sampling substrate. A good substrate must not only efficiently retain the sampled material prior to analysis; it must also permit it to be analyzed efficiently in the laboratory. It must either be possible to separate the sampled material from the substrate quantitatively, or it must be possible to perform the analysis in the presence of the substrate. A given substrate may be ideal for some analyses and entirely unsuitable for others where, for example, its presence may preclude the analysis of choice or where it may contain constituents that would interfere with the analysis.

Specificity and Interferences

One of the most difficult aspects of the analytical chemistry of environmental contaminants is that one is generally applying trace microanalytical techniques to samples of mixed composition. The presence of co-contaminants of unknown composition and concentration may either enhance or depress the response characteristic of the component of interest that is being analyzed. When the potential for interferences is known, it is generally desirable to achieve greater analytical specificity by performing chemical separations prior to the analyses.

Legal Requirements

When sampling is performed to demonstrate conformance with legal codes or standards, it may be necessary to use certain specified sampling and analytical procedures rather than, or in addition to, those best suited to scientific investigation.

MEASUREMENT OF FLUID FLOW

Characterization of chemical contamination in air or water generally requires data on either the fluid flowrate and duration of sample collection or on the sampled volume. In other words, when extractive sampling techniques are used, the determination of concentration requires measurement of both the sampled volume as well as the amount of chemical contaminant in the sample. Furthermore, the concentration at the sampling location may be different from that at other locations, and the determination of contaminant transport in a surface stream or airborne plume may require measurements of flow and concentration at many points across the flow cross-section.

Instrumentation that assesses fluid flow (i.e., flowmeters) can be divided into three basic groups on the basis of the type of measurement made: integral-volume meters, flowrate meters, and velocity meters. In volume meters and flowrate meters, the whole fluid stream passes through the instrument. In this respect, they differ from velocity meters, which measure the velocity at a particular point at the flow cross-section. Since the flow profile is rarely uniform, the measured velocity will almost always differ from the average velocity, and it may be necessary to make many measurements in order to determine the average value. However, when the flow field is very large, velocity sensors may be the only indicators than can be used.

Integral-Volume Meters

Some integral-volume meters are used exclusively for measurements in air; these include the wet-test meter and dry-gas meter. Others, which operate by positive displacement, are available for both air and liquid flow metering.

A wet-test meter (Fig. 9–2) consists of a partitioned drum half submerged in a liquid, usually water, with openings at the center and periphery of each radial chamber. Air or gas enters at the center and flows into an individual compartment causing it to rise, thereby producing rotation. This rotation is indicated by a dial on the face of the instrument. The volume measured will be dependent on the fluid level in the meter, since the liquid is displaced by air.

The dry-gas meter, shown in Figure 9–3, is very similar to the domestic gas meter. It consists of two bags interconnected by mechanical valves and a cycle-counting device. The air or gas fills one bag while the other bag empties. When the cycle is completed, the valves are switched and the second bag fills while the first one empties. The alternate filling of two chambers as the basis for volume measurements is also used in twin-cylinder piston meters. Such piston meters can also be classified as positive displacement devices. These meters consist of a tight-fitting moving element with individual volume compartments that fill at an inlet and discharge at an outlet. Another type of multicompartment continuous rotary meter uses interlocking gears.

Volumetric Flowrate Meters

The integral-volume meters discussed previously are all based upon the principle of conservation of mass, specifically, the transfer of a fluid volume from one location to another. On the other hand, the flowrate meters described in this section operate upon the principle of the conservation of energy; they are based upon

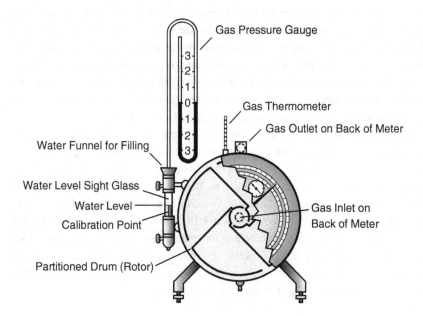

FIGURE 9–2. Schematic diagram of a wet-test meter.

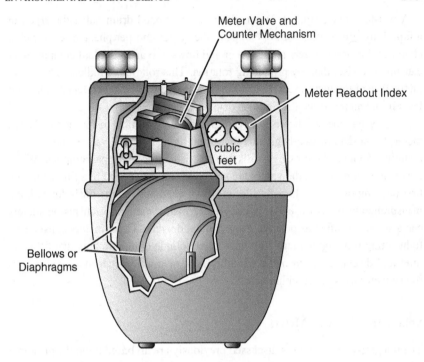

FIGURE 9–3. Schematic diagram of a dry-gas meter.

Bernoulli's theorem for the exchange of potential energy for kinetic energy and/ or frictional heat. Each consists of a flow restriction within a closed conduit. The restriction causes an increase in the fluid velocity and, therefore, an increase in kinetic energy, which requires a corresponding decrease in potential energy (i.e., static pressure). The flowrate can be calculated from knowledge of the pressure drop, the flow cross-section at the constriction, the density of the fluid, and the coefficient of discharge, which is the ratio of actual flow to theoretical flow and makes allowance for stream contraction and frictional effects.

Flowmeters that operate upon the conservation of energy principle can be divided into two groups. One type, which includes orifice meters, venturi meters, and flow nozzles, consists of devices that have a fixed restriction; these instruments are known as variable-head meters, because the differential pressure head varies with flow. A special subclass of this group includes weirs and flumes specifically used to measure flowrates of water. In these, the flow channel is only partially filled with water, and the height of the water column in the restriction varies, providing an indication of the flowrate. The other, smaller group, which includes rotameters, are known as variable-area meters, because a constant pressure differential is maintained by varying the flow cross-section.

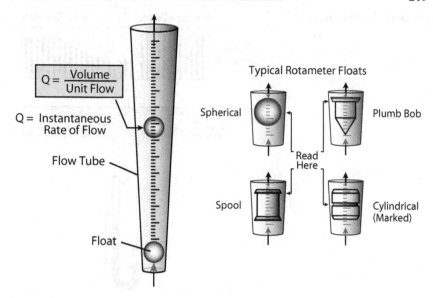

FIGURE 9–4. Typical rotameter and floats.

A rotameter (Fig. 9–4) contains a "float," which is free to move up and down within a vertical tapered tube that is larger at the top than at the bottom. The fluid flows upward, causing the float to rise until the pressure drop across the annular area between the float and the tube wall is just sufficient to support the float. The tapered tube is usually made of glass or clear plastic and has a flowrate scale etched directly on it. The height of the float indicates the flowrate. Floats of various configurations are used, as indicated in Figure 9–4. The term "rotameter" was first used to describe variable-area meters with spinning floats but now is generally used for all types of tapered metering tubes including those with a spherical float.

The simplest form of variable-head meter is the square-edged, or sharp-edged, orifice illustrated in Figure 9–5. It is also the most widely used because of its ease of installation and low cost. While the square-edged orifice can provide accurate flow measurements at low cost, it is inefficient with respect to energy loss; the permanent pressure loss for an orifice meter due to turbulence will often exceed 80%.

Venturi meters have converging and diverging angles of about 25° and 7°, respectively. They have high pressure recoveries (i.e., the potential energy), which is converted to kinetic energy at the throat and is reconverted to potential energy at the discharge, with an overall permanent loss of only about 10%. The characteristics of various other types of variable head flowmeters, for example, flow nozzles (Fig. 9–6), centrifugal flow elements, and so on are similar in most respects to those of either the orifice meter, venturi meter, or both.

FIGURE 9–5. Pipe-line orifice.

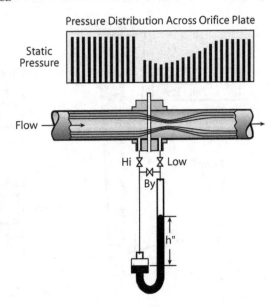

For water flow in open channels, the most common flow measurement techniques utilize weirs or flumes. A weir is essentially a dam or obstruction placed in the stream. It generally consists of a vertical plate with a sharp crest; the top of the plate can be straight, notched, or rectangular shaped (Fig. 9–7). The water level at a given distance upstream from the weir is proportional to flow rate.

There are a number of different types of flumes. The Parshall flume (Fig. 9–8) is a device that is similar to a venturi meter in that it consists of a converging section, a throat, and a diverging section. The level of the floor in the converging section is higher than the floor in the throat and diverging section. The head of the water surface in the converging section is a measure of the velocity through the flume and, therefore, of flowrate.

The flowrate in a pipe or tube may be strongly dependent on the flow resistance, and flowmeters with a very low resistance may be bulky and/or expensive. A metering element used in such cases is the bypass rotameter, which actually

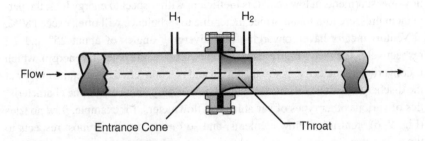

FIGURE 9–6. Flow nozzle (in a pipe).

FIGURE 9–7. Common types of weirs.

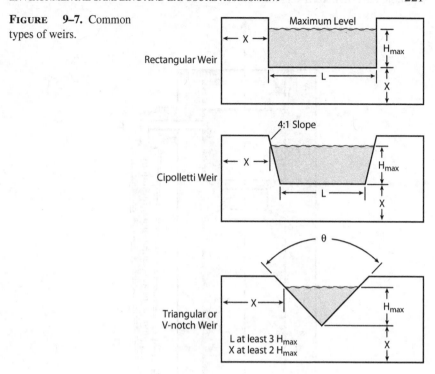

meters only a small fraction of the total flow; however, this fraction is proportional to the total flow. As shown schematically in Figure 9–9, a bypass flowmeter contains both variable-head and variable-area elements. The pressure drop across the fixed orifice or flow restrictor creates a proportionate flow through the

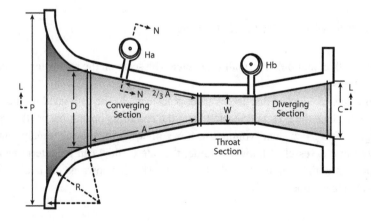

FIGURE 9–8. Parshall flume.

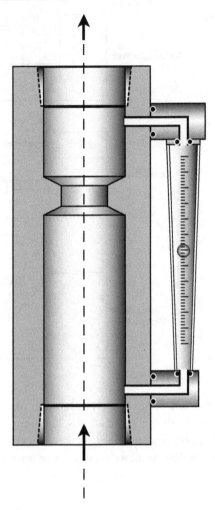

FIGURE 9–9. Schematic of a bypass flowmeter.

parallel path containing the small rotameter. The scale on the rotameter indicates the total flow.

Flowrates of aqueous discharges from pipes can also be estimated by measuring the distance the flow is projected from the end of an open pipe, as illustrated in Figure 9–10. For small volumetric flows of water, where the stream can be diverted into a vessel of known volume, the rate of flow can be determined by measuring the time required to fill the vessel. This is known as the bucket and-stopwatch technique.

Flow Velocity Meters

One type of flow velocity meter, the Pitot tube (Fig. 9–11), is often used as a reference instrument for measuring velocity and, if carefully made, needs no

Open-Pipe Flow Measurement

Requires Two Dimensions that Locate the Surface of Stream after it Leaves the Pipe

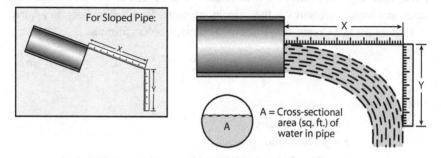

Measurement of Velocity and Discharge from a Pipe

When Y = 1ft.

Velocity (V) = 4.0 X

Discharge in GPM = 450 AV

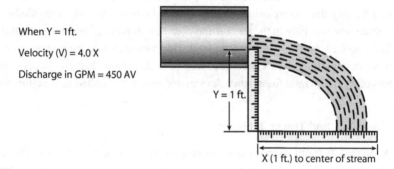

A = Cross-sectional area (sq. ft.) of water in pipe

X (1 ft.) to center of stream

FIGURE 9–10. Flowrate measurements based upon the distance a stream is projected from the end of an open pipe.

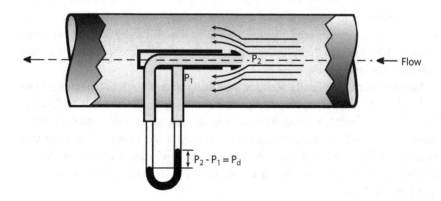

$P_2 - P_1 = P_d$

FIGURE 9–11. Pitot tube.

calibration. It consists of an impact tube whose opening faces axially into the flow and a concentric static pressure tube with eight holes spaced equally around it in a plane that is eight diameters from the impact opening. The difference between the static and impact pressures is the velocity pressure. Bernoulli's theorem applied to a Pitot tube in a stream simplifies to the dimensionless formula:

$$V = \sqrt{2g\,P_v} \qquad\qquad (9-1)$$

where

V = linear velocity
g = gravitational constant
P_v = pressure head of flowing fluid (velocity pressure)

There are several other ways beside the Pitot tube to utilize the kinetic energy of a flowing fluid to measure velocity. One technique is to align a jeweled-bearing turbine wheel axially in the stream and count the number of rotations per unit time. Such devices are known as rotating-vane flowmeters. Some are very small and are used as velocity probes. Others are sized to fit the whole duct and become indicators of total flowrate; these latter devices are sometimes called turbine flowmeters.

Mass Flow and Tracer Techniques

A thermal meter measures mass flowrate with negligible pressure loss. It consists of a heating element in a pipe or duct section between two points at which the temperature of the stream is measured. The temperature difference between the two points is dependent on the mass rate of flow and the heat input.

The principle of mixture metering is similar to that of thermal metering. However, instead of adding heat and measuring temperature difference, a tracer is added and its increase in concentration is measured, or clean fluid is added and the reduction in concentration is measured. The measuring device may react to some physical property, such as thermal conductivity or vapor pressure.

Mass flow may also be obtained using an ion-flow meter. In this device, ions are generated from a central disc and flow radially toward a collector surface where the ionic charges are collected, providing a basis for the flowrate measurement. Airflow through the cylinder causes an axial displacement of the ion stream in direct proportion to the mass flow.

An instrument specifically used for measuring water flow is the magnetic flowmeter (Fig. 9–12). This apparatus operates according to Faraday's law of induction, that is, the voltage induced by a conductor moving at right angles through a magnetic field will be proportional to the velocity of movement of the conductor through the field. In this device, the water is the conductor, and a set of electromagnetic coils in the meter produces the field. The induced voltage is measured to obtain the flowrate.

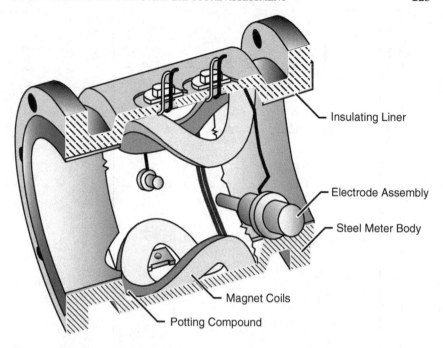

Insulating Liner

Electrode Assembly

Steel Meter Body

Magnet Coils

Potting Compound

FIGURE 9–12. Magnetic flowmeter.

The surface velocity of water may be measured by placing floating objects (e.g., wood, cork, etc.) in the stream and measuring the time required for the object to traverse a measured distance between two points. The velocity within a given length of channel may also be obtained by the use of dyes.

Factors to consider in the selection of flow-metering devices for aqueous streams are outlined in Table 9–1.

AIR SAMPLING

Instrumentation

A description of available instrumentation for air sampling can easily occupy a large reference volume. The discussion in this chapter is limited to the more fundamental considerations in the selection of instruments and sampling techniques for air contaminant evaluations.

Elements of an Air-Sampling Train

A number of essential elements comprise an air sampling system, known as a sampling train. These may all be incorporated into a single compact housing

TABLE 9–1. Selection Chart for Water Flow Measuring Devices

	ORIFICE		VENTURI	NOZZLE	PITOT	ELBOW	LO-LOSS TUBE	MAGNETIC FLOW METER[a]
	CONCENTRIC	SEGMENTAL OR ECCENTRIC						
Accuracy, and amount of empirical data	E	F	G	G	*	P	G	E
Differential for given flow and size	E	E	G	G	F	P	E	None
Pressure recovery	P	P	G	P	E	E	E	E
Use on dirty service	P	F	E	G	VP	P	G	E
For liquids containing vapors	E[b]	E	E	G	F	F	G	E
For vapors containing condensate	E[c]	E	E	G	P	F	G	None
For viscous flows	F	U	G	G	†	U	F	E
First cost small size	E	G	P	F	G	E	P	P
First cost large size	E	G	P	F	G	E	P	P
Ease of changing capacity	E	G	P	F	VP	VP	P	E
Convenience of installation	G	G	F[d]	F	E	E	F[d]	F[e]

All ratings are relative: E, excellent; G, good; F, fair; P, poor; VP, very poor; U, unknown.
*For measuring velocity at one point in conduit, the well-designed pitot tube is reliable. For measuring total flow, accuracy depends on velocity traverse.
†Requires a velocity traverse.
[a]Restricted to conducting liquids.
[b]Excellent in vertical line if flow is upward.
[c]Excellent in vertical line if flow is downward.
[d]Both flange type and insert type available.
[e]Requires pipe reducers if meter size is different from pipe size.
Source: US Environmental Protection Agency.

or portable instrument, or they may be separate elements. A sampling probe is needed when sampling in a moving stream, such as in a duct or stack and, when used, is the first element in the train. It is not needed when sampling from relatively still air in the ambient atmosphere or at the breathing zone of a worker. In such cases, it should not be used, since it would present some opportunity for sample losses without any corresponding benefit.

In sampling quiescent air, the inlet configuration of the sampler is a significant element of the train, and one whose influence is not always recognized, especially when sampling for particulate matter (PM). The inlet size and configuration, in conjunction with the sampling flowrate, establish an actual upper size cutoff for particle acceptance; particles larger than this size will not be aspirated into the inlet.

The next element in the train is the sample collector. It should precede other elements, such as flow-measuring devices and pumps, which could remove or add contaminants to the sample stream. The selection of the type of sample collector depends on the nature of the sample to be collected.

A static pressure sensor should be located downstream of the sample collector, preceding the inlet to the flow-metering element. There will usually be a significant pressure drop between the inlet to the probe or sample collector and the inlet to the flowmeter due to the flow resistance of the sample collector and the flow path itself. Most flowmeters are calibrated for atmospheric pressure inlets and therefore require a correction when used with a reduced inlet pressure. The necessity and magnitude of this correction can be determined by a static pressure measurement at this point.

The accuracy of the concentration measurement is equally dependent on the sampled volume as on the mass of the collected sample. The sampled volume can be determined either directly by an integrating flowmeter or from the product of the flowrate and the sampling interval. Flowrate measurements are made more frequently than are integrated volume measurements, because the meters are generally smaller, lighter, and less expensive. The precision of many flowmeters used in air sampling is, unfortunately, frequently poor and may act to limit the overall precision of the concentration determination. Care should be taken to avoid leakage between the inlet of the sampler and the inlet of the flowmeter, so that the volume measured represents only sampled air. Leakage downstream of the flowmeter can reduce the sampling rate but would not affect the accuracy of the measurement. The most commonly used flowrate meters in air samplers are rotameters and orifice meters.

The final element of the train is the air mover or suction source. It can be an air pump, a blower or fan, or an ejector. An ejector is a device that uses a stream of high-pressure fluid flowing through a jet to create a secondary stream of sampled air. The flow-induction fluid can be compressed air, steam, or water. The selection of air-mover type and size is dependent on the choice of sampling rate and the pressure drop to be overcome in maintaining the desired flow.

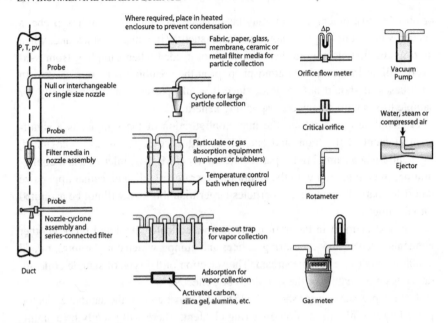

FIGURE 9–13. Sampling system components.

The sequence of the train elements is important but not sacred. Circumstances sometimes lead to the selection of a different configuration, and this is acceptable when it can be demonstrated that excessive losses or errors are not introduced. Some typical sampling-train elements and their normal sequence are illustrated in Figure 9–13.

Sample Collectors for Gases and Vapors

Gases and vapors can be extracted from an airstream by absorption, adsorption, or condensation. The most common absorption technique involves intimate contact between gas bubbles and an absorbing liquid. Gas washers in various configurations are illustrated in Figure 9–14. Table 9–2 presents examples of some gases commonly collected in liquid sorbents.

The airstream may be broken down into bubbles by a submerged jet, as in impingers (Fig. 9–14, gas washer A) and simple bubblers (Fig. 9–14, gas washer B). Finer bubbles, and hence greater gas-liquid contact and removal efficiency, can be obtained by passing the air through a porous plug or frit rather than a single orifice jet. Such a device, known as a fritted bubbler, is illustrated in Figure 9–14, gas washer D.

Absorption can also take place into a liquid film on a solid support. The support can be a screen or filter that has been coated with a reagent chemical. The airstream can be drawn through the screen or filter, or these devices can be used as passive samplers. In the latter, the gas or vapor molecules to be trapped reach

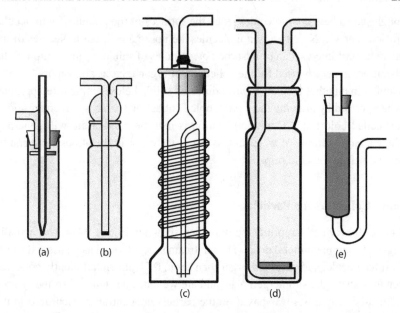

FIGURE 9–14. Various gas washers: (a) Midget impinger; (b) simple gas washer; (c) spiral absorber; (d) fritted bubbler; (e) glass-bead column.

the collection surface by molecular diffusion. The nitrogen dioxide (NO_2) sampler illustrated in Figure 9–15 utilizes an orifice as a diffusion barrier of known characteristics in order to eliminate the effects of bulk fluid motion on the transport rate of NO_2 from the ambient source to the collection surface at the screen.

Gases and vapors can also be captured by adsorption onto surfaces. The most commonly used adsorption collectors are cylinders filled with granules of an efficient adsorbent, such as activated charcoal or silica gel. As a gas stream is drawn

TABLE 9–2. Some Gases and Vapors Collected in Liquid Sorbents

CHEMICAL	SAMPLE COLLECTOR	SORPTION MEDIUM
Ammonia	Impinger	Sulfuric acid
Carbon dioxide	Fritted bubbler	Barium hydroxide
Formaldehyde	Fritted bubbler	Sodium bisulfite
Hydrogen sulfide	Impinger	Cadmium sulfate
Methyl mercury	Impinger	Iodine monochloride in hydrochloric acid
Ozone	Impinger	Potassium iodide in potassium hydroxide
Sulfur dioxide	Impinger	Sodium tetrachloromercurate

through such a bed, the molecules reach the surfaces of the granules by molecular diffusion. The O_2, N_2, and argon molecules are not retained, but molecules of the contaminant of interest can be adsorbed by granules of suitable composition. Solid sorbents are generally used for the collection of organic gases and vapors.

Contaminants that condense into liquids or solids below ambient temperature can be removed by drawing them past cooled collection surfaces, which are sometimes called cold traps. One complication is that the condensate will generally include a large volume of water unless the water vapor in the atmosphere can be removed before the cold trap.

Sample Collectors for Particles

While useful for gas sampling, the impinger, shown in Figure 9–14, was actually designed to sample mineral dusts. The airstream is accelerated as it passes through the orifice nozzle, and particles larger than about 0.75 μm are efficiently collected when they strike the submerged plate and are wetted. The standard formerly used for silica-bearing dusts was based on the particle concentration measured in the suspension within the flask. However, more recently, almost all particle sampling has been performed using dry-collection techniques. These include other inertial samplers, such as impactors, cyclones, and centrifuges, as well as filtration, electrostatic precipitation, and thermal precipitation.

Inertial (impactor) collectors

An impactor is very similar to an impinger but does not use a trapping liquid. The airstream is accelerated through a nozzle and directed at a collection surface. Particles larger than a certain size, termed cut-size, strike the surface and may be retained; smaller particles, having less momentum, are carried away with the carrier flow. The cut-size depends on the velocity in the jet. At ambient pressures, the

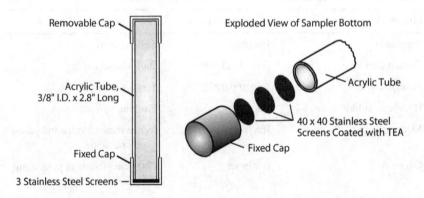

FIGURE 9–15. Schematic diagram of personal NO_2 sampler (Palmes tube).

smallest cut-size readily attainable is about 0.5 μm diameter; lower cut-sizes can be attained by operating the sampler within a vacuum chamber. If the particles and the collection surface are both dry, the particles are more likely to bounce and escape rather than be retained. Impactors can only be used successfully for such applications when the collection plate is coated with an adhesive layer. However, even with the use of an adhesive, total collection must be limited to avoid particle bounce. The adhesive can rapidly become saturated, and particles striking those already collected are poorly retained.

In a series, or cascade, impactor (Fig. 9–16), the same volumetric flow is passed through a series of impaction jets, with each successive stage having a smaller jet cross-section and particle cut-size. It therefore sorts the aerosol into a series of fractions on the basis of aerodynamic particle size. Knowing the proportions of the sample mass on each stage, and the stage constants (i.e., the cut-sizes of each stage), the overall size distribution of the sampled aerosol may be estimated.

Another widely used type of inertial sampler is the cyclone. In the cyclone configuration illustrated in Figure 9–17, the sampled air is drawn into a tangential

FIGURE 9–16. Diagrammatic cross-sectional view of a cascade impactor while sampling. The particle size is exaggerated.

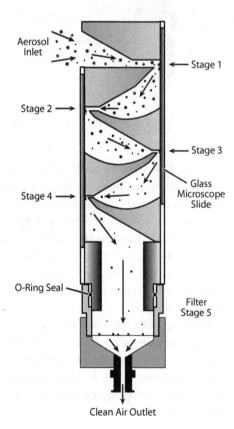

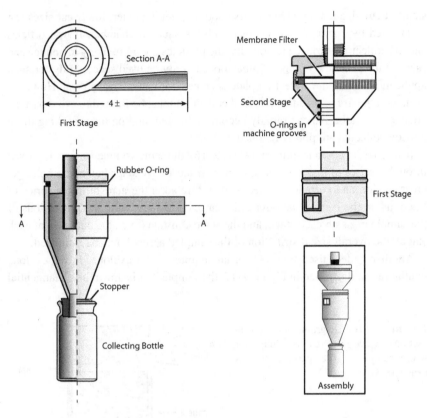

FIGURE 9–17. A miniature cyclone-filter unit for two-stage sampling of aerosols.

inlet. The flow follows a spiral path downward along the outer wall and then forms a tighter spiral as it flows up the axial exit pipe. Particles larger than the cut-size strike the outer wall and either remain there or migrate slowly down the wall into the conical section at the bottom. The major application of cyclones as samplers is as the first stage of a two-stage collector. They can also be used for aerodynamic size distribution evaluations in a multicyclone array, with each cyclone having a different cut-size. Their advantage over cascade impactors for such applications lies in the much larger sample masses that can be collected without artifacts due to particle bounce.

More precise separations and collection capabilities for smaller particles can be achieved using aerosol centrifuges. The particles enter at one side of a laminar flow channel and are deposited gently along a collection foil according to their aerodynamic diameter. However, aerosol centrifuges are basically laboratory instruments, their sampling flowrates are very low, and they are relatively large and expensive.

Filtration

Filters are currently the most widely used type of aerosol sampler. While no single type of filter is suitable for all applications, filters are available in a wide variety of types, ratings, and sizes, and there is generally at least one that is well suited for each particular application. Regardless of filter type, they all have inherent advantages, namely minimal equipment, low cost, and convenience with respect to sample handling. The major mechanisms of particle collection operative in air filters are inertial impaction, Brownian diffusion, and interception. In some cases, where the particles or the filter material have high levels of electrical charges, electrostatic precipitation may contribute significantly to the overall collection efficiency.

There are four basic types of filters: fibrous filters, consisting of a mat of randomly oriented fibers of cellulose, glass, asbestos, polystyrene, and so on; membrane filters, consisting of a plastic material with a gel structure with interconnecting pores; polycarbonate pore filters, which are a solid sheet with uniform parallel holes; and granular beds. The basic structures of polycarbonate pore and fiber filters are illustrated schematically in Figure 9–18.

In selecting a filter for a specific application, certain factors should be considered, including physical size, composition, and structure of the filter and whether these would limit the analyses to be performed; collection efficiency and flow resistance and how they may be expected to change during the sampling interval (e.g., with loading); and the characteristics of the available sampling pumps and whether they can provide the desired suction capacity and sampling rate throughout the entire sampling interval.

While filters may appear to be very simple devices, the mechanism by which they capture airborne particles is frequently misunderstood. The most common misconception is that they function like the sieves or screen collectors used to separate large particles from liquid streams and that particles smaller than the pore or void size will penetrate through the filter. This is generally not the case in air

FIGURE   9–18. Cross-sectional comparisons of Nuclepore and cellulose-fiber filter thicknesses.

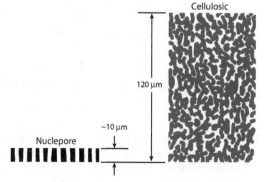

filtration, where particles much smaller than the void size can be collected on the filter with very high efficiency.

The flow path through polycarbonate pore filters is relatively simple, while flow through all other types is quite complex. The air undergoes accelerations, decelerations, and numerous branchings and directional changes in negotiating its way through the complex web of pores or voids. As the aerodynamic particle size and flow velocity increase, the probability of particle retention by impaction increases, while with decreasing particle size and flow velocity, the particles' Brownian displacement and retention time within the filter structure increase, increasing the percentage of retention by this mechanism.

The particles most likely to penetrate a sampling filter are those that have both minimal Brownian displacement and minimal momentum. Under most conditions, maximum penetration occurs for particle diameters of about 0.3 to 0.5 μm. For some widely used sampling filters, such as most membrane filters with pore sizes below 3 μm and most glassfiber filters, the maximum penetration will be less than 1%. On the other hand, some commonly used cellulose fiber filters may have particle penetrations as high as 70%, and polycarbonate pore filters may have even greater penetrations. These considerations are illustrated in Figure 9–19. Thus, the collection efficiency of filters with characteristic penetration minima can be

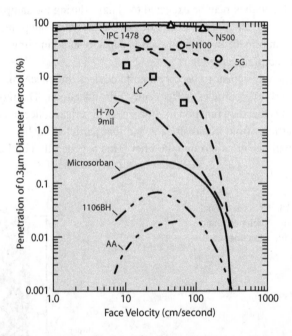

FIGURE 9–19. Penetration characteristics for various filters: AA-cellulose membrane, 0.8 μm pores; 1106BH-glass fiber; Microsorban-polystyrene fiber; H-70-cellulose-asbestos mixed fiber; W-41-Cellulose fiber; LC-teflon membrane, 10 μm pores; 5G-thin glass fiber; IPC 1478-thin cellulose-fiber mat; N100-Nuclepore, 1 μm pores; N500-Nuclepore, 5 μm pores.

greatly improved when they are operated at face velocities (the velocity normal to the filter face) that are greater or lower than those at which the minima occur.

In fibrous and granular filters, there is defense in depth; that is, the particles are collected by impaction on and diffusion to the granular or fiber surfaces throughout the depth of the filter. In order to analyze the material collected, one must either extract it from the filter or analyze it on the filter itself. On the other hand, in membrane and polycarbonate pore filters, which are much thinner, the particles are collected on or close to the upstream surface. This makes it possible to perform some kinds of analyses without removing the sample from the surface. These include electron or optical microscopic examinations of the particles, counting of α-particle emissions without excessive α-particle absorption within the filters, and X-ray fluorescence analysis of elements with low-energy X-ray emissions.

Thermal precipitation

Particles can be extracted from a sample stream by thermophoresis in a device known as a thermal precipitator. Prior to the mid-1950s, these were widely used in Europe to collect particles for microscopic determinations of particle number concentration and size distribution. In this device, a heated wire causes an unequal bombardment of particles by gas molecules, creating a net migration of the particles away from the wire and, in effect, a dust-free zone around the wire. When the temperature is high enough and the sampling flow-rate is low enough, the diameter of the dust-free zone can be larger than the width of the flow channel. In this situation, essentially all of the particles are collected on a linear trace, as illustrated in Figure 9–20.

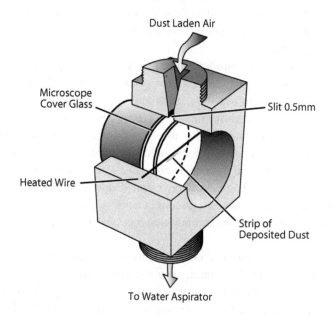

FIGURE 9–20. Sampling head of a thermal precipitator.

Electrostatic precipitation

When an airstream containing particles carrying electrical charges passes through a flow channel having a potential gradient normal to the flow, the charged particles will be deflected toward the electrode closer to ground potential. The efficiency of collection increases with the number of charges on the particle and the potential gradient. Very high collection efficiencies can be achieved in electrostatic precipitator samplers.

Electrostatic-precipitator samplers were widely used for particle sampling in the 1930s and 1940s because of their high collection efficiency and their low and constant flow resistance. However, precautions are needed to avoid unintended high-voltage discharges, and they cannot be used in atmospheres containing combustible vapors. With the development of improved filters for air sampling, which were simpler to use and much less expensive, electrostatic precipitation has since become limited to special sampling applications.

Special Considerations for Air Sampling

Ambient air concentrations of chemical contaminants are greatly influenced by meteorological factors, and their interpretation is, therefore, facilitated by knowledge of wind speed and direction, mixing layer depth, temperature profile, humidity, and so on. It is generally desirable to have appropriate weather instrumentation that can represent ambient conditions at the sampling site. Furthermore, as indicated previously, the selection of sampling sites requires a compromise between an ideal of a relatively large number of sites at strategically located positions and practical limitations imposed by fiscal constraints, availability of sites secure from malicious mischief, accessibility for sample changing and maintenance, and availability of power.

There are some special considerations in the selection of specific instrumentation for ambient air sampling. For airborne gases and vapors, the trend in recent years has been to use direct-reading continuous instruments with rapid-response gas-phase sensing, with each instrument being specific for a particular contaminant. Direct-reading instruments have also been favored for measurement of particle size distributions. For aerosol composition determinations, the trend has been toward sample collection in discrete aerodynamic size fractions. For example, a size cut is made at about 2.5 µm in aerodynamic diameter when sampling for fine particles that derive primarily from gaseous precursors and at 10 µm to capture the larger particles that derive from mechanical processes and are emitted as discrete entities.

Air sampling in occupational environments is generally much more highly focused than is ambient air sampling. It is usually designed to determine exposure levels for specific individuals or groups of individuals engaged in a common activity. It is also concerned with only one or, at most, a limited number of specific air contaminants at each work site.

There are two basic approaches available for determining an individual's average exposure. The more traditional one involves dividing the workday into a number of specific subfractions during which the air concentration is reasonably constant or reproducible. Samples are then collected in sufficient numbers in each activity interval to permit a reliable characterization of the exposure in that activity. The product of the exposure concentration at a given activity and the time devoted to that activity during each workday represents the contribution of that activity toward the overall exposure. The time-weighted average exposure is the sum of all the products of concentration and time, divided by the total daily work time. A representative determination of a time-weighted average occupational exposure is illustrated in Table 9–3.

TABLE 9–3. Sample Determination of a Time-Weighted Average Occupational Exposure

Given: Workers perform three different operations that involve potential exposure to airborne Pb during each workday. Breathing zone (BZ) air samples were collected at each of the operations, and general air (GA) samples were collected that represent the workers' exposure during the balance of the workday.

Sampling rate for all samplers: 15 L/min
Sampling interval for BZ samples: 10 min
Sampling interval for GA samples: 60 min

Sample collections:	Operation #1: 20.6, 24.3, and 18.2 µg
	Operation #2: 35.2, 33.5, and 39.1 µg
	Operation #3: 6.5, 9.7, and 7.8 µg
	General air: 3.7, 1.9, and 5.1 µg
Working time/day:	Operation #1: 60 min
	Operation #2: 90 min
	Operation #3: 90 min
	General air: 4 hr (240 min)

CALCULATION

OPERATION	AVG. SAMPLE (µg Pb)	SAMPLED VOLUME (LITERS)	PB CONCENTRATION (C) (µg/m³)	EXPOSURE TIME (T) (MIN)	C × T (µg-MIN/m³)
1	21.0	150	140.2	60	8,412
2	35.9	150	239.6	90	21,564
3	8.0	150	53.3	90	4,797
GA	3.6	900	4.0	240	960
				Totals 480	35,733

Time-weighted average exposure = 35,733/480 = 74.4 µg/m³

Note: Threshold limit value (for 2000) = 50 µg/m³, so the workers in this case are overexposed. Operation #2 accounts for 60% of the daily exposure. If, by application of engineering controls, the concentration at Operation #2 was reduced to 25 µg/m³, the time-weighted average would drop to 34.2 µg/m³.

An alternate approach to the determination of average daily exposure involves lightweight, self-contained samplers and battery-powered pumps worn by individual workers without restricting their mobility. The sampling rates are low, but the sampled volume accumulated over a workday is usually large enough to permit an accurate determination of the average exposure for the day. Although there are fewer samples to analyze using this approach, it provides no information on peak levels of exposure or which of the activities during the day may have accounted for the bulk of overall exposure.

The trend in recent years has been toward a greater reliance on personal samplers for routine monitoring of exposures. In sampling airborne PM, there has been a growing utilization of size-selective samplers, where the PM is aerodynamically separated into sample fractions on the basis of its presumed fate in the respiratory tract. One application of these is for dusts that produce pneumoconioses following deposition in the alveolar region of the lungs. The PM that penetrates to this region is simulated by the second stage of a two-stage sampler with a first-stage collector removing the fraction that is expected to be deposited in the upper respiratory tract and the tracheobronchial tree and a second stage that collects all of the PM penetrating the first stage.

There are a number of special considerations related specifically to source sampling for air contaminants. The air flowing in discharge ducts and stacks generally has much higher concentrations of contaminants than does the ambient air or workroom air. As a result, it is important to use sampling substrates that can retain relatively large sample masses without losses due to resuspension in the sample stream and that do not exhibit changes in sampling rate due to clogging.

When the flow channel is large and/or where there are branch entries or elbows near the sampling port, the flow pattern and contaminant concentration across the channel are likely to be far from uniform. Thus, in order to determine the average concentration or total mass discharge rate, it is necessary to measure both the flowrate and concentration at numerous points across the channel. Such a series of measurements is known as a traverse; standardized traverse locations for circular and rectangular cross-sections are available and are shown in Figure 9–21.

It is important that the stream drawn into the sampler probe at each traverse point be representative of the stream at that point. No special precautions are needed when sampling most gases or PM with diameters smaller than approximately 2 μm. However, for larger particles moving in high-velocity airstreams, the particle momentum may lead to a biased sample. The bias may be either for or against the larger particles, depending on the velocity and orientation of the sampling probe with respect to the stream flow. Representative samples of aerosols containing larger particles can only be taken using the technique of isokinetic sampling, where the probe inlet faces into the airstream and the velocity in the probe is equivalent to that of the stream flowing past it. Isokinetic sampling is illustrated in Figure 9–22.

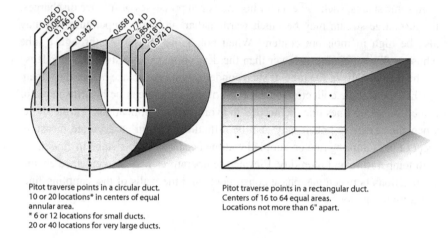

Pitot traverse points in a circular duct.
10 or 20 locations* in centers of equal
annular area.
* 6 or 12 locations for small ducts.
20 or 40 locations for very large ducts.

Pitot traverse points in a rectangular duct.
Centers of 16 to 64 equal areas.
Locations not more than 6" apart.

FIGURE 9–21. Traverse points for flow measurements in ducts.

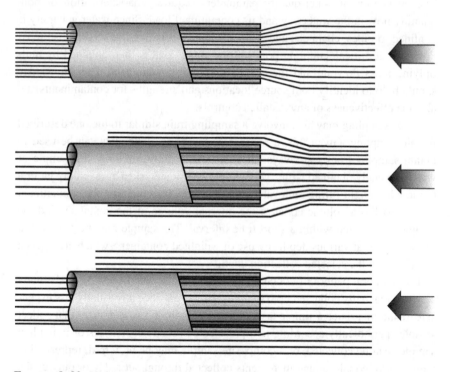

FIGURE 9–22. Various types of aerosol sampling. (top) Isokinetic sampling; (middle) Flowrate less than isokinetic rate. Flow streamlines diverge outward near the edge of the probe, causing larger particles to diverge into probe, and smaller particles to follow streamlines; thus, sample has too many large particles. (bottom) Flowrate greater than isokinetic rate. Streamlines diverge inward; sample has too few large particles.

In some stacks, such as vents for hot industrial processes or furnace discharges, the discharge stream may be much warmer than ambient temperature and may also be high in moisture content. When hot, moist air passes through a tube whose wall temperature is lower than the dew point, the moisture will condense on the inside walls of the tube. If such condensation takes place before the stream reaches the sample collector some, or all, of the sample can be lost. Insulated or heated sampling lines may be needed to avoid such complications. Furthermore, when the stream is hot, it may also be difficult to obtain an accurate measurement of the sampling rate or sampled volume because of the change in air density with temperature and humidity. Elevated temperatures may also enhance chemical reactions between gas-phase constituents and the walls of the sampling line, resulting in line losses.

WATER SAMPLING

The evaluation of water-quality parameters requires characterization of both quantity flow in the waterway and its contaminant load. Since water generally is confined to pipes, channels, and defined streams, it is possible to establish mass flowrates at critical sampling points. Total waste load is then obtained by multiplying mass flowrate and the concentration of contaminant. This knowledge should help to identify both source locations and strengths for contaminants and also the effectiveness of any installed controls.

Water sampling may also involve a sampling train similar to the one described for air sampling. Given this, a number of sampling techniques can be used to obtain samples from waterways. In one, termed grab sampling, the sample of water is usually obtained with a very simple instrument, such as an empty jar or a bottle with a tight-fitting removable cap. The sample containing the contaminant of interest has a volume equal to the internal volume of the container, and the sample is collected within a short time interval. The sample may be obtained at the surface or at various depths by use of weighted containers, which are opened and closed at the selected collection depth.

The temporal pattern of contaminant levels can be determined by analyzing a series of manually collected grab samples. However, since it is usually less expensive to automate such a simple task, it is more common to use devices that collect samples periodically according to a preselected schedule. A series of individual samples can be collected, or a composite sample may be preferred, representing a mixture of equal-volume increments collected throughout a day or duty cycle. However, intermittent sampling at fixed time intervals may miss surges in contaminant discharge and, therefore, it may be desirable to draw the sample continuously. Continuous samplers are generally operated at a very low flowrate to avoid processing larger volumes than those needed for the laboratory analyses.

Samples of constant volume collected on a fixed time schedule and samples collected continuously at a constant sampling rate can be used to determine the flux of contaminant directly only when the stream being sampled moves at a constant rate. When the stream flow varies, both the flowrate and the concentration must be determined in order to obtain the average or total mass flux of contaminant. Alternatively, the samples can be collected at variable rates that are proportional to the flow.

There are two basic types of proportional samplers. One type collects a definite volume at irregular time intervals; the other collects a variable volume at equally spaced time intervals. Both are flow dependent. In one, flow dictates the time interval, while in the other, flow regulates the sample volume. In most of these, the pressure differential that is needed to extract the sample is provided by the moving stream itself, ensuring the required proportionality. A schematic diagram of a proportional-sampler control system is presented in Figure 9–23.

A variety of approaches have been used in flow proportional sampling. In some, a constant sampling flow is pumped through a pipe. After a predetermined volume has passed through the flowmeter, a diverter is activated and a sample is taken. An air-lift automatic sampler can be used to obtain samples when a pump cannot be used. When the compressed air supply is shut off (the air-control valve is connected to a timing relay from the flow-measuring device), a spring in the sampler raises a piston, normally kept closed by the air, which opens an inlet so that water enters the sampler and goes to the sample container. The air valve is then opened and the piston is forced down, closing the inlet. The air passes

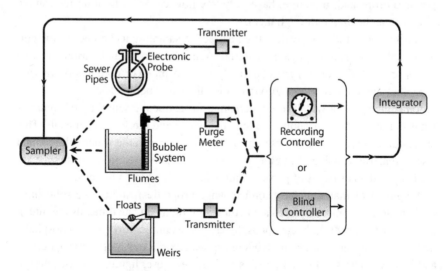

FIGURE 9–23. "Flow-proportional" sample-control systems.

through the air-escape port into the main chamber of the sampler and forces the liquid up the sample line into the collecting container. The cycle is then repeated.

In a vacuum-type sampler, which is very often used in water with high suspended-solids content or corrosive properties, a signal from a flowmeter activates the vacuum system, which lifts liquid through a suction line into the sample chamber. When the chamber is filled, the vacuum line is automatically closed. The pump then turns off, and the sample is drawn into the sample container. A float check prevents any liquid from reaching the pump. Another type of automatic sampler consists of a cup or dipper on a chain. When activated by a signal, the cup is carried down through the stream and returns filled with a sample, which is emptied into a container. In theory, most automatic-type samplers may be connected to flow-measuring devices in order to obtain proportionate samples.

Another approach to water sampling involves extractive techniques. These have many advantages for water sampling, corresponding to those previously discussed for air sampling. The samplers are much lighter, more compact, and easier to preserve and ship, and they allow the concentration of contaminants from large volumes of water in a practical manner. However, effective and reliable extractive techniques have not been developed for a large number of water contaminants of interest, including many soluble gases and salts and most colloidal solids.

Effective extractive techniques are available for suspended solids, many organics, and some of the toxic metallic ions. Membrane and glass-fiber filters are widely used for collecting suspended solids. In liquid filtration, sieving is an important mechanism of collection, and a major fraction of the particles smaller than the void or pore size can penetrate the filter. Membrane filters are available with pore sizes as small as 0.01 μm, but the smaller-pore-size filters have large pressure drops and, therefore, have very low flow capacity, limiting the sample size that can be drawn through them.

Adsorption is the most widely used extractive sampling technique for the collection of water-soluble organic material. Activated carbon and various organic resins (e.g., XAD, Tenax) are commonly used in pipes or columns of varying capacities and configurations through which the water percolates.

Ion exchange is a widely used technique for extractive sampling of contaminants in ionic form; it is applicable to both organic and inorganic chemicals. This technique involves the reversible transfer of ions between the water solution and a solid material capable of binding the ions. The resins used in ion exchange are graded powders or beads of porous compounds, packed into columns.

A number of materials have ion-exchange properties. Among the solid inorganic materials commonly used are natural aluminum silicate minerals (zeolites); hydrous TiO_2 or ZrO_2; metal phosphates, for example, zirconium phosphate, microcrystalline ammonium molybdophosphate; oxides of Al(III), Si(IV), Fe(III), and Mn(IV). The solid organic resins include cellulose, lignin, cation-exchange resins such as natural clay, chelating resins (with amine carboxylates as functional groups), and anion resins (usually in OH⁻ or Cl⁻ forms). The selectivity for specific ions depends largely upon the chemical structure of the resin and its physical configuration (e.g., degree of cross-linking).

Special Considerations in Water Sampling

It is especially difficult to collect representative samples of water in a free-flowing stream or river. The velocity profile in a stream is highly variable, depending as it does on the cross-section for flow, the volumetric rate of flow, the disturbances introduced upstream by bends in the stream, tributary streams, surface winds, stream bed obstructions, and so on. The stream flow varies in the short term with the recent history of precipitation and in the longer term with seasonal variations in runoff. Variations in flow affect both the levels of contaminants attributable to surface runoff and the resuspension of contaminants from the stream sediments.

Since no one point in the stream cross-section can be considered representative of the overall stream, it follows that contaminant flux can only be determined accurately from an analysis of samples collected at a series of representative locations across the flow cross-section. Unfortunately, the appropriate traverse locations are not as easily defined as those for the circular and rectangular cross-sections of Figure 9–21; it may be necessary to sample at only one or a few points and to accept the uncertainties introduced by the circumstances. Although ideally the sample should be taken from mid-stream where the velocity is greatest and the possibility that any solids have settled is minimal, in practice, most samples are collected at the edge of the stream and near the upper surface. One potential error introduced by such a practice is illustrated in Figure 9–24.

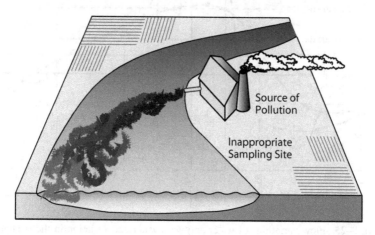

FIGURE 9–24. An illustration of a potential problem encountered by sampling near the edge of a stream.

The collection of representative samples of effluent wastewaters is relatively simple in comparison with sampling of process lines or free-flowing streams. In most cases, the number of outfalls of discharge points will be limited, their volumetric rate of flow will generally be known, or at least definable, and the flow will be relatively well mixed. The major variables will usually be in the temporal aspects of the waste discharges. Industrial wastes may be expected to vary greatly in both composition and volume with time, reflecting the batch nature of many production operations, the periodicity of routine cleanup and maintenance operations, and the consequences of accidental, unanticipated releases.

The first step in designing an effluent sampling program is to define all of the significant discharge points. The next step is to determine their likely variability in discharge and to select the best means of characterizing the discharges from each. If the total amounts discharged are of primary concern, it may be satisfactory to continuously collect an integral sample whose volume is proportional to the overall flow. On the other hand, if the objectives include identifying the contributions of the component sources to the overall discharge, it may be necessary to collect a series of effluent samples at appropriate times throughout the production or other cycle. Some of these considerations are illustrated in Figure 9–25.

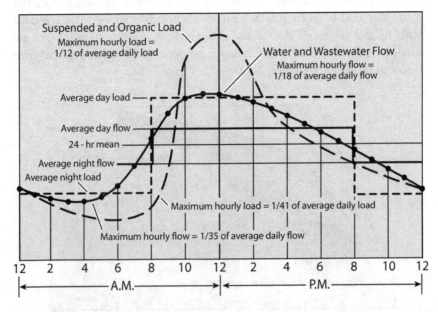

FIGURE 9–25. Flow variations of water and wastewater and variation in the strength of wastewater.

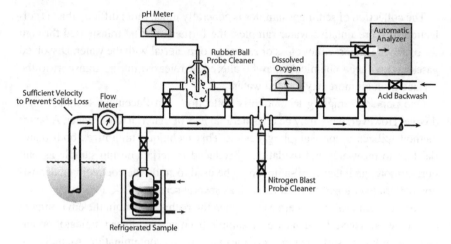

FIGURE 9–26. Examples of possible precautions needed in a sampling train to prevent changes in composition in the water sample.

Some special precautions are necessary in the handling of water samples. Water that is contaminated with chemicals is often contaminated with microorganisms as well. Since the metabolism of these microorganisms may add or subtract materials of interest between the time of collection and the subsequent laboratory analyses, water samples are often refrigerated to minimize such complicating effects.

Many other time-dependent changes may occur in water samples after their collection. These may include changes in pH, loss of some dissolved gases and absorption of others from the atmosphere, change in valence state of metal ions via oxidation-reduction reactions, and precipitation of metal cations due to chemical reactions. It is important that the procedures used to stabilize one contaminant of interest do not affect the determination of another. Some precautions which may be needed in a water-sampling train to preserve contaminants of interest are illustrated in Figure 9–26.

BIOTA, SEDIMENT, AND SOIL SAMPLING

A complete evaluation of the extent and significance of environmental contamination by a given material will frequently require the analysis of soil and aquatic sediments and biota, since they can act as temporary or permanent sinks for the material. As long as the contaminant is available for ingestion by mobile life forms, absorption by roots or leaves of plants, or chemical leaching into the surrounding medium, it retains a potential for producing adverse effects.

The collection of sediment samples is generally much more difficult than is collection of representative water samples; the former must be transported through the overlying water without loss or exchange of material with the water. Dissolved gases and reduced chemicals may be altered or released during transport to the surface through more oxygenated waters.

One popular sampling technique is collection of a flocculent sample with a dredge. However, there may be considerable loss of material to the water. A better method utilizes a core-sampling device. This technique minimizes both material loss to the water and oxidation of reduced materials during sample ascent. Furthermore, an intact core sample may be used to study the contaminant deposition pattern over a period of time, such as several seasons.

Contaminant evaluations are easier when the pathways within the environment are well established. In such cases, a sample from an established indicator organism may provide a sufficient index of the extent of the contamination. An indicator organism is a plant, animal, or microbe that provides an index of environmental conditions by its presence, absence, or specific characteristics. The narrower the tolerance of the indicator organism for a specific condition, the greater is its utility for describing ecologic conditions. For example, presence of an aquatic organism that can only survive in a pH from 7.0 to 8.0 is an accurate field monitor of mild alkalinity in a waterway.

Indicator plants may be used in evaluating the extent of chemical contamination of the ambient air by characteristic pathological changes. For example, intercostal markings on the leaves of violets indicate high sulfur dioxide (SO_2) levels; banding on snowstorm petunia leaves indicate aldehydes; white flecks on tobacco leaves indicate high ozone (O_3) levels. Other plants may accumulate chemicals and thus serve to provide an index of ambient levels. For example, samples of Spanish Moss growing in the southeastern US on trees downwind of highways were collected in order to determine their lead (Pb) content, on the basis that this content was proportional to traffic density. Certain plants are associated with the presence of particular chemicals in soil. For example, the growth of Eastern columbine in the eastern US is indicative of high soil calcium carbonate content, and plants of the genus *Astragalus* in the western US grow in association with selenium (Se) in soil.

Some of the best biological indicators of chemical contamination can be found in the aquatic environment. Some filter-feeding crustacea can extract organic contaminants and metallic ions very efficiently from water. Aquatic plants may also greatly concentrate some of the trace metals that accumulate in the bottom sediment of streams and lakes. Certain algae, such as the green and blue-green types, are indicators of high-nutrient content in waterways. Samples of aquatic biota may be obtained with nets, dredges, or an artificial substrate, such as glass slides for microorganisms.

FOOD SAMPLING

Concern about the extent of chemical contamination of food may lead to several kinds of sampling programs. One involves identification of specific batches of contaminated, adulterated, or spoiled food products that may reach the market. The random- and spot-sampling programs of the Food and Drug Administration and various state agencies fall into this category. When the samples indicate that the quality of the food is substandard, the batch of food product represented by the sample will be destroyed or diverted to another usage where its quality would be acceptable.

A second type of program focuses on the definition of pathways and accumulation rates for materials of interest that do not present immediate hazards but may be worthy of concern and continued observation. One example is the worldwide food-sampling program and analysis for ^{90}Sr content during and after the period of atmospheric testing of thermonuclear weapons in the 1950s and 1960s. Another is the continuing monitoring of pesticides in foods.

In much of this large-scale food sampling, the extent of the problem is determined from a relatively limited number of composite samples rather than from large numbers of analyses of individual foods. The approach is generally known as market-basket sampling. In it, a representative diet is selected and the food products needed to prepare this diet are purchased from regular retail outlets. The foods are prepared for cooking as they would be in the home (i.e., wrappings, bones, trimmings, etc., are discarded). The parts that would be consumed are then analyzed for their content of the contaminant(s) of interest.

Market-basket sampling has several important advantages for routine monitoring of overall contaminant levels and their trends. These are fewer samples that need to be analyzed; the samples are relatively large, avoiding problems of too limited analytic sensitivity; and the results are directly relatable to an average population exposure. On the other hand, there are important limitations in that there is no information on the major sources within the overall diet, and there is no information on maximum levels of exposure within the population attributable to unusual dietary patterns.

BIOMARKERS OF EXPOSURE

The sampling and analysis of ambient air, drinking water, and foods can indicate the contaminants to which people are exposed. However, environmental data cannot define the fractions retained after, for example, inhalation or ingestion either initially or at later times. For most toxicants, the amount retained is very variable among individuals and with time in a given individual. Furthermore, the basic metabolic pathways are poorly defined for many contaminants. For these reasons, it is usually a good practice to use biological samples as supplements to environmental samples in evaluating the health significance of exposures to environmental contaminants.

Biological samples have, in the past, been most widely used in occupational health evaluations. Biological exposure indices have been used for various chemicals, their metabolites, or the alterations they produce in biological constituents or physiological functions. The kinds and sizes of the samples needed for such analyses depend on the contaminant and its metabolism and the sensitivity of the assay.

The easiest kinds of samples to collect are urine and exhaled air. Hair and fingernails may also be useful and can also be collected fairly readily in many cases. It is usually more difficult to obtain the cooperation needed for the collection of blood or fecal samples, but they frequently are needed and can be successfully collected. Table 9–4 lists some biological indices of contaminant exposure.

An ever-present problem in the collection and handling of all of these types of samples is contamination, and extreme caution must be exercised in the selection and handling of the sample container. The material to be analyzed must not

TABLE 9–4. Some Biological Samples Used as Exposure Indices

CONTAMINANT	SPECIMEN FOR EVALUATION	CHEMICAL INDICATOR
Arsenic	Urine, blood, hair	Arsenic
Benzene	Urine	Phenol
	Blood, exhaled breath	Benzene
Cadmium	Hair, nails	Cadmium
Carbon monoxide	Blood	Carboxyhemoglobin
Cyanide	Blood	Cyanmethemoglobin
Fluoride	Urine, hair	Fluoride
Hydrogen sulfide	Blood	Sulfhemoglobin
Lead	Hair, blood	Lead
	Urine	Lead, delta aminolevulinic acid
Nitrite	Blood	Methemoglobin
Organic mercury	Urine, hair	Mercury
Parathion	Urine	P-nitrophenol
Polycyclic aromatic hydrocarbons	Urine	Analysis for parent compound
Vinyl chloride	Exhaled breath	Vinyl chloride

be extractable from the container itself, and anyone handling the sample must be aware of all the possible means of introducing artifacts and avoid them. For example, industrial workers usually have much more of the chemical of interest on their hands than in their urine and, if they have not first washed thoroughly, can easily contaminate a urine bottle while filling it.

Another problem is the collection of a representative sample. Most biological samples are short-interval (grab) samples containing a material of interest whose body concentration varies considerably with time. Proper interpretation may depend on the availability of supplementary information, such as the times of exposure relative to the times of sampling and the characteristic metabolic time constants.

For some contaminants, especially radioactive materials, the overall retention in the body and within various organs can sometimes be determined by noninvasive, external measurements, in this case, of emitted radiation. Estimates of individual and population burdens of contaminants can also be made by analysis of tissues taken at autopsy from accident victims. However, the great difficulties associated with obtaining adequately sized samples of this type generally precludes the acquisition of a significant body of data through this approach.

METHODOLOGY FOR SAMPLE ANALYSIS

Concentrations of contaminants in environmental samples may be measured by laboratory analysis following sample collection or by use of direct-reading instrumentation in the field. Many of the instruments used in the field are merely more compact and/or rugged versions of those used in the laboratory.

Sample collection refers to the collection of a defined weight or volume of sampled material for a subsequent laboratory analysis or series of analyses. The analyses are performed at a later time, and the results are not generally available until long after the sampling has been completed. In other words, there is no possibility of feedback to guide further sampling or to initiate immediate reduction of exposure.

In direct measurements, on the other hand, the sampling and analysis are performed in the same instrument. This technique allows feedback during the sampling period at the sampling site, so that the site and/or frequency of sampling may be modified so as to permit a more thorough evaluation of exposure and its determinants or to justify a prompt shutdown or evacuation when exposures are acutely hazardous. In using any direct-reading instrumentation, accurate calibration is essential to the correct interpretation of instrument response.

Some direct reading instruments are known as continuous analyzers. In these, an intrinsic property of the contaminant is measured as the sampling stream passes through a sensing zone. There may be no sample collection, and an uninterrupted output response is a direct function of the concentration of the contaminant of interest. The output can be recorded and can also be integrated to yield average concentrations over specified time intervals.

Other types of direct measurements are based upon extractive sample collection and immediate analysis of the sample. These instruments perform collection and analysis within the same housing and are known as semicontinuous analyzers. A representative fraction of the contaminant is obtained and analyzed, and the process then repeats itself. The measurement phase is generally performed on a liquid, after controlled dosage of a reagent chemical, using electrochemical or colorimetric techniques. In air sampling, the contaminant is first extracted by a collecting liquid in a gas washer or scrubber. Ideally, the analyzing period is sufficiently short so that no significant chemical changes occur before another sample is measured. However, because of the finite time lag between the sampling and measurement steps, which is generally on the order of a few minutes, and as a result of the mixing and diffusion within the liquid stream, a monitor of this type has a slower response than does a continuous gas-phase analyzer.

There are a number of advantages to direct measurements: both peak and average concentrations can be determined; results are immediately available, permitting one to identify sources and initiate corrective actions as appropriate; continuous and unattended operation is possible under some circumstances; and continuous recordings of concentrations may be useful in retrospective data evaluations and, in industry, as legal documentation of employee exposures for compliance with established standards.

On the other hand, there are many distinct advantages in sample collection with subsequent laboratory evaluations. For example, sample collection is a much simpler operation in the field than is quantitative chemical analysis, and the hardware needed for collection is less expensive, smaller, and less sensitive to mechanical and thermal stresses than that of sensitive analytic equipment. Because it is smaller and more portable in most cases, the sampling equipment is easier to transport and operate in the field and can be set up in locations not suitable for continuous monitors. The analyses performed in the laboratory on field-collected samples can be more sensitive, accurate, and reliable. The samples can be analyzed by highly specialized equipment that cannot be incorporated into field monitors. Furthermore, a series of different analyses can be performed on each sample, and the effects of co-contaminants can be eliminated by preliminary chemical separations.

Laboratory Analysis of Environmental Samples

All of the samples of air contaminants, water, soil, biota, food, and biological materials brought to the analytical facility can be evaluated by similar techniques.

While they may differ considerably in their nature, such samples generally have a number of common characteristics. These include the likely presence of a variety of cocontaminants at concentrations equal to or greater than the contaminant of interest and the presence of a collection substrate or matrix from which the contaminant must be extracted prior to analysis. Thus, the first task in any laboratory evaluation is generally to separate the contaminant of interest from its sampling substrate matrix and any cocontaminants so that it can be further analyzed unambiguously.

Separation is often combined with preconcentration procedures. Preconcentration improves the sensitivity of the analytical method and increases the precision and accuracy of analysis. The necessity to concentrate prior to analysis is dependent upon the specific analytical method that is to be used and the lower limit of detection of the procedure. For example, organic constituents of water are sufficiently numerous and usually present at such low concentrations that they require preconcentration and separation. Many inorganics are present at concentrations above the sensitivity limits of analytical methods; however, numerous interferences can arise from other agents that should be removed prior to analysis. Many of the current analytical methods are subject to well-characterized interferences, which are noted in published standardized procedures for contaminant analysis.

Contaminants collected by extractive sampling techniques generally need to be separated from the collection matrix. Following adsorption on activated carbon, the organic chemicals are extracted from the carbon using a chloroform wash, sometimes followed by methanol. Following evaporation of the solvent, the weights of the organic extract residues are expressed in units of µg/L as carbon chloroform–extract and carbon alcohol–extract. These extracts may be further separated into groups, for example, bases, weak acids, strong acids, and so on, via differential solubility techniques.

Recovery of contaminants from ion-exchange resins is performed by elution with a solvent appropriate for the particular material being analyzed. Ion exchange is a particularly useful separation technique for selective removal of ions that may interfere with the analysis of other ions.

A number of specific separation techniques are used in contaminant analyses. These are presented in Table 9–5. Often, more than one procedure is required prior to final analysis. Many of these separation techniques can also be used for the preconcentration of contaminants.

The selection of the best analytical techniques in a specific situation is often quite difficult. Factors to consider in any choice are any sample preprocessing required, specificity with regard to the chemical being analyzed, influence of the matrix, amounts needed for determination of the chemical accuracy for the concentration ranges expected, and time and cost. The full

TABLE 9–5. Some Common Chemical Separation Techniques

TECHNIQUE	PRINCIPLE	TYPICAL SEPARATION APPLICATIONS
Coprecipitation	Removal of an ion from solution via adsorption on a carrier precipitate	Trace metals
Liquid-liquid extraction (solvent extraction)	Physical separation based upon the selective distribution of a substance in two immiscible solvents (e.g., water and a water-immiscible organic solvent)	Many organic and inorganic compounds, especially water-immiscible organics and metal ions
Freeze concentration	Water in sample is frozen, yielding pure ice crystals and leaving water-soluble impurities in a liquid phase with a reduced volume	Wide range of inorganic and organic solutes
Adsorptive bubble separation	Passage of gas bubbles through a solution or suspension in a vertical column; material of interest may be adsorbed at bubble-solution interface or onto bubble surface	Surface active agents; colloids and some other particles
Distillation, evaporation, sublimation	Separation from water via removal of a component as a gas or vapor	Organic compounds; dissolved inorganic gases
Centrifugation	Decrease time required for particles to sediment by increasing the gravitational forces affecting them	Macromolecules; suspended and colloidal particles
Chromatography	Based upon the differential migration of substrates due to their selective retention in a fixed phase while subjected to movement by a flowing bulk phase (gas or liquid)	Wide applicability; generally used for organics
Gel filtration	Lodging of colloidal particles and high-molecular-weight organic compounds in pores of a cross-linked gel packed in columns	Colloids; organics
Chelation separation	Complexing of metals by certain reagents	Metals
Liquid-anion exchange (ion-pair formation)	Separation of metal complexes by formation of an ion pair with an onium cation (e.g., trialkyl ammonium) formed by use of a strongly basic solvent	Metals
Dry ashing	Combustion of material (e.g., filter, biological tissues) and recovery of contaminants of interest in unburned residue or in gaseous emissions	Metals
Wet ashing	Oxidation of material in acid to separate contaminant of interest from an organic matrix	Metals

range of analytical instrumentation that can be used for the qualitative and quantitative analyses of environmental samples could not be adequately described in a whole volume, let alone a small part of this chapter. An overview of some of the more commonly used techniques and their typical applications are presented in Table 9–6.

Direct Measurement of Environmental Concentrations

As mentioned earlier, many instruments may be used for direct measurements of contaminants in the field. This discussion of equipment for the direct determination of air and water concentrations is brief, limited to an illustration of some of the more common or innovative approaches to the general task. Note that similar instrumentation may be used for both water and air sampling. The determining factor is the specific state in which the sample must be introduced to the instrument.

Direct-Reading Instrumentation for Airborne Contaminants

Gases and vapors

Any characteristic physical property of a chemical can be utilized in its detection and measurement. A variety of instrumental principles can be employed to maximize sensitivity, specificity, and precision. For example, a widely used instrument for monitoring airborne CO is based on its capacity to absorb specific wavelengths of infrared radiation (IR). Thus, CO molecules in the path between an IR source and an IR sensor will reduce the incident flux at the sensor. A higher concentration or longer path length will increase the attenuation. Other gases and vapors present in relatively high concentrations in the air, including methane and water vapor, will also absorb IR. However, they do not absorb at the same wavelength as does the CO and, by using an IR beam of an appropriate wavelength, the instrument response can be limited to CO. It also follows that the same basic instrument can be used to make specific measurements of the concentrations of other airborne molecules using appropriate IR wavelengths.

A schematic diagram of an IR instrument is shown in Figure 9–27. It uses both a sensing cell and a sealed reference cell, and the signal is proportional to the difference in the heat absorbed by the two detectors which, in turn, varies with the content within the sample cell. In this configuration, small variations in IR output by the source affect both detectors equally and do not significantly affect the output signal.

Instruments have also been developed for airborne contaminants, such as mercury vapor and benzene, which strongly absorb ultraviolet (UV) light. Other

TABLE 9–6. Some Common Methods of Contaminant Analysis

METHOD	PRINCIPLE OF OPERATION	INSTRUMENTATION	SAMPLE[A]	SPECIFICITY[B]	SENSITIVITY[C]	TYPICAL APPLICATIONS
Gravimetric	Isolation of component of interest in the form of one of its compounds which shows insolubility under test conditions and has a known chemical composition; weigh the product	Conventional lab equipment	SLG	Good	1–10 μg	Numerous analyses for inorganic and organic materials
Titrimetric (e.g., acid-base; oxidation-reduction; precipitation)	Reaction of material in predictable manner with a standard solution of known concentration	Conventional lab equipment	L	Good	10^{-6}–10^{-7} M in solution	Numerous analyses, e.g., acidity, alkalinity, water hardness (Ca^{-2}, Mg^{-2}), Cl^-, S^{-2}, transition metals, DO
Absorption spectrophotometry						
Visible	Measure of selective absorption or transmission of visible light, usually following reaction with a reagent	Colorimeter; spectrophotometer	SLG	Fair	0.005 ppm	Metals (e.g., Fe, Cu); nutrients (NO_2^-, NO_3^-, PO_4^{-3}), NH_3, phenol SO_2, NO_x, oxidant gases (O_3), COD, TOC

Method	Principle	Instrument		Rating	Detection limit	Applications
Ultraviolet	Absorption of UV light	UV spectrophotometer	SLG	Fair	0.005 ppm	Organic compounds, some inorganic gases, e.g., SO_2, O_3, NO_2
Infrared	Absorption of IR (from heated filament)	IR spectrophotometric	SLG	Fair	1 ppm	Organic compounds, some inorganic gases, e.g., CO, CO_2, SO_2, NO_x, NH_3, O_3, TOC
Emission spectroscopy						
Flame photometry	Spectral-emission analysis following excitation by flame (arcs or sparks are also used)	Flame photometer; spectrograph (arc)	SL	Good Excellent	0.001–0.1 ppm	Metals, halogens, phosphorus, sulfur (depending on specific method)
X-ray fluorescence spectrometry	Spectral-emission analysis following X-ray excitation	XRF spectrometer	SL	Good	10 ppm	All elements having atomic numbers above 11
Spectrofluorimetry	Reemission of radiation absorbed by dissolved molecules	Recording spectro-fluorimeter	SL	Good	0.001 ppm	Organic compounds; some inorganic constituents

(*continued*)

TABLE 9–6. (*continued*)

METHOD	PRINCIPLE OF OPERATION	INSTRUMENTATION	SAMPLE[A]	SPECIFICITY[B]	SENSITIVITY[C]	TYPICAL APPLICATIONS
Chemiluminescence	Emission of spectral radiation resulting from a chemical reaction	Chemiluminescence analyzer	G	Excellent	ppb	O_3, NO_x
Atomic absorption spectrophotometry	Absorption of radiation by free atoms in a vapor state	AA spectrometer	SL	Excellent	0.001 ppm	Most metals
Chromatography (gas; liquid; thin-layer)	Differential migration of agents due to selective retention in a fixed phase while subjected to movement by a flowing bulk phase	Gas chromatograph, liquid chromatograph	LG (gas chr) SL (liquid chr) SL (thin-layer)	Excellent (gas) Good (liquid)	ppb–ppm (TLC)	Organic compounds, trace metals, anions
Electrochemical						
Conductivity	Measurement of electrical conductivity following absorption of an agent through a solution	Conductivity meter	LG	Poor	0.01 ppm	Any gas which forms electrolytes in aqueous solution, e.g., SO_2, NO_2, Cl_2, H_2S, NH_3
Anodic stripping voltammetry	Electrolysis of solution and electrodeposition followed by stripping at various potentials	DC-pulse polarograph	L	Good	0.001 ppm	Some metals, e.g., Cd, Cu, Fe, Pb, Zn

Method	Principle	Instrument	State	Quality	Sensitivity	Applications
Coulometry	Quantitative electrochemical conversion from one oxidation state to another	Coulometer	L	Good	1 ppm	SO_2, NO_2, O_3, H_2S, olefins, F^-, mercaptans
Polarography	Electrolysis of solution and measurement of current-voltage relation	Polarograph	L	Good	0.1 ppm	Some organic compounds and some metals, e.g., Sb, As, Cd, Cr, Ni, Zn
X-ray diffraction	Recording of scattered X-rays from sample subject to X-ray beam	X-ray diffractometer	S	Good	—	Solids of crystalline structure, e.g., asbestos, quartz
Microscopy (polarizing; fluorescence)	Optical analysis of enlarged images	Various types of microscopes	S	Excellent	—	Nature and size of particles, e.g., fibers, dusts
Mass spectrometry	Determination of material by creation of ionic current via electrostatic or electromagnetic field and separation of the ions according to their mass. Generally used in conjunction with gas chromatographs	Mass spectrometer	SLG	Excellent	ppb	Organic compounds

(*continued*)

TABLE 9–6. (*Continued*)

METHOD	PRINCIPLE OF OPERATION	INSTRUMENTATION	SAMPLE[a]	SPECIFICITY[b]	SENSITIVITY[c]	TYPICAL APPLICATIONS
Neutron activation analysis	Measurement of emitted ionizing radiation following production of radionuclides by neutron bombardment	—	SL	Excellent	ppb to ppm	Hg, As, Pb, Fe, and about 66 other elements which can become radioactive when bombarded with neutrons

[a]Sample: Most usual form of sample suited to the specific method; S = solid; L = liquid; G = gas.

[b]Specificity: This is an indication of the general ability of the method to measure the material in the presence of matrix interferences. The classification are very broad and use of preconcentration or separation methods may increase specificity.

[c]Sensitivity: Typical lower limits of detection are presented; these may differ depending upon the specific chemicals analyzed and the background concentrations present. DO, dissolved oxygen; UV, ultraviolet; IR, infrared radiation; TOC, total organic carbon; XRF, X-ray fluorescence; AA, atomic absorption; TLC, thin-layer chromatography; DC, direct current.

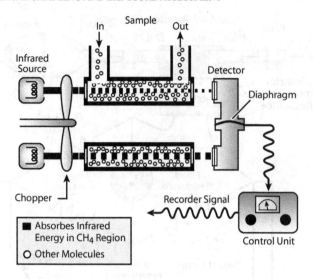

FIGURE 9–27. Schematic of an infrared analyzer.

instruments can detect molecules that fluoresce upon excitation by a radiant source. In this case, the detector is tuned to the fluorescence frequency, rather than the source frequency. Instruments of this type are currently being used for CO and SO_2 analyses.

Of still greater complexity are instruments that analyze patterns in the radiant spectra of excited molecules, rather than just total absorption over a given wavelength band. These second derivative spectrometers can be used to monitor a wide variety of gases and vapors, including SO_2, NH_3, NO_2, and benzene.

Another approach to continuous measurement of gas and vapor contaminants is to react them with other chemicals in the stream, or at surfaces in contact with the stream, and to measure the rate of reaction or the presence of reaction products. One of the earliest applications of this approach, and one that is still widely used, is the combustible gas indicator. A gas stream is passed over a catalytic filament at which combustible vapors burn. The heat of combustion raises the temperature of the filament, reducing its electrical resistance. The change in resistance provides an indication of the concentration of combustible material. A schematic diagram of this type of instrument, in which the catalytic filament is one leg of a Wheatstone bridge, is shown in Figure 9–28.

A slightly more sophisticated approach, which has also been used for many years, is utilized in the halide meter. It employs the "Beilstein" reaction, in which halogen atoms in contact with a copper element at high temperature emit an intense green light. In the halide meter, halogen-containing hydrocarbons are burned in a copper arc, and the intensity of the spectral emission indicates the concentration of the vapor.

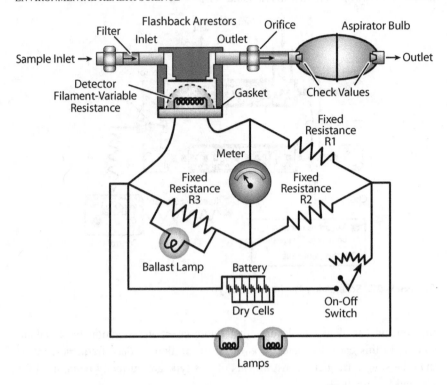

FIGURE 9–28. Schematic of a combustible-gas indicator flow system with circuit diagram.

Neither the combustible gas meter nor the halide meter can provide accurate or specific determinations when there are mixtures of air contaminants that influence the signal strength. The former responds to all combustible vapors, and its response depends on the heat of combustion. The halide meter responds only to halogens but has a different degree of response for different halogen compounds.

Some instruments utilizing gas-phase conversions achieve specificity by employing highly individual chemical reactions. For example, the O_3 monitor illustrated in Figure 9–29 utilizes the chemiluminescent reaction between O_3 and ethylene. An excess of ethylene is used, and the luminescence is limited by the O_3 concentration. None of the other airborne gases react with ethylene, so there are no significant interferences.

Particles
There are only a few methods for direct measurement of particle number concentrations in airstreams without actual sample collection, and almost all of these utilize light-scattering properties of the particles. Some measure the integral scatter of an aerosol. Since the amount of light scattered by each particle varies greatly with particle size, these instruments have little value unless the size distribution is known and constant. Other instruments measure the scattered light

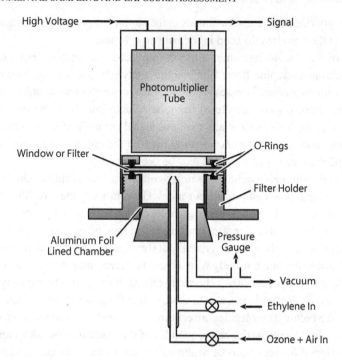

FIGURE 9–29. Schematic of ozone monitor utilizing ozone-ethylene chemiluminescent reaction and photometric detection.

from individual particles as they pass through the sensing zone and sort the pulses by size. Thus, they can accumulate number concentrations within defined size intervals for particles larger than about 0.3 μm in diameter. They are calibrated for reflective spherical particles and therefore may not be accurate for size analyses of particles with other shapes and optical properties.

Air Sample Collection with Online Measurements

In online systems, the analysis instrumentation is combined with the sampling instrumentation. A sample is collected as a grab, composite, or continuous sample and then analyzed by a direct-reading instrument. Most sample-collecting instruments with direct-measurement capabilities combine one of the extractive-sampling techniques previously described with a simplified version of a laboratory-type analytical instrument.

Many of the older online instruments used wet chemical techniques following extraction of the air contaminant by a bubbler or gas washer. A once widely used SO_2 instrument passed the scrubbing solution through an electrical conductivity cell. However, it was not specific for SO_2, since other ionizable contaminants in the air would also contribute to the conductivity. Other two-stage instruments used colorimetric reagents to collect the contaminants and could give more specific analyses, provided that cocontaminants in the air did not produce or quench color changes at the wavelengths used in the built-in colorimeter.

Most of the current instruments use dry-collection techniques. For example, the gas chromatographic flame photometric analyzer illustrated in Figure 9–30 collects various sulfur (S) gases simultaneously on a chromatographic column. When the sampled gases are eluted from the column into the flame photometric detector, they each give essentially the same signal per mole of S. However, since each compound has a different retention time on the column, they give distinct peaks, and the quantity of each can thus be determined.

Perhaps the simplest example of sample collection with a rapid readout of concentration is that provided by detector tubes. Glass tubes of the type illustrated in Figure 9–31 are filled with adsorbing granules coated with a colorimetric chemical reagent that changes its color after reacting with a specific gas or compound class.

A specific small volume of air, typically $100 \, cm^3$, is drawn through the tube from one end toward the other, usually with a hand-powered pump; the contaminant of interest reacts with the adsorbed chemical, changing its color. The capacity of the granules to take up the chemical is limited, so that the penetration of contaminant into the tube before its molecules are captured is dependent on concentration. The length of color change is, therefore, an index of the concentration. Detector tubes were developed for occupational health and safety evaluations and are currently available for about 200 gases and vapors. However, their accuracy is generally not very good; only 39 tube types for 14 gases have been demonstrated to have

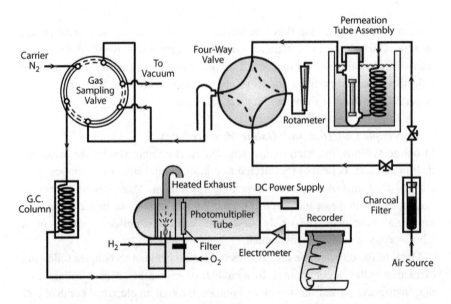

FIGURE 9–30. Schematic of an automated gas chromatograph/flame photometric detector system for analysis of SO_2 and other S compounds.

FIGURE 9–31. A detector tube.

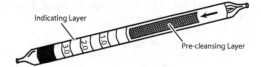

accuracies equal to or better than about 25%, according to tests conducted by the National Institute of Occupational Safety and Health (NIOSH). Some detector tubes have been calibrated for long-period sampling using mechanical sampling pumps and can therefore be used to determine lower airborne concentrations in industrial and community environments.

While there are many direct-reading instruments that are specific for gases and vapors, there are virtually none for airborne particles.

Direct-Reading Instrumentation for Waterborne Contaminants

Direct-reading instruments are commonly used for the measurement of water-quality parameters other than contaminant concentrations, such as temperature, pH, and dissolved oxygen. Examples of these instruments are presented in Table 9–7.

One of the more commonly used methods for the direct measurement of dissolved contaminants is the ion-selective electrode. A variety of electrodes are available that respond rather selectively to specific ions; their response depends upon the potential of a measuring electrode versus a reference electrode. Among the contaminants for which selective electrodes exist are NH_4^+, CN^-, SO_4^{-2}, F^-, Na^+, and Ag^+. Spectrophotometric methods utilizing the IR- or UV-absorption characteristics of specific contaminants are also widely used. Some of their applications are for measurement of total organic carbon and certain dissolved inorganic gases. Electrochemical methods, such as coulometry, potentiometry, and polarography, are used for the measurement of a wide range of metals in water.

Water Sample Collection with Online Measurements

The water-quality instrumentation that measures a rate or product of reaction between a contaminant and a reagent is essentially the same as that described earlier for air contaminants, except that a reactor vessel replaces the scrubber and the sampled stream contributes water as well as contaminant to the mixture in the reactor.

QUALITY ASSURANCE

The accurate determination of the concentrations of trace contaminants in environmental media is no small or simple task, as should be evident from the preceding discussion. Fortunately, there has been a much greater awareness of the

TABLE 9–7. Direct Measurements for Some Nonspecific Water-Quality Parameters

PARAMETER MEASURED	TYPICAL INSTRUMENTATION	PRINCIPLE OF OPERATION
pH	pH meter	Voltage difference between a measuring electrode (a glass tube having a special membrane which responds to H+ at one end) and a reference electrode
DO	DO electrode or DO probe (gas membrane electrode)	Gas dissolved in the water diffuses through a semipermeable membrane into an electrochemical cell compartment, which is comprised of a sensing electrode, reference electrode, and supporting electrolyte. DO reacts at sensing electrode to produce a current flow
Turbidity	Turbidimeter	Measures amount of light passing directly through the water
	Nephelometer (low turbidity ranges)	Measures light-scatter in water
Temperature	Thermometer	Expansion and contraction of a liquid
	Thermocouple	Measures voltage generated at junction of two wires made of different metals
	Thermistor	Electrical resistance varies with temperature in special resistors

DO, dissolved oxygen.

problems of quality assurance of environmental data in recent years, and significant progress has been made in the availability of standard reference materials and calibration instrumentation and techniques for analysis of specific contaminants. There are an increasing variety of laboratory certification programs and programs for interlaboratory evaluation of analytical methods. Some of these important activities are summarized in Table 9–8.

EXPOSURE DATA RESOURCES AND MANAGEMENT

To be useful, repositories of exposure data should contain not only measurements of contaminant concentrations in relevant media but also information about the circumstances that give rise to the concentrations and to potential exposures to those media (i.e., simultaneous measurements of the activities of populations of interest). This section discusses some salient aspects of current

TABLE 9–8. Organizations Publishing Recommended or Standard Methods and/or Test Protocols Applicable to Air Sampling Instrument Calibration

ABBREVIATION	FULL NAME AND ADDRESS
ANSI	American National Standards Institute, Inc. 11 W. 42nd Street, 13 Floor New York, NY 10036
ASTM	American Society for Testing and Materials D-22 Committee on Sampling and Analysis of Atmospheres and E-34 Committee on Occupational Health and Safety 100 Barr Harbor Drive West Conshohocken, PA 19428
AWMA	Air and Waste Management Association 1 Gateway Center, Third Floor Pittsburgh, PA 15222
EPA/NERL	US Environmental Protection Agency National Exposure Research Laboratory Quality Assurance Division Research Triangle Park, NC 27711
NIOSH	National Institute of Occupational Safety and Health Editor, NIOSH Manual of Analytical Methods (NMAM®) Division of Physical Sciences and Engineering 4676 Columbia Parkway Cincinnati, OH 45226

exposure databases and the manner in which current assumptions and practices in exposure monitoring affect the nature and usefulness of such data.

The past three decades have seen a change in the availability of methods for sampling and analysis of environmental exposures, as well as in the means to store and manipulate data through ever cheaper and more efficient computers. Coupled with the regulatory legislation of the same period, these changes have triggered an explosion in the volume of exposure-related data collected and stored. Substantial financial, human, and technological resources are devoted to this task throughout the nation. However, the quality, availability, and usefulness of the resulting data resources often leave much to be desired for their application to the characterization of long-term exposures and of temporal changes in exposure. In many cases, present data collection approaches are neither comprehensive nor cost effective, and the collected data have not necessarily been optimally used.

Government agencies, particularly the Environmental Protection Agency (EPA) and the Food and Drug Administration (FDA), are the major collectors of human exposure data, with the exception of the occupational environment, where the Occupational Safety and Health Administration (OSHA), NIOSH, and the Mine Safety and Health Administration have primary domain. The EPA monitoring requirements typically involve contaminants in source emissions (i.e., industrial stacks or discharge pipes) in environmental media and, much less frequently, in human tissue or blood. The data, however, are frequently inadequate for estimating exposure in various exposure assessment applications. The monitoring is largely driven by regulatory or legal mandates related to compliance with source emission or environmental standards. Due to the current regulatory structure, the measurement of environmental contaminants is highly compartmentalized among and within various government agencies (OSHA, EPA, FDA, etc.), frequently resulting in artificial barriers that limit the ability of any one agency to assess total human exposure and develop effective mitigation strategies.

Regulatory efforts frequently separate contaminants by whether they are found in the different media; whether they are ingested, inhaled, or absorbed through the skin; and by the environment in which the exposure occurs (e.g., workplace or home). Contaminants in multienvironmental media and multihuman exposure pathways, such as pesticides, automotive fuels, polycyclic aromatic hydrocarbons, and heavy metals, are not monitored in a systematic way, so the media and pathway of exposures cannot be assessed or actual exposures quantified.

A primary justification for human exposure assessment is to support actions that protect public health and welfare. Seldom, however, are environmental exposure data gathered in combination with measures of dose or effects. The National Health and Nutrition Examination Surveys of the National Center for Health Statistics comprise one of the few national health surveys that have gathered biomarker indicators of exposure in combination with health outcome data. These studies, however, have relied mostly on questionnaires to assess exposures, with some use of environmental monitoring data that were collected independently. Exposure monitoring needs to be linked to indicators of dose (i.e., biomarkers) and health or comfort outcomes, as well as to information about the sources and circumstances that gave risk to the exposures.

Since environmental contaminants, whether found in the air, water, food, or soil, are generally found as part of a complex mix, current efforts to monitor contaminants that focus on single compounds may not reflect the complex nature of the mix or the toxicity from interactions among the components. If monitoring efforts are to be tied to effects, they must take into account the complex nature of contaminant mixtures.

Environmental monitoring networks are designed to assess environmental concentrations of regulated contaminants in areas of suspected high concentrations and media where the potential for large-scale exposure exists (i.e., ambient air, soil, and water quality monitoring networks). While these monitoring networks may meet mandated regulatory needs, they usually do not provide data for the direct measurement of human exposure or data necessary to estimate human exposure, characterize exposure distributions, or identify high-risk groups. Monitoring locations may not have been selected to be statistically representative of populations or geographic areas. Population time-activity patterns may not have been measured or considered in site selections or in the interpretation of data.

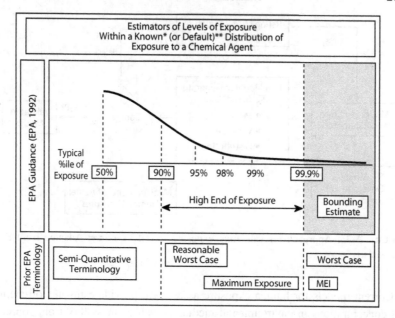

FIGURE 9–32. EPA Science Advisory Board recommendation on exposure terminology.

When applying data for risk assessment, the distribution of the exposures needs to be described. Figure 9–32 outlines the EPA's current recommendations for describing such a distribution.

Each database has associated with it potential sources of bias, error, and inconsistencies resulting from all aspects of the data gathering effort (e.g., site selection, frequency of sampling, sampling and analytical methods used, etc.), and quality control and quality assurance procedures vary greatly among these databases. In many cases, documentation for the data presented and the limitations and inconsistencies are not clearly identified, resulting in considerable difficulties in combining existing monitoring data for the estimation of human exposures.

MODELING OF EXPOSURES

Distributions of direct measurements of exposures based on actual measurements are seldom available for risk assessments of populations of interest. Therefore risk assessors generally rely on exposure models. They often begin with source characterization (Chapter 3); extend through dispersion modeling in environmental media (Chapter 4); assume standard patterns of ventilation, ingestion, and activity sites and levels; and correct, to the extent possible, for population distributions of age, sex, and socioeconomic variables.

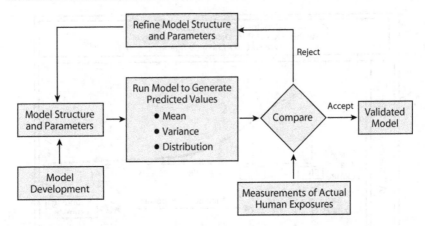

Figure 9–33. Approach for validating exposure models. (*Source*: Adapted from EPA materials.)

Current models for human exposure assessment provide estimates of chemical concentrations in environmental media, as well as mass flux from sources, through environmental compartments, to human receptor target organs, tissues, or cells. Most of the modeling effort has been focused on the solution of chemical fate and transport equations for the prediction of concentrations in air, food, water, and soil. Common pathways of concern include drinking water, incidental ingestion of soil, inhalation of vapors and PM, and ingestion of fish and garden vegetables.

Many of the current exposure models are undergoing in-depth evaluations and refinement. Regardless of the model used, validation is essential to determine how the relevance of the model to the actual situation of concern. Since there are numerous data gaps that confront the risk assessor when using exposure models, conservative assumptions (default values) are generally built into the exposure models to provide some assurance that exposures will not be underestimated in terms of either median levels or upper-bound exposures. The basic approach to model validation is illustrated in Figure 9–33.

Exposure assessment for application in risks assessments is always performed using a number of assumptions. Some of these, and their shortcomings, are summarized next.

Time-Activity Patterns

Consideration of the time spent by individuals in different locations and their activities have been incorporated to a limited extent in many exposure assessments conducted for risk assessment purposes. Current practice is generally based on certain assumptions:

Time-activity patterns are invariant during an individual's lifetime; there are no significant differences in time-activity patterns of the population as a function of age, gender, socioeconomic status, or ethnic origin that may affect exposures; and there are no significant differences in time-activity patterns of the population in relation to regional variability, as well as urban, suburban, or rural place of residence.

Time-activity patterns at any given age also change significantly over time due to societal trends. For example, children used to remain at home until school age, but today more infants and preschoolers spend most of their day in childcare settings. There is also a trend toward lengthening the school year, which could result in children spending less time in the home or outdoors or in changed times of the year for school vacations. Children and adolescents are also increasingly involved in structured, competitive sports activities, either indoors or outdoors, for significant amounts of time. The duration and place where these activities are undertaken could impact children's exposures and their inhalation doses associated with elevated ventilation rates. The trend toward telecommuting and the increased number of home-based businesses increases the amount of time spent at home, as compared to a separate workplace, for at least some fraction of the adult population. A large segment of the adult population also engages in sports or exercise activities, both indoors and outdoors, during significant amounts of time, for health maintenance or cosmetic reasons. As adults have more free time, they are likely to engage in less strenuous, longer duration exercise and leisure activities (e.g., daily walks, golf) performed outdoors, resulting in less time spent indoors. Consideration of these societal trends means that activity patterns cannot be considered fixed in time.

Although time and location budgets in the US and abroad are rather consistent (i.e., approximately 90% of the time is spent indoors), there is often inadequate consideration of time-activity pattern variability as a function of socioeconomic status, gender, ethnic origin, or location of residence. For example, population subgroups in lower socioeconomic strata with significant levels of unemployment would be expected to have very different time-activity patterns than those in the middle or upper socioeconomic strata. They are also less likely to engage in the health maintenance and leisure activities previously described. The type of employment also can influence the outdoors versus indoors time budget (e.g., construction, lawn maintenance, etc.). Gender also affects individual time-activity patterns. Even when over half the women with school-age children work away from the home, a significant number still remain within the home environment most of the day. There is also limited information on differences in activity patterns that may affect exposures as a function of the region of the country or urban versus suburban versus rural populations due, at least in part, to the lack of activity in behavioral science research.

Exposure Pathways

Although exposure assessments, as applied to risk assessment, have incorporated the consideration that certain exposure sources and pathways may be more important for some population subgroups (e.g., ingestion of contaminated soil by children), current practice is, by and large, based on the following assumptions:

- source emissions and concentrations of contaminants in environmental media, foodstuffs, and consumer products to which an individual may be exposed are invariant during an average lifetime of 70 years.
- the relative importance of different exposure pathways does not change over an individual's lifetime (exception: soil ingestion by children).
- there are no significant differences in relevant exposure pathways as a function of gender, socioeconomic status, or ethnic origin.

Reformulation of consumer products, for example, cosmetics, cleansing agents, personal care products, occurs continuously, both in terms of the types and relative concentrations of the chemical constituents of such products. Emissions from mobile sources in particular, but those from industrial point sources as well, have shown a consistent trend of reductions during the past half century, largely as a result of improvements in technology and regulatory pressures. The types of emissions have also changed in many cases. For example, benzene and other aromatic compounds have replaced tetraethyl lead as engine antiknock additives, with the consequent changes in airborne Pb and organic chemical concentrations, as well as the exposures associated with mobile sources. Changes in the types and relative consumption of different fuels, as well as improvements in combustion technology, have also resulted in reductions of airborne concentrations of some pollutants such as SO_2 and PM. As a result, exposures have changed both qualitatively and quantitatively well within the generally assumed 70-year life span.

The types and relative proportions of different foodstuffs in the diet and, consequently, food consumption–related exposures are generally assumed to be constant over a lifetime after infancy. However, food preferences can vary significantly with the age of the individual, as well as with societal trends. Young children may consume proportionally larger amounts of fruits and milk in their diet as compared to adults. Adolescents may consume significantly larger amounts of "fast" or "junk" foods. As adults, they may become more aware of the importance of a healthy diet and reduce their intakes of such foods, fat, and meat and revert to an increased consumption of fruits and vegetables. Advice regarding what constitutes a healthy diet also changes over time, therefore affecting exposures as well. In addition, a significant fraction of the population at any given time may be on special diets.

Dietary sources of exposure also vary over time as a result of changes in agriculture and food production practices. For example, pesticides, herbicides, and other chemicals used in the production, preservation, and processing of foods vary as a result of regulatory and market pressures. New food products are introduced into the marketplace constantly, affecting dietary exposures. The levels of contaminants present in locally grown foods may not be a good indicator of dietary exposures for a particular population. Foods grown in one area of the country are distributed nationally; food imports have also increased significantly as a result of increased trade. The types and relative amounts of foods consumed also vary according to socioeconomic status, ethnic origin, and gender.

Lifestyle and Other Personal Factors

Some exposure assessment models assume that, in general, lifestyle and/or other personal factors do not impact exposures significantly and that lifestyle and/or other personal factors that might impact exposures do not vary significantly throughout an individual's lifetime or across the population. However, lifestyle factors do significantly affect exposures. The most notorious behavioral factor with a strong influence on exposures to a wide range of contaminants is smoking. But other lifestyle factors are also potentially important. Dietary intakes of, for example, benzo(a)pyrene may not only be strongly affected by food preferences but can also be strongly affected by the method used to cook the food.

Residential Mobility

Some assessments assume that individuals reside in the same location throughout their lifetime. The reality is, however, that the US population is highly mobile, with the average family changing residences every five to seven years on average. There may also be differences in mobility associated with socioeconomic status.

Reliability of Predictive Models Based on Sources and Transport

Exposure assessment can be based on the assumption that exposures can be adequately assessed by using source emissions or concentrations of contaminants in media and environmental transport models and a limited number of exposure scenarios. The sophistication and complexity of models used for exposure assessment have increased notably and have led to the development of new models that attempt to link exposure to the internal dose of a chemical at a critical target site. However, there is often a significant lack of validation of such models with actual data. All models are based on explicit and implicit assumptions about how contaminants move into, through, and across environmental compartments, as well as on some conceptual framework about the most significant routes of exposure. The weakness of this approach is that, frequently, neither the assumptions nor the model results are validated with actual exposure data. The EPA's total exposure assessment methodology studies are an excellent example, in which measurements of volatile organic compounds in outdoor, indoor, and personal air demonstrated that the long-held assumption that populations living near industrial sources experience higher exposures to these compounds than those living in less industrialized areas was incorrect. The study showed that indoor sources dominate exposures to many volatile organic compounds and that exposures to the study populations were similar and not dominated by the presence of nearby sources. Another example is the assumption in some models that ingestion is the dominant route of exposure for chemicals in potable water. In many cases, major contributions to internal dose derive from dermal contact via bathing or inhalation of vapors and aerosols released in showers and sprays.

Temporal Variability at Specific Locations

Although some models take into consideration long-term accumulation of contaminants in environmental compartments, more often it is assumed that the concentration of contaminants in media remains constant over time. As a result of this very assumption, exposures associated with those contaminants and compartments are also assumed to be invariant over time. To a large extent, this assumption is related to the heavy reliance on environmental transport and fate models that are generally based on steady-state condition assumptions. Concentrations of contaminants in media can either increase or decrease over time due to changes in environmental conditions or human activities that may result in either enhancing or diminishing the release and bioavailability of the contaminant in specific compartments.

Infiltration of Outdoor Air into Indoor Spaces

Although there has been some recognition that indoor environments are more significant contributors to some chemical exposures than outdoor air, exposure assessment practice is still largely based on the following assumptions: buildings do not significantly affect exposures to outdoor airborne pollutants and that buildings are static and indoor concentrations of airborne contaminants can be modeled using steady state assumptions.

As noted, most of the population spends about 90% of their time indoors. As a result, indoor environments are typically more important in determining some contaminant exposures than are exposures in outdoor air. Indoor exposures to airborne pollutants are the sum of the fraction of the outdoor concentration that penetrates into a building plus the contribution from indoor sources. The extent of infiltration of outdoor air and contaminants into a structure depends on a number of variables related to the building itself (e.g., construction, ventilation system), environmental conditions (e.g., season temperature differences between outdoors and indoors), the type of contaminant (e.g., reactive or nonreactive, small or large particles), and human factors (e.g., opening of windows, obstruction of air intakes to maintain higher temperatures indoors, use of the space for purposes other than the original design).

Population Distributions of Exposure

Another common assumption in exposure assessment is that population exposure to any contaminant can be described by one of the well-characterized statistical distributions (e.g., log-normal) and that the distribution of exposures in the population is not affected by factors such as socioeconomic status, gender, ethnicity, and lifestyle factors. These assume that population exposure distributions behave similarly because concentrations of contaminants in media generally approach one of the simpler statistical distributions, that is, typically a unimodal, log-normal distribution. The idea that, for risk assessment purposes, an acceptable exposure can be established at a certain percentile in the upper tail of such distribution is also based on the assumption of a "well-behaved" statistical distribution of exposures.

Figure 9–32 shows the current terminology for describing exposures at the upper end of the population distribution. When total exposures occur via multiple media, and especially when indirect exposures to contaminants emitted into and transported through the atmosphere are involved, the situation can become quite complex. While data on population exposures and their distributions are very limited, there is some evidence that population exposure distributions are often skewed; that is, they may be multimodal because the overall distribution of population exposures is a juxtaposition of different population subgroups' exposures.

10

Risk Management

Risk management is a process by which action is taken to reduce or eliminate risk factors. It is carried out at many different levels. Individuals make risk management decisions every day through their dietary choices, use of medications, personal habits, types and levels of exercise, use of personal and mass forms of transportation, crafts and hobbies, household maintenance, and, to some extent, the ways that they perform their tasks at work. Their choices in each of these activities are likely to involve exposures to chemical agents that could affect their health. However, individuals have only limited control over some aspects of factors that affect their exposures, dependent as they are on the contents of the foods and beverages they ingest, the consumer products that they buy and use, and the concentrations of some of the pollutants in the air they breathe or the water they drink. However, through their attention to product labels and other forms of public health and consumer product information, they can make informed choices that can minimize their exposure.

Employers have more clearly defined obligations and legal requirements to minimize employee exposures to workplace stressors. Workplace health and safety standards, if followed, should ensure a low order of risk to employees. Legally mandated effluent emission limits, if followed, should be protective of the public living downwind or downstream of such effluents or living near waste storage sites or repositories.

Manufacturers are responsible for the safety of the products they sell. Product liability lawsuits and the costs of worker's compensation claims awarded are further stimulants to industry to maintain safe workplaces and products. Unfortunately, there have been many cases of irresponsibility, as documented by continuing cases of occupational disease and the public and corporate expenses devoted to the clean-up of abandoned waste repositories and contaminated work sites.

Various federal, state, and local governmental agencies are responsible for the effective enforcement of public statutes enacted to protect workers and the public from exposures to chemical agents, for developing a knowledge base for identifying and evaluating risk management options, and for providing information to the public about the residual risks that remain from exposures resulting from uncontrolled sources of toxicants. In the United States (US), the Environmental Protection Agency (EPA) has, as its primary mission, the management of environmental risks. The EPA's process for risk management is depicted in Figure 10–1. It begins with risk assessment and extends into techni-cal means of control, regulatory options, and risk communication to concerned segments of the public.

Educational institutions and the media can affect the management of risks by informing the public about the nature and extent of the risks and what can be done to reduce them both by actions at the individual citizen's level and by actions

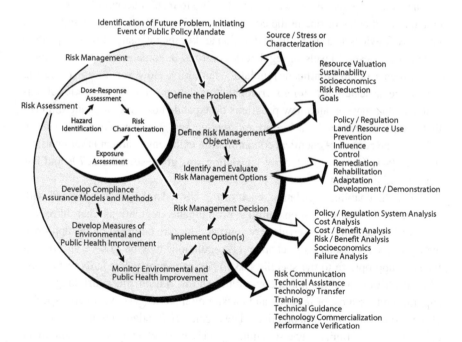

FIGURE 10–1. Risk management/risk assessment paradigm.

taken by the executive and legislative arms of government. Examples of options for risk reduction are presented in Table 10–1.

This chapter discusses means of controls that are or can be applied to minimize emissions from sources and exposures to people and also discusses legislative and regulatory actions that can have a major impact on exposures.

CONTROL APPROACHES

When a decision has been made that an imminent or serious potential hazard exists due to the presence of a chemical contaminant in the environment and that there is a need to control its further release or usage, choices must be made concerning the means by which control is to be achieved.

Regulatory Control

One approach is regulatory control. This can involve outright bans of specific chemicals, specification of permissible usages, permissible times for operation and discharge of effluents, and permissible environmental concentrations and/or amounts of effluent discharges. Another regulatory approach, which is largely restricted to the occupational environment, is to control access to potentially hazardous environments. Management may be able to restrict access of personnel to contaminated areas or obtain the cooperation of its employees in using personal protective devices so as to limit the duration and/or the intensity of their exposure.

There are several levels of regulatory controls. In some areas, state and local governments retain significant regulatory authority. However, to an increasing extent, the basic authority for governmental regulation comes from a series of federal acts that have created new programs or expanded existing ones and directed that certain objectives be achieved and maintained. The key federal legislation affecting the control of chemical contamination in the environment is summarized in Box 10–1. The pace of federal legislation was greatest in the 1970s and has tapered off since then.

Under the enabling legislation, each of the US federal agencies charged with environmental-control responsibilities promulgates regulations that have the force of law and, unless overruled by court actions, are enforced by the appropriate agencies. As discussed in chapter 11, regulations may be of several different types as appropriate to the specific task. They can specify standards or permissible levels of contamination in community and workroom air, in drinking water supplies and streams, or in foods and packaging materials. They can also specify effluent standards and work practices. The regulatory standards involving numerical concentration limits may also require specific types of analytical measurement techniques.

TABLE 10–1. Examples of Risk Reduction Activities[a]

	INDIVIDUALS	GROUPS[b]	INDUSTRY	OTHER INSTITUTIONS[c]
Preventing pollutant generation	Energy and water conservation	Car pooling	Raw material substitution	Purchase of biodegradable products
	Purchase nonhazardous household products	Integrated past management	Process redesign	Purchase of recycled products
	Organic gardening	Land acquisition for environmental protection	Product redesign	Zoning to protect critical resources
Recycling and reuse	Reuse of paint cleaners	Community solid waste recycling	Solvent reclamation	Paper recycling
	Trade in used car batteries	Oil recycling	Use of scrap iron in steel making	Commercial glass recycling
	Donate unused paint to school art department	Community hazardous waste recycling	Kraft process for chemical and energy recovery in pulp making	Use composted yard waste for fertilizer
Treatment and control	Asbestos removal	Water supply treatment	Solid and hazardous waste incineration	Co-composting sludge and solid waste
	Auto inspection and maintenance	Community composting	Air pollution control devices	Waste-water treatment
	Heating system maintenance	Landfill wood chipping	Accident prevention programs	Chemical inventory, audit and control systems
Reduce residual exposure	Home ventilation for radon, gas stoves	Proper sanitary landfill	Secure chemical landfill	Smoke-free work areas
	Home water filtration devices	Land use planning	Pollutant dispersion technologies	Proper building ventilation
	Do not fish in polluted waters	Household hazardous waste collection	Controlled pesticide application	Purchase bottled water

[a]Many of these strategies, e.g., energy conservation, can be employed by all.
[b]Communities, community groups.
[c]Federal and state government, academia, health-care institutions, commercial business, etc.

Box 10–1 Major Federal Legislation Pertaining to Control of Chemicals

Air Pollution Control Research and Technical Assistance Act (1955)

Provided temporary authority and funding for five years for a federal program of research in air contamination and technical assistance to local governments. (Extended in 1959 for an additional four years.)

Clean Air Act (1963)

Granted permanent authority for federal air pollution control activities. Provided for federal grants to state and local air pollution control agencies to establish and improve their control programs; federal action to abate interstate air contamination through a system of hearings, conferences, and court actions; and an expanded federal research and development program with particular emphasis on motor-vehicle exhaust and sulfur oxide emissions from coal and fuel-oil combustion.

Amendments to Clean Air Act (1965)

Provided for the promulgation of national standards relating to motor-vehicle exhaust and cooperation with Canada and Mexico to abate international air contamination.

Air Quality Act (1967; amending the 1963 Clean Air Act)

Declared a national policy of air-quality enhancement and provided a procedure for designation of air-quality control regions and setting of standards by cooperation between federal and state governments; provided for registration of fuel additives.

Amendments to Clean Air Act (1969)

Extended authorization for research on low-emission motor vehicles and cleaner fuels.

Amendments to Clean Air Act (1970)

Provided for the establishment of national ambient air-quality standards and their achievement through the implementation plans of air-quality control regions and states; a 90% reduction from 1970 levels of hydrocarbon and carbon monoxide emissions from automobiles by the 1975-model year and 90% reduction from 1971 levels in nitrogen oxide emissions by the 1976-model year (with one-year extensions if necessary) and studies of aircraft emissions.

Amendments to Clean Air Act (1977)

Required the Environmental Protection Agency (EPA) to apply different levels of stringency to emission controls on new sources for criteria pollutants in areas meeting existing National Ambient Air Quality Standards (NAAQS), that is, best available central technology, than in areas exceeding the NAAQS. For these nonattainment areas, existing sources were required to use reasonably available control technology. Also mandated the EPA to increase the promulgation of National Emissions Standards for Hazardous Air Pollutants, to re-examine the adequacy of the NAAQS every five years, and to establish a Clean Air Scientific Advisory Committee to conduct public reviews of the criteria to be used as bases for NAAQS revisions and to advise the EPA administrator on research needs for ambient air pollutants.

Amendments to Clean Air Act (1990)

Required a phase in of 50% reductions in emissions of sulfur dioxide and nitrogen oxides from stationary sources; required technology based emission controls for 189 specific hazardous air pollutants and assessments of residual risks after the application of such controls; required tighter emission controls for motor vehicles; required a production phase-out of stratospheric ozone-depleting chemicals.

"Refuse Act of 1899" (part of River and Harbor Act of 1899)
Prohibited discharges into navigable rivers.

Public Health Service Act (1912)
Authorized investigation of water contamination in relation to human diseases.

Oil Pollution Act (1924)
Prohibited discharge of oil by any means, except in emergency or by accident, into navigable waters of the United States.

Water Pollution Control Act (1948)
Provided five-year authorization to fund research studies, low-interest loans for construction of sewage and water treatment works, and a Federal Water Pollution Control Advisory Board. Authorized the Department of Justice to bring suits against individuals or firms but only after notice, hearing, and consent of the state involved. (Extended in 1952 for an additional three years.)

Amendments to Water Pollution Control Act (1956)
Provided for permanent authority, grants for construction of sewage-treatment plants, and abatement of interstate water contamination by federal enforcement through a conference public-hearing court-action procedure.

Federal Water Pollution Control Act (1961)
Permitted the Secretary of Health, Education and Welfare (HEW), through the Department of Justice, to bring court suits to stop contamination of interstate waters without seeking permission of the state. Extended pollution abatement procedures to navigable intrastate and coastal waters, with permission of the state. Authorized seven regional laboratories for research and development in improved methods of sewage treatment and control. Authorized funds for grants to local communities for sewage-treatment plants.

Oil Pollution Act (1961)
Enacted to implement provisions of the International Convention for the Prevention of the Pollution of the Sea by Oil, 1954.

Water Quality Act (1965)
Declared a national policy of water-quality enhancement. Established the Federal Water Pollution Control Administration (FWPCA). Provided for the states to adopt water-quality standards for interstate waters and plans for implementation and enforcement, to be submitted to the Secretary of HEW (later to Secretary of Interior after FWPCA was transferred to the Department of the Interior) for approval as federal standards; authorized the Secretary to initiate federal actions to establish standards if the state criteria were inadequate. Authorized grants for research and development to control storm water and combined sewer overflows and authorized funds for sewage treatment plant grants.

Clean Water Restoration Act (1966)

Provided for grants for research and development of advanced waste-treatment methods for municipal and industrial wastes. Authorized grants for construction of treatment plants. Amended the Oil Pollution Act of 1924 by transferring responsibility to Secretary of the Interior and provided for suits against "grossly negligent, or willful spilling, leaking, pumping, pouring, emitting or emptying of oil."

Water Quality Improvement Act (1970)

Strengthened federal authority to deal with sewage discharges from vessels, hazardous contaminants, and contamination from federal and federally related activities. Provided for liability for oil spills from onshore- and offshore-drilling facilities and from vessels.

Marine Protection, Research, and Sanctuaries Act (Ocean Dumping Act) (1972)

Regulated ocean dumping of materials that may be hazardous to human health and welfare.

Ports and Waterways Safety Act (1972)

Regulations relating to prevention or mitigation of damage to the marine environment due to transport of oil and other hazardous materials.

Water Pollution Control Act (1972, Clean Water Act)

Extended national water-pollution program to intrastate waters. Established a system of national effluent limitations with goals of best practicable water-pollution control technology by mid-1977 and best available technology by mid-1983; national performance standards for new industrial and publicly owned waste-treatment plants; and a national system of permits for discharge of contaminants. Authorized funds for treatment facilities.

Safe Drinking Water Act (1974, amendment to Public Health Service Act)

Authorized the EPA to establish federal drinking-water standards for protection from all harmful contaminants, applicable to all public water supplies in the United States. Established a joint federal–state system for compliance with standards and protection of underground sources of drinking water.

Amendments to Water Pollution Control Act of 1972 (1977)

Delayed installation of best available control technology until 1984. Continued federal grants for sewage-treatment plants. Added 65 chemicals and classes of chemicals to list of toxic agents.

National Ocean Pollution Planning Act (1978)

Established comprehensive five-year plan for federal ocean pollution research and development and monitoring programs and coordinated research on Great Lakes and estuaries of national importance.

Safe Drinking Water Act and Amendments (1986)

Provision of maximum limits for contaminants in public drinking water and techniques for their removal.

Ocean Dumping Ban Act (1988)

Prohibited issuance of new permits for dumping of sewage sludge or industrial waste into ocean waters; phased out existing permits by December 31, 1991.

Safe Drinking Water Act and Amendments (1996)

Required specific analyses for emerging contaminants, such as Cryptosporidium; provided a revolving fund to help state and local authorities improve drinking-water systems.

SOLID WASTES

Solid Waste Disposal Act (1965)

Began a national research, development, and demonstration program for solid wastes. Provided financial assistance to interstate, state, and local agencies for planning and establishing solid-waste disposal programs.

Resource Recovery Act (1970)

Provided for research into new and improved methods to recover, recycle, and reuse wastes, and financial assistance to states in the construction of solid-waste disposal facilities.

Resource Conservation and Recovery Act (1976)

Promulgated regulations for treatment, storage, and ultimate disposal of hazardous solid wastes.

Comprehensive Environmental Response, Compensation and Liability Act ("Superfund")/Superfund Amendments and Reauthorization (1980)

Imposed notification and public disclosure requirements on persons who handle, store, or dispose of hazardous substances: established national hazardous substance response plan, including methods of determining priorities among releases and provided for liability for response costs incurred by government; imposed financial liability on companies for direct damage to individuals and natural resources.

PESTICIDES AND FOOD ADDITIVES

Federal Food, Drug, and Cosmetic Act (1938)

Provided for the establishment of tolerances for pesticide residues in food.

Federal Insecticide, Fungicide, and Rodenticide Act (1947)

Provided for registration of "economic poisons" by the US Department of Agriculture for products marketed in interstate commerce and seizures of adulterated, misbranded, unregistered, or insufficiently labeled pesticides.

Miller Amendment to the Federal Food, Drug, and Cosmetic Act (1954)

Provided for condemnation of raw agricultural commodities containing pesticide residues in excess of tolerances as fixed by the Secretary of HEW.

Federal Insecticide, Fungicide, and Rodenticide Act (1972)

Gave the EPA authority to regulate uses of pesticides so as to prevent environmental damage.

Food Quality Protection Act (1996)

Repealed Delaney clause; eased regulation of processed foods and tightened regulation of raw foods.

Walsh-Healey Act (1936)

Enabled federal government to set standards for safety and health in workplaces engaged in activities relating to federal contracts.

Coal Mine Health and Safety Act (1969)

Directed the Secretary of Labor to set standards for exposure to coal-mine dust and to compensate miners for coal workers pneumoconiosis resulting from past exposures.

Occupational Safety and Health Act (1970)

Established National Institute for Occupational Safety and Health to conduct research and training, develop exposure standards, and make inspections relevant to the protection of the health and safety of workers.

Mine Safety and Health Act (1973)

Created the Mine Safety and Health Administration within the Labor Department to establish standards for occupational exposures of workers in metal and nonmetal mines and associated ore processing mills.

RADIOACTIVE WASTES

Atomic Energy Act (plus amendments) (1954)

Regulated release of radioactive wastes into the environment.

Uranium Mill Tailings Radiation Control Act (1978)

Required remediation of former uranium mill processing and disposal sites.

Nuclear Waste Policy Act (1982)

Provided requirements for disposal of high-level radioactive wastes.

Low-Level Radioactive Waste Policy Amendments Act (1985)

Provided requirements for disposal of low-level radioactive wastes.

GENERAL

Freedom of Information Act (1966)

Provided public access to government-held information.

National Environmental Policy Act (1969)

Declared a national policy for the environment to encourage harmony between humans and their environment, to promote efforts that will prevent or eliminate damage to the environment, and to enrich the understanding of ecological systems and natural resources important to the nation. Established a three-member Council on Environmental Quality, in the Executive Office of the President, charged with making studies and recommendations to the president and with preparing an annual Environmental Quality Report. Directed that all agencies of the federal government include in every recommendation or report on proposals for legislation and other major federal actions significantly affecting the quality of the human environment a detailed statement on the environmental impact of the proposed action; any adverse environmental effects that cannot be avoided should the proposal be implemented; alternatives to the proposed action; the relationship between local short-term uses of the environment and the maintenance and enhancement of long-term productivity; and any irreversible and irretrievable commitments of resources that would be involved in the proposed action should it be implemented.

Environmental Quality Improvement Act (1970)

Established national policy of enhancement of environmental quality though state and local governments, encouraged and supported by the federal government.

Coastal Zone Management Act (1972)

Made federal funds available to encourage states to develop comprehensive management programs to increase effective management, beneficial use, protection, and development of coastal zones.

Hazardous Materials Transportation Act (1975)

Regulated handling and transportation in commerce of hazardous materials (amended in 1976).

Energy Policy and Conservation Act (1975)

Required states to prepare energy conservation plans in order to receive financial assistance for eligible programs; set energy conservation standards for appliances; authorized Secretary of Energy to make grants for energy inspections and audits of, and instruction of energy conservation improvements in, schools and hospitals.

Toxic Substances Control Act (1976)

Provided for regulation of new chemicals or new uses for existing chemicals to prevent adverse effects upon human health or the environment. Provided for chemical testing, as well as authority to delay use of or ban specific chemicals or uses of chemicals.

Emergency Planning and Community Right-to-Know Act (Title III of 1986 Superfund amendments) (1986)

Required owners and operators of facilities containing extremely hazardous substances to notify state emergency response commission and required joint preparation of emergency plans.

Pollution Prevention Act (1990)

Established waste-management hierarchy, with preference for source reduction.

Energy Policy Act (1992)

Provided requirements for improvements in the energy efficiency of a wide range of items, including transportation vehicles and industrial and home appliances, and promotion of the use of renewable resources.

Federal regulations that establish numerical standards provide a uniform frame of reference and goals to be achieved and maintained, putting all parties involved on an equal basis and giving them due notice of what is required. These regulations relieve industry and local authorities from the need to make technical judgements as to the levels that are toxic or may produce environmental damage. Such judgements, as previously discussed, can be very difficult and, when made for a particular situation on a local basis, are likely to differ substantially from place to place and time to time.

Governmental regulations may also specify requirements for work practices, for the maintenance of records on effluent discharges, and for the exposures and medical histories of employees. Such regulations may therefore affect the manner in which industry achieves control of exposures through both administrative and technical controls.

Administrative Control in Occupational Environments

Administrative controls of contaminant emissions and exposures depend upon the ability of management to motivate and control the personnel whose actions affect contaminant release, dispersion, and accessibility of themselves and others to contaminated areas. The most effective means of administrative controls are those achieved through programs of education and motivation. Personnel whose actions affect their own exposures, or the extent of contaminant releases to the community environment, should be aware of the potential consequences of the resulting exposures and of the effects that their own work practices have on the extent of the releases. They should be thoroughly familiar with the intended functions, proper operating parameters, and maintenance procedures for the process equipment in their unit and for any emission-control devices associated with its operation, so that substandard equipment performance may be recognized and controlled. These people should also be trained to cope with malfunctions and emergencies, including safe procedures for shutting down process operations, and be thoroughly familiar with evacuation routes and procedures.

Personnel training programs are also needed for the proper use of personal protective devices for exposure prevention, such as gloves, aprons, coveralls, eye protectors, hearing protectors, and respiratory protective devices. These programs should include the criteria for the selection of the particular devices provided, when and how they should be worn or used, the degree of protection that they provide, their useful life and replacement schedules, and the conditions under which they should be maintained and stored between periods of use.

Management's role goes beyond the education and encouragement of employee participation. Other aspects include human engineering of equipment, incentives for positive efforts, and disincentives or penalties for poor performance. In terms of human engineering, known as ergonomics, the provision of process controls and protective equipment within easy reach, which can be readily operated with natural motions, will usually increase effective performance. On the other hand, the lure of special incentives or the threats of penalties are not usually very effective in achieving a uniformly high level of employee performance.

Management also has a responsibility to limit the access of personnel to contaminated areas when unlimited access would result in overexposures. This type of control can take several forms. One involves the designation of personnel who rightfully have access to certain areas, as well as those who do not have such access. The former category may include employees who have been fitted with personal protective devices. It may also include personnel without protective equipment but whose work time within the contaminated area will be sufficiently brief that their cumulative exposures will still be well within the permissible daily limits.

Among those for whom management is obligated to restrict access to contaminated areas are people who may be hypersensitive, have preexisting conditions, or have accumulated prior exposures that would place them at special risk. The information needed for the identification of such personnel can be obtained at preemployment and periodic medical examinations and, for some conditions and materials, by special screening tests.

Technological Control

Implicit in previous discussions is that ways and means exist to eliminate or control contaminant sources when exposures that they cause are deemed to be excessive. One may expect that the administrative and technical skills and ingenuity displayed in creating the technologies that generate contaminants could also be employed in controlling them. By and large, this expectation is realistic. When the motivation is provided, whether by enlightened self-interest or legislative mandate, it has been possible to achieve substantial control over ambient levels of chemical contaminants. One of the major benefits of the developing consciousness of the impact of environmental contamination on human health and welfare, exemplified by the substantial body of environmental legislation (Box 10–1), has been a major impetus toward the application of existing technology, as well as the development of new and improved technologies, for the prevention and/or control of contaminant effluents.

The technological approach for sources is known as engineering control. In this approach, controls are built into the industrial production process itself in order to reduce the formation and/or release of chemicals into the workplace and outside environment. This may be achieved through appropriate design and selection of materials, processes, and equipment; the intent is to depend as little as possible on the individual actions of plant personnel or consumers. Engineering controls can be subdivided into two broad types, namely process controls and source or effluent controls. The former emphasize the reduction or diminution of contaminant formation, while the latter act to reduce or prevent the release to the environment of any contaminants that are formed. The basic approaches to engineering control are discussed next.

Material substitution

Substitution of materials can ensure against exposure and contamination due to a particular chemical by eliminating it from the process entirely and replacing it with another chemical that is less toxic or offensive. This technique is usually practical where the material in question is used as a solvent, pigment, thermal or electrical insulator, or for other applications where there are a number of alternate materials with similar properties that may be used. Some notable examples of materials substitution that have substantially reduced occupational disease incidence and/or environmental contamination are presented in Table 10–2.

Process modification

The control of contaminant discharges by modification of process procedures and equipment involves using process elements that provide a greater degree of enclosure and/or the combination of a series of process operations into a single step. Both of these techniques are generally used in switching from batch-processing operations to continuous processing. The loading, unloading, and transfer of materials from one batch-processing vessel to another provides ample opportunities for uncontrolled releases, and these are minimized in continuous operations. Some notable examples of process substitutions that have substantially reduced contaminant release are also presented in Table 10–2.

Isolation

When the generation of contaminants cannot be completely eliminated, the next best approach is to prevent contact of the contaminants with people and the environment. This can be done by isolating the contaminant source or, occasionally, by isolating the people to be protected from the source. It is generally easier and less expensive to isolate the source; usually this is done by enclosing it within some structure that will eliminate or reduce the escape of the material. On the other hand, when the sources are numerous and the people to be protected are few, it may be less expensive to isolate the people and create a clean artificial environment for them. This is frequently done, for example, in foundries, where the crane operator works within a cab provided with clean breathing air; the operator moves, within the cab, through dusty air above the foundry operations.

Another approach to isolation from contaminants is use of personal protective devices. Gloves, aprons, face shields, goggles, and various kinds of protective

TABLE 10–2. Applications of Control by Substitution

SUBSTITUTION OF LESS-TOXIC MATERIALS		
ORIGINAL MATERIAL	LESS TOXIC SUBSTITUTE(S)	APPLICATIONS
Yellow phosphorus	Red phosphorus	Safety-match heads
Carbon tetrachloride	Perchloroethylene	Cleaning solvent
	Methylene chloride	Degreasing solvent
Benzene	Toluene, cyclohexane	Chemical raw material, solvent
Lead	Titanium, zinc	Interior paint pigments
Mercury nitrate	Various materials	Caroting fur for hats
Beryllium	Calcium phosphate	Phosphors for fluorescent lamps
Asbestos fibers	Glass fibers	Insulating materials
Sand	Silicon carbide	Abrasive cutting and shaping
Radium	Tritium	Luminous phosphors in dials

SUBSTITUTION OF EQUIPMENT OR PROCESSES THAT DISPERSE LESS CONTAMINANT		
ORIGINAL EQUIPMENT OR PROCESS	LESS CONTAMINATING EQUIPMENT OR PROCESS	APPLICATIONS
Spraying	Dipping	Painting of parts and products
Sandblasting	Hydroblast; chemical etching	Cleaning of parts
Soldering or welding	Riveting, crimping, or adhesive bonding	Joining of parts
Batch charging	Continuous feeding	Addition of materials
Diesel engines	Battery-powered electric engines	Fork-lift and delivery vehicles
Flanged piping	Welded pipe	Transfer lines
Mercury gauges	Mechanical gauges	Instrumentation

clothing can provide effective barriers against skin and eye contact, provided the proper materials are selected for their construction.

Respiratory protective devices provide defense against the inhalation of airborne contaminants. There are two basic types of devices: air-supplying and air-purifying respirators. In the former type, an excess of clean air is introduced into or around the nose or mouth. The air can be supplied by a portable tank under pressure, as in scuba diving, or by an air hose from a fixed tank or pump located out of the contaminated air. Portable tanks can only contain enough air or oxygen for brief periods, while air hoses limit mobility and freedom of action.

In air-purifying respirators, the contaminated air passes through a canister containing a filter and/or bed of adsorbing granules that extract the contaminants and pass the air. One type uses a battery-powered fan to push the air through the canister and thereby deliver it at positive pressure to the interior of a face mask or hood. It therefore resembles the air-supply type of device in that leakage is outward. On the other hand, in many other types of air-purifying respirators, the driving force to move the air through the canister is provided by the wearer's lungs, the pressure inside the mask is lower than that outside, and leakage is inward. Thus, even with complete contaminant removal from the air passing through the canister, the inhaled air will be partially contaminated if there is any leakage through or around the seal between the mask and the face. There is always resistance to flow through the canister, and this resistance generally increases with increasing collection efficiency and increasing flow.

For maximum protection, the correct air canister must be used, that is, one that is rated for the particular contaminant in the exposure atmosphere and within its concentration and time limits. If it is used at higher concentrations, or if its adsorptive capacity is exceeded by prolonged usage, it will fail to be protective. With a well-fitted mask, the appropriate air-purifying canister, and the motivation to use it properly, an individual can be protected from the inhalation of contaminated air, and respirators are, therefore, very useful for protection during infrequent brief operations and during unanticipated emergencies when used by well-trained and motivated workers.

Source and Effluent Control

When processes generate contaminants that must be controlled, the most economical and effective controls are those that exert their action closest to the source. It becomes more difficult and expensive to collect a chemical from a waste stream as process volume increases and the concentration decreases, such as following dilution by ambient air or water, and as the number of other constituents increases. Co-contaminants may produce interferences in the separation procedures and may also increase the cost of, or limit the freedom for, ultimate disposal of the extracted material. The use of process controls may act to concentrate contaminants in a small volume of air or water, thus decreasing the cost of any source cleaning devices, as well as increasing their efficiency.

Effluent controls include the removal of contaminants from waste streams and recycling of the extracted materials back into the process or disposal in an environmentally acceptable manner. When recycling or sale of the extracted material is feasible, it will usually be necessary to segregate waste streams according to their content and to treat some separately in order to avoid mixing and dilution of materials. On the other hand, where the material is simply to be disposed, it may be possible to combine various waste streams in ways favoring the neutralization of one waste with another. For example, an acidic liquid waste can be mixed with an alkaline waste to produce a neutral salt. One example is to use boiler-flue (smokestack) gases containing CO_2 to neutralize alkaline liquid wastes.

Control by dilution

In the past, it was often stated that "dilution is the solution to pollution." As time went on, this old adage became harder to justify. There will, however, always be some suitable application of control by dilution, such as for contaminants with short residence times in the environment and for those whose degradation products are innocuous. For contaminants that persist or that are degraded in the environment into persistent chemicals, control by dilution is inappropriate. Also, what was acceptable when there were few sources may become unacceptable when the number of sources increases.

In the air environment, toxic chemicals from power plants and industrial operations may be vented through high stacks. Such stacks can penetrate normal inversion layers and, as discussed in chapter 4, the maximum ground-level concentrations decrease approximately as the square of the effective stack height. Some power plants and industrial operations depend on dilution via tall stacks most of the time but switch fuels or process operations to control effluent levels on those days when stack dispersion will not be adequate.

In the water environment, dilution rates depend on stream flow, and industrial operations frequently utilize detention basins on site to hold some of the wastes for periodic discharge during times of high stream flow.

In the occupational environment, dilution ventilation is sometimes used to reduce air concentrations from levels slightly above to slightly below acceptable limits, especially when there are multiple small sources and where the material is considered more of a nuisance or housekeeping problem than a toxic hazard. Where the sources are few and well defined, or where toxicity is involved, source control by local exhaust ventilation is usually preferred over dilution ventilation.

Local exhaust ventilation is one of the key techniques for controlling air-contaminant levels within the occupational environment. The principle is to create sufficient suction in the zone of active contaminant release to draw essentially all of the contaminant into a hood enclosure, along with a minimum volume of workroom air. The contaminant stream is then generally guided through a duct system to an air-cleaning device, fan, and discharge stack Examples of some local exhaust hood designs are illustrated in Figures 10–2 and 10–3.

A special consideration in ventilation control of air contaminant concentrations in industry is the cost of supply make-up air. For each volume of contaminated air exhausted, it is necessary to supply an equivalent volume of clean air, and the latter usually must be cooled or heated to maintain the desired temperature in the workroom. With the increasing costs of energy, the relatively large volume requirements of dilution ventilation make it increasingly less attractive as a contaminant control technique. This is, however, still a major technique used in underground mines.

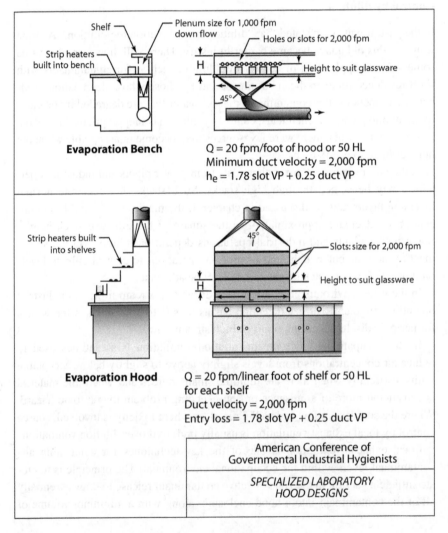

Evaporation Bench

Q = 20 fpm/foot of hood or 50 HL
Minimum duct velocity = 2,000 fpm
h_e = 1.78 slot VP + 0.25 duct VP

Evaporation Hood

Q = 20 fpm/linear foot of shelf or 50 HL for each shelf
Duct velocity = 2,000 fpm
Entry loss = 1.78 slot VP + 0.25 duct VP

American Conference of
Governmental Industrial Hygienists

SPECIALIZED LABORATORY HOOD DESIGNS

FIGURE 10–2. Specialized laboratory exhaust-hood designs.

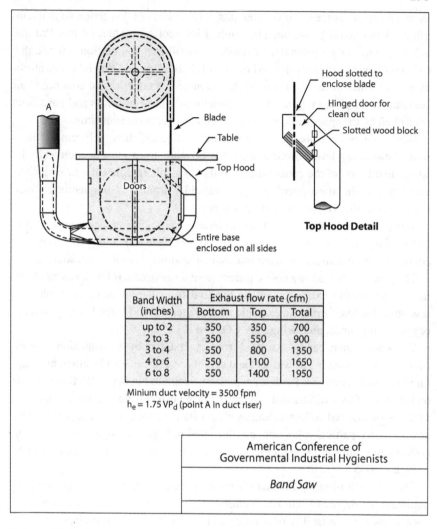

Top Hood Detail

Band Width	Exhaust flow rate (cfm)		
(inches)	Bottom	Top	Total
up to 2	350	350	700
2 to 3	350	550	900
3 to 4	550	800	1350
4 to 6	550	1100	1650
6 to 8	550	1400	1950

Minium duct velocity = 3500 fpm
$h_e = 1.75\, VP_d$ (point A in duct riser)

American Conference of
Governmental Industrial Hygienists

Band Saw

FIGURE 10–3. Hood design for a band saw.

A widely used technique for source control of industrial dusts is to wet the dust generated in operations such as crushing, grinding, cutting, and drilling. Dust generation may be greatly reduced by spraying a water mist or a stream of water onto the working face of the machine or tool.

CONTROL OF AIR CONTAMINANTS

The effective control of air quality ultimately depends upon the control of source emissions. For the criteria air contaminants (i.e., particulate matter [PM], Pb, SO_2, CO, NO_2, and O_3), the first step in their control was to construct an inventory of sources, source strengths, and patterns of dispersion so that control priorities could be set up. The original inventories established that PM and SO_2 were primarily attributable to stationary fossil-fuel combustion sources, that CO and O_3 were primarily derived from mobile sources, and that NO_2 was attributable in similar measure to both of these source categories. Most noncriteria air contaminants that are regulated are primarily of industrial origin and have been controlled by the specification and enforcement of effluent standards.

While source controls are preferable in most cases, there still remain times when satisfactory levels of air quality can be maintained most economically by adequate dilution of the contaminants in the atmosphere. Since the mixing characteristics of the atmosphere are quite variable, dilution is frequently coupled with specifications on intermittent discharge.

Source controls are discussed first in terms of air-cleaning techniques. This is followed by a discussion of the application of available technology to the specific control of air contamination from mobile and stationary combustion sources.

The process of removing contaminants from an airstream is known as air cleaning. It can involve extraction of the material from the air or its conversion to less objectionable forms. An example of the latter method is the burning of toxic organic compounds, producing only CO_2 and H_2O.

Gases and vapors can be extracted from an airstream by condensation, adsorption, and absorption in gas washers and scrubbers. PM can also be efficiently captured in some scrubbers. Particles can also be captured in dry collection systems on the basis of their diffusional and inertial displacements during passage through filters or specialized collectors and by electrostatic precipitation. The basic mechanisms of dry-particle collection are illustrated in Figure 10–4, and those used to collect gases and vapors, and in some cases particles by scrubbing, are illustrated in Figure 10–5.

The selection of any particular cleaning method depends on the specific contaminant, the degree of removal required, the volumetric rate of flow to be processed, and the availability of suitable and economical means of disposal of the materials extracted from the air or of the secondary reaction products formed during the control process.

Air Cleaning Techniques

Combustion

The simplest, and generally the most economical, way to dispose of organic wastes or other combustible contaminants in air is to burn them. If combustion is complete, all wastes are converted into CO_2 and H_2O. The burning of so-called "sour" waste gases (i.e., containing H_2S and organic S compounds) in refineries results in SO_2 in the effluent, but this SO_2 is generally much less of a problem

Brownian Separator

The Brownian Separator removes particles in the size range from 0.01 to 0.05 μm. This device consists of filaments, usually glass, arranged so that the space between them is less than the mean free path of the particle to be collected. Thus, the particle must collide with one of the filaments, where it is captured.

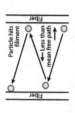

Filtration

The panel-type filter is limited to relatively light dust loadings because cleaning is time-consuming and costly. The baghouse, on the other hand, can be used with heavy loadings. These units are made up of a large number of cylindrical bag filters, which may be cleaned automatically to provide continuous operation. The baghouse should not be used for oily, hygroscopic or explosive dusts. Gas temperature is limited by the composition of the filter medium.

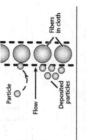

Electrostatic Precipitator

In this common collecting device, a high voltage is imposed on a relatively slow-moving gas stream. Charged particles are attracted to the grounded electrode, where they are periodically removed by rapping, vibration or washing. Conventional precipitators are quite bulky because of low air velocities. They have the advantage of providing dry collection, generally down to the 0.2 to 0.5 μm range. Theoretically, there is no minimum limit to the size of particles that can be collected.

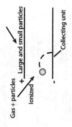

Centrifugal Collectors

Available in a number of different designs, cyclone separators rely on centrifugal force to drive particles to the wall of the chamber where they drop out of the gas steam into a collector. They are useful for relatively coarse separations. A large-diameter cyclone will remove particles 15 μm and larger. Small-diameter units usually connected in parallel are effective down to the 10-μm range.

Inertial Separators

A variety of dry mechanical collectors are based on inertial impaction of particles on baffles arranged in the gas stream. Because there is a practical limit on how closely together the baffles can be placed, these devices are best for separations above 20 μm. However, some designs can achieve 5 to 8 μm separations.

Gravity Settling

The oldest and simplest means of particle separation is the gravity setting chamber, in which particles fall onto collecting plates. These are useful for removing coarse dusts above 40 μm, and may be placed ahead of other separation equipment.

FIGURE 10-4. Basic methods for dry-particle collection.

Spray Towers

In spray towers the gas stream passes at low velocity through water sprays created by pressure nozzles. These units are simple, but only moderately effective. They will remove particles down to the 10 to 20 μm range and absorb very soluble gases.

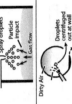

Wet Cyclones

In this version of the cyclone separator, swirling gas flows through water sprays. Droplets containing dusts and absorbed gas are separated by centrifugal force and collected at the bottom of the chamber. Wet cyclones are more effective than spray towers. They will absorb fairly soluble gases and remove dusts down to the 3 to 5 μm range.

Venturi Collectors

The venturi design relies on high gas velocities on the order of 100 to 500 ft/sec. through a constriction where water is added. The impact breaks the water into droplets, with the fineness of spray determined by gas velocity. Venturis collect particles to the 0.1 μm level and recover soluble gases.

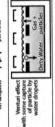

Perforated Impingement Trays

Here, gas flows through small orifices at 40 to 80 ft/sec. and hits a liquid layer to form spray. The liquid captures particles and gases, and is collected on impingement baffles. Performance is relatively good with particle collection possible down to 1 μm and absorption of fairly soluble gases.

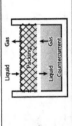

Packed Towers

Packed towers were primarily created for gas absorption. With some new designs, they can also be used for dust removal. Gas-liquid flow may be concurrent, cocurrent or crossflow. Two kinds of mass and particulate transfer are possible, depending on the packing. Extended surface packings provide absorption by spreading the liquid surface, and collect particles by cyclonic action. Recovery is limited to 10-μm sizes and larger. With filament packing, absorption by surface renewal is twice as effective and particles to 3 μm are separated by inertial impact. A cross-flow tower gives the highest solids handling capability with the lowest pressure drop.

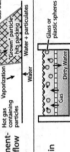

Nucleation Scrubber

This process grows submicron particles, down to 0.01 μm, by condensation. it collects grown particles on filament-type packing by inertial impact at low energy. At the same time, absorption occurs as in a conventional cross-flow packed tower.

Turbulent Contactor

With this device gas flows through spherical packing made up of 0.05 to 2-in. balls, which oscillate or bounce in liquid. Particles down to the 1 to 5 μm range collect by baffle impact on the spheres. Gases will absorb in the turbulent liquid and absorption efficiency will depend on the number of stages.

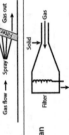

Wet Filters

This is an open filter whose efficiency is improved by spraying with water. Its absorption ability is limited to those gases that are very soluble.

Chromatographic Recovery

The chromatographic unit contains inexpensive extended surface solids with a mono-to-termolecular coating of an active reagent. Thus the solute gas need only diffuse to the surface of the solid where it reacts either by ionic or molecular reaction. Where the reaction is ionic, the absorption is generally irreversible. Where the reaction is molecular, the absorption can be reversible, and the solute gas can be stripped off in essentially pure form.

FIGURE 10-5. Basic equipment that may be used for both particle and gas collection.

than the original compounds. In most cases, the amounts of SO_2 and NO_x in the effluent of a waste-gas combustion system will be negligible in comparison to the other regional anthropogenic sources of these contaminants.

Two methods are widely used for the combustion of effluents: the flame method and the catalytic method. In the former, an external gas supply is generally needed to maintain a flame temperature sufficiently high to ensure complete combustion. An example is a petroleum refinery flare tower used to dispose of hydrocarbon vapors. The other method is based on catalytic oxidation and is generally used for low-concentration streams. Such devices are used on post-1975 cars to reduce the CO and hydrocarbon content of exhaust gases.

Adsorption and Condensation

Contaminant gases and vapors can be removed by passing the airstream through a bed of granules whose surfaces will adsorb and retain the contaminants that reach them by diffusion. For example, activated carbon is often used to adsorb organic vapors. Since the surfaces eventually become saturated, the adsorption bed must either be periodically replaced or reconditioned, and provisions must be made for the disposal of the bed or the materials collected on and stripped from the bed during its reconditioning. Some gas-phase contaminants may be removed by use of cold surfaces within or lining the flow path. Similar considerations concerning reconditioning apply here as well. Organic vapors are often controlled using this technique.

Adsorbants may be impregnated with various materials (Table 10–3) to convert the contaminant to less-harmful forms (e.g., CO into CO_2) or to promote a reaction that results in a form that may be more readily collected.

Scrubbing

Liquid streams and droplets can be used to extract gases and vapors from airstreams in devices known as scrubbers. The gases and vapors are removed by absorption into the liquid phase. The removal efficiency depends on both the solubility and the degree of saturation within the liquid and also on the contact time between the contaminant in the gas phase and the surfaces of the liquid phase. Table 10–4 summarizes contaminants commonly removed by scrubbing and the scrubbing media used.

Scrubbers can also be used to collect PM. Particles that contact a liquid surface are collected, with removal due to impaction, diffusion, and, in some cases, electrostatic precipitation. Not surprisingly, collection efficiency is strongly dependent on particle and droplet sizes.

Scrubbers, such as the Venturi type, which break up the liquid into very small droplets, may have relatively high collection efficiencies for small particles. However, these devices consume much more energy in the process than do gas washers, such as the spray tower or packed tower.

TABLE 10–3. Some Adsorbent Impregnations for Gaseous Contaminants

ADSORBENT	IMPREGNANT	CONTAMINANT	ACTION
Activated alumina	Potassium permanganate	Easily oxidizable gases, e.g., formaldebyde	Oxidation
	Sodium carbonate or bicarbonate	Acidic gases	Neutralization
Activated carbon	Bromine	Alkenes, e.g., ethylene	Conversion to dibromide
	Iodine	Mercury	Conversion to HgI_2
	Lead acetate	Hydrogen sulfide	Conversion to PbS
	Oxides of Cu, Cr, V, etc.; noble metals (Pd, Pt)	Oxidizable gases, including H_2S and mercaptans	Catalysis of air oxidation
	Phosphoric acid	Ammonia; amines	Neutralization
	Sodium carbonate or bicarbonate	Acidic vapors	Neutralization
	Sodium silicate	Hydrogen fluoride	Conversion to fluorosilicates
	Sodium sulfite	Formaldehyde	Conversion to addition product
	Sulfur	Mercury	Conversion to HgS

The contaminants contained within the scrubbing liquids will generally require special handling in terms of treatment and disposal. Considerations in the handling of such contaminated liquids are discussed later in this chapter, under water treatment techniques.

TABLE 10–4. Some Widely Used Scrubbing Media for Gaseous Contaminants

CONTAMINANT	SCRUBBING MEDIA
Chlorine (Cl_2)	H_2O, NaOH, NH_3
Hydrogen chloride (HCl)	H_2O, NaOH, NH_3, $Ca(OH)_2$, $Mg(OH)_2$
Hydrogen sulfide (H_2S)	NaOH, Na_2CO_3, KOH, K_2CO_3
Nitrogen oxides (NO_x)	$Ca(OH)_2$, $Mg(OH)_2$, Na_2SO_3 – $NaHSO_3$, urea
Sulfur dioxide (SO_2)	NaOH, Na_2CO_3, Na_2SO_3, KOH, K_2CO_3, K_2SO_3, CaO, $Ca(OH)_2$, $CaCO_3$, NH_3

Bag filters

Fibrous-filter media formed into cylindrical sleeves or bags are the most widely used type of dry-particle collectors for air cleaning. By appropriate selection of fiber type, diameter, packing density, fiber-mat thickness, and total surface, a high degree of collection efficiency may be achieved under a wide variety of operating and ambient conditions and for aerosols with a broad range of characteristics.

Filtration in bag filters is similar in some respects and different in others to that occurring in air-sampling filters. The face velocity across a bag filter is generally only a few centimeters per second, and, therefore, collection seldom depends on impaction deposition. Furthermore, the filtration occurs in the accumulated dust layer as well as in the filter mat itself. By contrast, in air sampling, the dust loadings are generally much lower, and a major portion of the sampled volume may be drawn through the filter before a significant dust layer accumulates. The clean filter must, therefore, be an efficient PM collector.

As dust accumulates on a bag filter, the flow must pass through the dust layer. Thus, the resistance to flow increases and will eventually impose an increasing energy burden on the air mover used. An analogous situation occurs in canister-type home vacuum cleaners that use a permanent-type felt bag, which must periodically be emptied. When the bag is emptied, some dust clings to the felt and, therefore, the pressure drop and collection efficiency never return to the original new-bag levels. The same considerations apply to industrial bag filters, which are emptied and reused many times before they develop tears or other defects that lead to their replacement.

Two typical configurations for bag-type industrial air cleaners are illustrated in Figure 10–6. In these devices, numerous identical bags hang vertically within an enclosure called a bag house. The bags are closed at one end and PM-laden air enters the housing at the top or bottom, then flows through the felt surfaces of the bags and out of the center of each bag into the clean-air plenum. The PM accumulates on the outsides (pressure-jet type) or the insides (traveling-ring type) of the bags, until the pressure drop across the bag reaches the preselected action level. In the pressure-jet device, the dust cake is periodically removed by a pressure pulse created by air-jet discharges affecting the bag surface. The pulse makes the bags expand, and the dust layer falls off and down into the collection hopper. Between cleaning cycles, the dust reaccumulates on the bags. In some bag filters, the cleaning is done by mechanical bag shaking. In the traveling-ring type, a doughnut-shaped ring with small air jets on its inner surface travels slowly up and down each bag to dislodge the dust.

Bag-filter control devices are available with an enormous variety of design features, numbers, sizes and shapes of bags, efficiency ratings, and operating conditions and can be used for many, but not all, air-cleaning applications. They cannot be used effectively at very high temperatures without special high-temperature fiber mats or for dusts that tend to gum or accumulate moisture excessively.

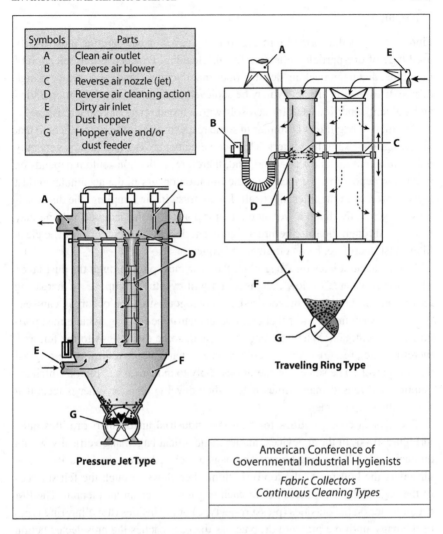

Symbols	Parts
A	Clean air outlet
B	Reverse air blower
C	Reverse air nozzle (jet)
D	Reverse air cleaning action
E	Dirty air inlet
F	Dust hopper
G	Hopper valve and/or dust feeder

Traveling Ring Type

Pressure Jet Type

American Conference of Governmental Industrial Hygienists
Fabric Collectors Continuous Cleaning Types

FIGURE 10–6. Two continuous-cleaning fabric collectors.

Electrostatic precipitators

Electrostatic precipitators used for air-cleaning applications are generally single-stage types, in which both the charging and collection of the particles take place in the same electrode configuration.

A negative high voltage is applied to the corona wires, creating a bipolar glow region around each wire and a unipolar stream of negative air ions that migrate toward the collection plates on each side. Particles in the airstream passing through are charged and accelerated toward the collection plates.

The collection plates must be periodically cleaned in order to maintain a high overall collection efficiency, since the resistivity of the dust cake affects the voltage gradient and corona current and, thereby, the performance of the precipitator; the resistance to air flow does not change appreciably with dust accumulation. The plates can be cleaned by any of several techniques, as appropriate to the precipitator and the dust. They can be mechanically shaken or periodically washed. Alternatively, the plates can be continually wetted by a falling film of water and thereby maintain a more constant operating characteristic.

High-voltage power supplies and electrodes for precipitators are very costly to install, and precipitators are generally not economically competitive with bag filters, except for very large installations. However, for high-temperature, high collection efficiency air-cleaning applications, precipitators may be the only types of air cleaners available. They are widely used to clean coal-fired power-plant effluents.

Inertial separators

When particles larger than approximately 1 to 2 μm in diameter are suspended in a moving stream of air, their motion will deviate less than that of the carrier air when the stream is turned or deflected. If particle displacement within the stream is sufficient to bring them into contact with a surface, they can be collected. The probability of collection increases rapidly with increasing particle size, air velocity, and sharpness of the directional change in the flow. Two types of inertial air cleaners, the baffle type and the cyclone, are illustrated in Figure 10–4.

In a cyclone, dusty air enters tangentially, and the particles are projected onto the outer walls as the stream follows a descending spiral. As dust accumulates in an air-cleaning cyclone, it migrates in flocs along the wall toward the collecting hopper at the bottom. Part of the way down the conical section, the descending air spiral containing the fine particle fraction turns inward and upward and leaves the axial exit pipe on a tighter spiral.

Their simplicity of construction and operation, and their relatively low power requirements, make inertial collectors relatively inexpensive to purchase and operate. However, they are of limited value for controlling the release of particles that have health significance because of their inability to capture particles smaller than approximately 5 μm. These devices are, however, frequently used to collect nuisance-type dusts that have large particle sizes. They are also used as precleaners upstream of more efficient and expensive particle collectors, such as bag filters, electrostatic precipitators, or Venturi scrubbers. When the mass median particle size is large, an inertial precollector can remove a substantial fraction of the total mass at low cost, reducing the loading on the more expensive second-stage air cleaner.

Mobile-Source Controls

The control of mobile-source (i.e., motor vehicle) emissions was mandated by the US Congress in stages. The sequence of steps was intended to permit adequate lead time for the development and testing of control technology. The entire industry adopted essentially the same technology for the control of evaporative losses of fuel and gaseous leakage around the pistons but chose several different approaches to the control of tailpipe emissions of CO and hydrocarbons. Most adopted the catalytic converter to more completely oxidize exhaust gases; others initially elected to achieve more complete combustion within the engine, some with electronic ignition systems and others with different combustion chamber designs such as the stratified-charge engine. As emission requirements became more stringent, essentially all auto manufacturers installed catalytic converters.

Stationary Combustion Source Controls

The effluents from industrial and utility boilers fired by coal and heavy oil can contain substantial quantities of ash, SO_2, and NO. Up until the 1960s, the only exhaust-gas cleaning that was commonly done was removal of most of the ash, using either electrostatic precipitators or bag filters. Neither type of air cleaner removed any significant amount of the gas-phase contaminants.

The initial focus of stationary-source control under the Clean Air Act of 1970 was the reduction of SO_2 emissions by reducing the average S content of the fuel burned or extracting SO_2 from the effluent gas before it leaves the stack. Most utilities with an adequate supply of low-S fuel chose the former approach. The increased fuel costs could be passed directly to the power consumers. The alternative was to incur the additional expenses required for land acquisition and plant construction and operation for a stack gas-cleaning system. Some utilities sought permission for a third alternative, that is, dilution via high stacks, but the EPA refused to approve control by dilution, and this position has been upheld in the courts. Since there was not enough low-S fuel to accommodate the demand, some utilities installed flue-gas desulfurization (FGD) processes to clean their effluents.

In the US, six types of FGD systems have been used for controlling full-scale utility boilers. Most of the operating capacity involved lime/limestone-slurry or sodium carbonate scrubbing techniques. These are known as "throwaway" processes, since the SO_2 is removed in a form that is discarded. On the other hand, the magnesium oxide and Wellman-Lord processes produced salable sulfur products, such as elemental S, H_2SO_4, or liquid SO_2.

In concept, lime- or limestone-slurry scrubbing processes are very simple. In practice, however, the chemistry and the system design for a full-scale operation can become quite complex.

The overall absorption reaction taking place in the scrubber and the hold tanks for a limestone-slurry system produces hydrated calcium sulfite:

$$CaCO_3 + SO_2 + 1/2\,H_2O \rightarrow CaSO_3 \cdot 1/2\,H_2O + CO_2 \qquad (10\text{--}1)$$

With a lime-slurry system, the overall reaction is similar but yields no CO_2:

$$CaO + SO_2 + 1/2\,H_2O \rightarrow CaSO_3 \cdot 1/2\,H_2O \qquad (10\text{--}2)$$

(The actual reactant in Eq. (10–2) is $Ca(OH)_2$, since CaO is slaked in the slurrying process.)

In practice, some of the absorbed SO_2 is oxidized by oxygen absorbed from the flue gas; the resultant product appears in the slurry as either gypsum ($CaSO_4 \cdot 2H_2O$) or as a calcium sulfite/sulfate mixed crystal. The slurry is recycled around the scrubber to obtain the high liquid-to-gas ratios required.

Lime- and limestone-slurry scrubbing systems can be engineered for almost any desired level of SO_2 removal. Commercial utility systems are generally designed for 80% to 95% removal; however, some systems have, at times, achieved removal efficiencies as high as 99%.

Some low-S coals contain large amounts of alkaline metal oxides (e.g., CaO, MgO) in the ash that results from combustion. These coals appear particularly suitable to SO_2 control by scrubbing the waste gases with a slurry of the alkaline fly ash. There are two ways to add the ash to the system: (a) by collecting the fly ash in an electrostatic precipitator upstream of the scrubber and then slurrying the dry fly ash with water so that it can be pumped into the scrubber circuit or (b) by scrubbing the fly ash directly from the flue gas by the circulating slurry of fly ash and water.

"Throwaway" FGD systems depend, of course, on having a suitable place to throw the waste away. The enormous amounts of impure gypsum slurry produced by lime- and limestone-process plants must be piped to large settling and evaporation ponds. These FGD processes are, therefore, economically feasible only where land for such ponds is available close by and at low cost. The sodium carbonate process produces a solution of Na_2SO_4, which has no commercial value and cannot be readily disposed of.

The magnesium oxide process differs from the previously discussed techniques in that it is a "regenerable" or "salable product" process; the SO_2 removed from the flue gas is concentrated and used to make marketable H_2SO_4 or elemental S. This process employs a slurry of MgO or $Mg(OH)_2$ to absorb SO_2 from the flue gas in a scrubber and yields magnesium sulfite and sulfate. When dried and calcined, the mixed sulfite/sulfate produces a concentrated stream (10%–15%) of SO_2 and regenerates MgO for recycle to the scrubber. Carbon added to the calcining step reduces any $MgSO_4$ to MgO and SO_2.

In commercial applications, the scrubbing and drying steps would normally take place at the power plant. The regeneration, and the S or H_2SO_4 production steps, might be performed at a conventional H_2SO_4 plant. Alternatively, a central processing plant could produce sulfur from mixed magnesium sulfite/sulfate brought in from other desulfurization locations.

The Wellman-Lord (W-L) process is also a regenerable system. When coupled with other processing steps, it can make salable liquid SO_2, H_2SO_4, or elemental S.

The W-L process employs a solution of Na_2SO_3 to absorb SO_2 from waste gases in a scrubber or absorber, converting the sulfite to bisulfite:

$$Na_2SO_3 + SO_2 + H_2O \rightarrow 2\ NaHSO_3 \tag{10-3}$$

Thermal decomposition of the bisulfite in an evaporative crystallizer can regenerate sodium sulfite for reuse as the absorbent:

$$2\ NaHSO_3 \rightarrow Na_2SO_3 + SO_2 + H_2O \tag{10-4}$$

The evaporative crystallizer produces a mixture of steam and SO_2 and a slurry, which contains sodium sulfite/sulfate plus some undecomposed $NaHSO_3$ in solution. As water condenses from the steam/SO_2 mixture, it leaves a wet SO_2-enriched gas stream to undergo further processing for recovery of salable S.

Dilution and Intermittent Discharge

Dilution is a control technique only in the sense that it depends on turbulent diffusion in the atmosphere to reduce the concentration of the effluents to acceptable levels. It may also depend on the ability of the normal processes in the atmosphere to degrade, transform, or remove the contaminants, so that they do not accumulate excessively within the atmosphere. Consideration of the acceptability of control by dilution may even extend to the effects of the accumulation of airborne contaminants on the land, in surface waters, and in the biosphere. The technical considerations in determining the extent of dilution of contaminants discharged into the atmosphere are discussed in chapter 4.

Intermittent discharge is, as previously discussed, a useful adjunct to dilution in the atmosphere. Since effective dilution varies with the atmospheric lapse and wind velocity, control by dilution is better justified when there are means of stopping or reducing effluent discharges during periods when dilution factors are unfavorable. The most notable example of voluntary effluent reductions during periods of poor atmospheric dispersion is the use of reserve supplies of natural gas or low-S fuel for industrial and utility boilers as a means of limiting the build-up of ambient SO_2 levels.

CONTROL OF SURFACE WATER QUALITY

By the beginning of the twentieth century, the concentration of people in metropolitan areas and the increasing use of sewer systems that discharged directly into surface waters resulted in the frequent appearance of floating fecal matter and attendant problems of odor and visual blight. Thus, until about mid-century, water pollution control was primarily concerned with sewage and other wastes with similar properties. The installation of primary treatment facilities was generally sufficient to alleviate the immediate problem; this stage of treatment involves removal of larger suspended and floating solids and oil and grease, generally by mechanical means.

Primary treatment, however, only removed part of the oxygen demand of the waste, and the continuing increase in the total amount of sewage led to frequent depletion of the DO levels in surface waters around major metropolitan areas. Oxygen depletion, in turn, led to fish kills and odors associated with putrefaction. The general solution to this problem was the installation, beginning on a large scale in midcentury, of secondary-treatment facilities to supplement the primary treatment. Secondary treatment generally is synonymous with biological treatment. In this process, the liquid waste is contacted with bacteria, which consume most of the organic material in the waste during their metabolism. The actual reduction in oxygen demand achieved depends upon the ability of the bacteria to use the organics in the waste and the contact time of the contaminants with the bacteria.

Depending upon their biodegradability, nonsewage chemical contaminants in waste streams may be partially removed by the bacteria in secondary-treatment facilities. Some chemical contaminants, however, present serious problems to sewage treatment plant operations in that they are toxic to the bacteria. If too many bacteria are killed, the operation of the treatment plant can be seriously disrupted, and it may take many days for the reestablishment of an effective bacterial colony in the facility.

Primary-treatment steps are frequently needed to modify the waste to the extent that it will be amenable to biological treatment. Conditions calling for such preliminary treatments are listed in Table 10–5.

In recent decades, concern has shifted from the oxygen demand to the toxicity of contaminants in surface waters. This concern has led to an acceleration in the development of a variety of advanced treatment techniques, also known as tertiary treatment, for the removal of many chemicals from waste streams. Much of this modern technology has been based on adaptations of the process technology of chemical engineering that has been used to recover materials from industrial-process waste streams. However, applications of these techniques to large-volume municipal waste streams of extremely heterogenous composition are a greater challenge.

TABLE 10–5. The Concentration of Contaminants That Make Primary Treatment Necessary

CONTAMINANT	LIMITING CONCENTRATION
Acidity	Free mineral acidity
Alkalinity	0.5 lb alkalinity as $CaCO_3$/lb BOD removed
Suspended solids	> 125 mg/L
Oil, grease	> 50 mg/L
Dissolved salts	> 16 gm/L
Organic-load variation	> 4:1
Trace metals	> 1 mg/L
Sulfides	> 100 mg/L
Phenols	> 70–160 mg/L
Ammonia	> 1.6 gm/L

BOD, biochemical oxygen demand.

The kinds of waste-water treatment processes used, and their sequence, are illustrated in Figure 10–7. The various treatment schemes accomplish different levels of removal, and as the degree of treatment increases, the cost also rises. Note that there is some overlap in techniques used in the different stages of water treatment. The contaminants for which the various processes are effective are presented in Table 10–6.

Control of Chemical Contamination from Point Sources

Stream Manipulation

When there are numerous waste streams, with many having similar characteristics, it may be desirable to have two or more separate waste systems, for example, one for oily wastes, one for inorganic chemicals, and one for storm drainage. In some cases, it may not be necessary to clean the storm drainage at all, and keeping it out of the chemical-waste system would help to avoid flow surges that can overwhelm the capacity of the waste-treatment facilities.

Under some circumstances, it is better to mix waste streams than to segregate them for separate treatment. If, for example, one stream is acidic and another is basic, one waste can neutralize the other, producing a salt solution that may require no further treatment. Even when the salt is insoluble and requires collection and disposal, the net effect may still be beneficial, and the

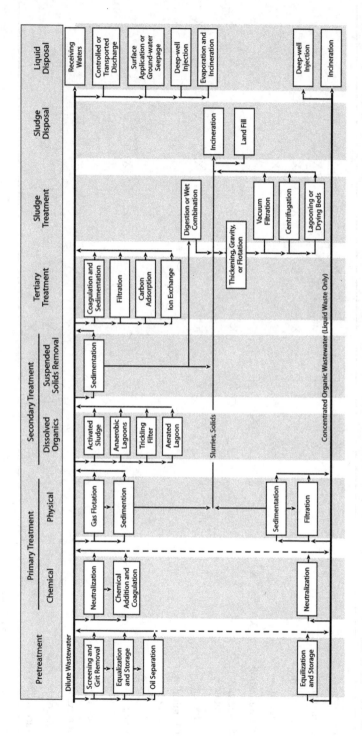

FIGURE 10-7. Waste water-treatment sequence. Very often, one process may be substituted for another, or a variety of different processes may be needed for treatment of specific contaminants. (Pretreatment is usually considered to be part of the primary treatment of wastewater.)

TABLE 10-6. Typical Applications of Wastewater Treatment Processes

			CONTAMINANT									
PROCESS	ACIDS	ALKALIES	TOTAL DISSOLVED SOLIDS	SUSPENDED SOLIDS	NITRATE	PHOSPHATE	TRACE METALS	ORGANIC MATERIAL (BOD)	OILS	PERSISTENT PESTICIDES	COLOR	TASTE AND ODOR
Filtration				X			X					
Sedimentation				X					X			
Clarification				X		X	X	X	X			
Flotation				X					X			
Neutralization	X	X										
Biological treatment					X	X		X			X	X
Lagooning				X	X	X		X	X		X	X
Chemical oxidation or reduction											X	X
Emulsion breaking									X			
Adsorption				X		X		X		X	X	X
Ion exchange			X		X	X	X					
Reverse osmosis			X	X	X	X		X				
Electrodialysis			X		X	X						

BOD, biochemical oxygen demand.

Source: Compiled from: Paulson, E.G. Water pollution control programs and systems. In H.F. Lung (ed.), *Industrial Pollution Control Handbook.* New York: McGraw-Hill, 1971; and Metcalf and Eddy, Inc. *Wastewater Engineering: Collection, Treatment, Disposal.* New York: McGraw-Hill, 1972, pp. 611–627.

use of chemicals for intentional neutralization can be minimized. The economic benefits of complementary wastes frequently lead neighboring industries to exchange waste streams or to share in the costs of operating a combined waste-treatment facility.

The mixing of streams can also be beneficial in terms of equalization of the strength of the waste, especially when one or more of the major sources are intermittent. The mixing of streams within the sewer system or preliminary-treatment steps may prevent concentration peaks that could adversely affect subsequent biological treatment or advanced chemical processes.

Reduction of Waste Volume

Reduction of waste volumes involves decreases in both the amount of contaminant and the amount of dilution water; both can generally be achieved by a good preventive maintenance program that detects and reduces leakage. Improvements in process design, equipment, and operating procedures may also make it possible to maintain or improve process output with reduced amounts of waste materials.

Water Treatment Techniques

Screening

Screening is a mechanical process that separates suspended particles on the basis of size. There are several types of devices that may have static, vibrating, or rotating screens. Openings in the screening surfaces range from several centimeters down to approximately 20 µm.

Screens are used for treating raw water, storm water, and waste waters from certain industries that produce large volumes of solids, such as food-process plants, textile mills, and pulp and paper mills. The amount of solids removal depends upon screen size and characteristics of the water or waste water being treated. For example, fine-mesh microstrainers, with screen apertures of approximately 20 µm, remove solids, such as various forms of plankton and general microscopic debris; coarse bar screens that have metal bars spaced approximately 25 to 50 mm apart are used to remove the large suspended solids. The screens are periodically cleaned, generally by automatic equipment. Comminutors, or large, circular garbage grinders, may be used to grind into smaller pieces those solids that penetrate the coarse screens.

Centrifugal separation

Hydrocyclones are used to remove relatively dense suspended solids from a liquid stream; a common application is for separating grit from stormwater or waste water. Figure 10–8 shows their operation. Fluid-pressure energy is used to create rotational fluid motion in a cyclone. This motion causes relative movement of

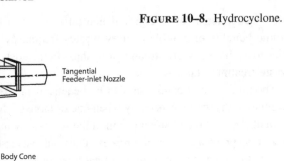

FIGURE 10–8. Hydrocyclone.

materials suspended in the fluid, and permits separation of these materials from the fluid. The bulk of the fluid is removed through a top outlet, and the separated solids are discharged through a second outlet located in an axial position at the bottom.

Smaller particles can be separated using centrifuges, but such separations are much more expensive, both in terms of capital outlay and operating costs.

Neutralization

When there is an excess of acid or base in the wastes, it may be necessary to treat the waste stream with sufficient basic or acidic chemicals to produce neutralization. For small volumes of acidic wastes, sodium hydroxide is used, and the resulting salts will usually not present any significant disposal problem. Lime and ground limestone are much less expensive per unit of basicity but produce insoluble neutralization products that will usually require separate disposal. Alkaline wastes can be neutralized with sulfuric or hydrochloric acids. Relatively inexpensive neutralizations can be achieved by bubbling flue gas containing CO_2 through the waste stream.

Precipitation

Precipitation reactions are often carried out in water and waste treatment. Lime is frequently used to precipitate calcium carbonate and magnesium hydroxide. In waste-treatment operations, the precipitation of metallic hydroxides from plating, metallurgical, and steel mill wastes and also from acid mine drainage is commonplace. Wastes that contain phosphates may also be treated to precipitate calcium or aluminum phosphate. Lime or limestone is used for the precipitation of calcium sulfite or sulfate from stack-gas scrubber water. Essentially, every reaction is carried out with the stoichiometric amount of precipitant and at the optimum pH. If the solubility of a compound exceeds the required effluent concentration, an excess of the precipitant may be added to reduce the solubility. Precipitation reactions are usually carried out in the presence of previously formed sludge. This minimizes supersaturation and also results in more rapidly settling precipitates because precipitation occurs on existing slurry particles.

Flocculation and coagulation

Flocculation is the process of bringing together fine particles, for example, colloids, so that they agglomerate, forming larger particles (flocs) that will settle more rapidly. Mechanical stirring or air injection may be used to cause the suspended particles to collide. Alternatively, a chemical coagulant may be used; these cause suspended solids to agglomerate by reducing colloidal charge levels. When a metallic cation such as aluminum, for example, as aluminum sulfate (alum), or iron, for example, as ferric chloride, is used, particulate metallic hydroxides are initially formed, which adsorb the colloids.

Synthetic polymers are often used as coagulants or as coagulant aids. Polymers may be cationic and adsorb a negatively charged particle; anionic, which permit bonding between a colloid and polymer; or nonionic, which adsorb flocculates by hydrogen bonding between the solid surfaces and the polar groups of the polymer.

Separation by density differences

Separation by density differences involves the removal of suspended solids or oils when the specific gravity difference from water causes settling or rising of the solids or oils during passage through a tank under suitably quiescent conditions. Static conditions that affect such separations are water temperature, which affects both viscosity and density of the water; specific gravity of the oil or suspended solids; size and shape of suspended oil droplets and of particles; and solids concentration, which results in free or hindered settling.

The removal of suspended solids via gravity settlement is termed sedimentation and is usually performed in tanks, known as settling tanks, that have a continuous flow and a detention time of several hours. The settling velocities of particles will change with time and depth, as particles agglomerate and form larger floc sizes.

Settling tanks are divided into four zones: inlet zone, to provide a smooth transition from the influent flow to a uniform steady flow desired in the settling zone; outlet zone, to provide a smooth transition from settling zone to the effluent flow; sludge zone, to receive and remove settled material and prevent it from interfering with the sedimentation of particles in the settling zone; and settling zone, to provide tank volume for settling free of interference from the other three zones. Settling tanks that follow preliminary treatment such as screening are known as primary clarifiers. Typical primary-clarifier units are shown in Figure 10–9.

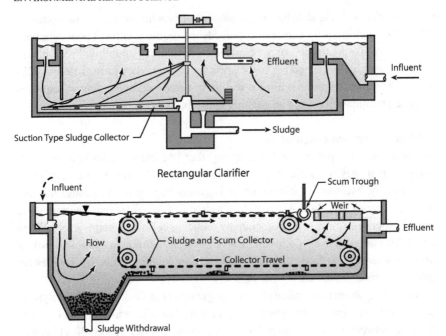

FIGURE 10–9. Typical clarifier units.

The surface area of a settling tank is one of the most important factors that influence sedimentation. The tank should be designed so as to produce a clarified effluent at minimum water temperature and to allow for separating floc particles that may not be of maximum size or density. This latter problem may be due to either partial deflocculation in the inlet zone or to partially ineffective coagulation. Tanks are generally approximately 30 m in diameter and approximately 4 to 5 m deep.

If the release of entrained air might cause floating floc, or if floating solids may be present, a scum baffle plus a scum skimmer mechanism are usually installed. Sludge removal from the tank bottom may be performed manually or by a continuous belt-drive system. An oil/water separator is used to separate free oil from refinery waste water. This type of unit will not separate soluble substances or break emulsions.

A corrugated-plate interceptor for oil is shown in Figure 10–10. Waste water enters a separator bay and flows downward through corrugated plates arranged at an angle of 45° to the horizontal. Oil collects on the underside of the plates and rises to the surface where it is skimmed. Solids settle into a sludge compartment, with the now-clarified waste discharging into an outlet channel.

In dissolved air flotation, illustrated in Figure 10–11, air is intimately contacted with an aqueous stream at high pressure, dissolving the air. The pressure on the

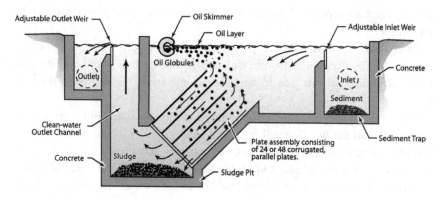

FIGURE 10–10. Corrugated-plate interceptor separator.

liquid is reduced through a backpressure valve, thereby releasing micrometer-sized bubbles that sweep suspended solids and oil from the contaminated stream to the surface of the air-flotation unit. Applications of this technique include treating effluents from oil refineries, metal-finishing processes, pulp and paper mills, cold-rolling mills, poultry processing, grease recovery in meat-packing plants, and cooking-oil separation from French-fry potato processing. An increasingly important application is the thickening of sludge.

The attachment of gas bubbles to suspended solids or oily materials occurs by several mechanisms. The suspended-solids/gas mixture is carried to the vessel surface after precipitation of air on the particle; collision of a rising bubble with a suspended particle; trapping of gas bubbles as they rise, under a floc

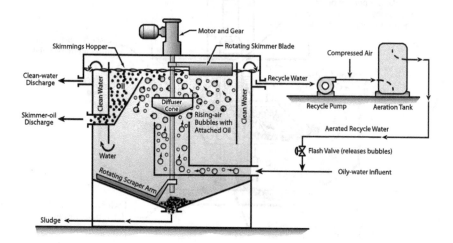

FIGURE 10–11. Dissolved-air flotation unit.

particle; and adsorption of the gas by a floc formed or precipitated around the air bubble.

Flocculants, such as synthetic polymers, may be used to improve the effectiveness of dissolved-air flotation. Also, coagulants, such as alum, may be used to break emulsified oils and to agglomerate materials for improved flotation recovery.

Flotation can also be achieved by induced air, as illustrated in Figure 10–12. Air drawn into the flotation cell by action of the rotor is mixed with the water and transformed into minute bubbles. Oil particles and suspended solids attach themselves to the gas bubbles and are borne to the surface of the water. Skimmer paddles push the contaminated froth from the top of the cell into collection launders. The tank is designed with sloping sides at the lower part to recirculate the water for continuous purification. The conical shape of the dispenser hood, with its perforated surface, acts to quiet the liquid surface.

The efficiency of contaminant removal in induced-air devices is often improved with chemical aids injected in the water, upstream in the flotation cell. In addition to breaking oil-water emulsions, they also make the air bubbles more stable and promote formation of froth on the surface of the water. Properly conditioned wastes leave the device nearly 100% oil-free, and the suspended solid load is also significantly reduced.

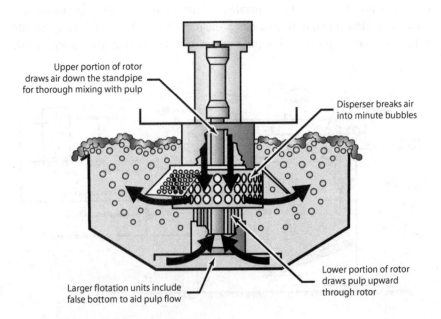

FIGURE 10–12. Induced-air flotation unit.

Filtration

Granular-media filtration is a liquid–solids separation method that uses flow through porous media, such as sand, to remove PM. To be readily filterable, the suspended solids must either be naturally flocculent or be made so by chemical coagulation. Granular media filters are used for removal of solids and oil from refinery waste waters, for effluents from activated-sludge treatment plants (see next section), and for effluents from pulp and paper plants. Heavy metal precipitates may be recovered in granular filters, and process streams are clarified by similar units.

The heart of any granular-media filter is the filter bed. The size and depth of the filter medium are the most important design parameters. As an example of granular-media filtration, the operation of a shallow-bed, gravity, granular-media filter is shown in Figure 10–13. The filter is completely automatic and is on a controlled cycle. Several filters are customarily used in parallel to avoid interruption of the flow stream when the filter backwashes. During filtration, waste water is passed through the media, and solids are removed. As the cycle progresses, the pressure drop across the filter increases to a preset maximum, indicating that the bed is full of suspended solids, and a backwash-cycle controller is energized. Alternately, a timed cycle or monitoring of the suspended solids may be used to terminate filtration.

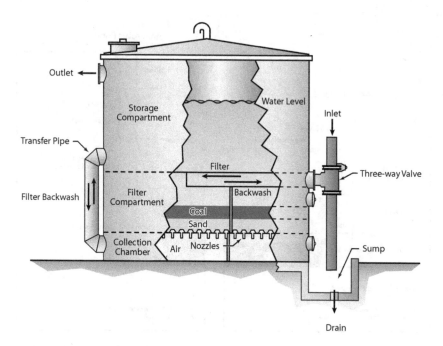

FIGURE 10–13. Granular-media filter (shallow bed, gravity type).

Backwash water flows by gravity through the collection chamber and the filter compartment to remove the suspended solids that were fluidized by the air wash. The backwash discharges by gravity to settling basins, where all of the suspended solids settle out and water is recovered. The addition of polymers to the backwash water accelerates the settling rate of the solids.

Biological treatment

Biological oxidation is the major process for the removal of oxygen demand in sewage wastes; it also finds extensive use for the treatment of industrial wastes containing biodegradable organics. In biological oxidative processes, microorganisms break down organic matter. These microorganisms are broadly classified as aerobic, facultative, or anaerobic. Aerobic organisms require O_2 for metabolism, anaerobes function in the absence of O_2, and facultative microbes may function in either an aerobic or anaerobic environment. Where land is relatively stable and inexpensive, cost-effective treatment can be done in simple ponds. Figure 10–14 shows the general characteristics of such ponds.

The predominant species used in biological systems are known as heterotrophic microorganisms; these require an organic carbon source for both energy and cell synthesis. Autotrophic organisms, in contrast, use an inorganic carbon source, such as carbon dioxide or carbonate. The autotrophs may derive energy for cell

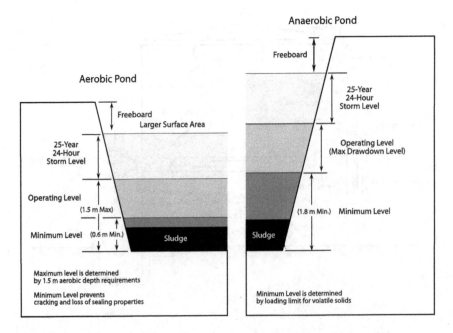

FIGURE 10–14. Characteristics of ponds for biological treatment of wastewaters.

synthesis from the oxidation of inorganic compounds of N or S (chemosynthetic bacteria) or from the sun (photosynthetic bacteria).

Basically, biological oxidative processes involve either of two mechanisms to accumulate and store the microbes, as a flocculated suspension of biological growth, known as activated sludge, which is mixed with the waste waters, or as a biological film fixed to an inert medium over which the waste waters pass.

Biological processes may not be applicable to many specific industrial waste waters. For example, cyclic organics, especially if halogenated, are resistant to biological degradation. Organochlorine pesticides may prove toxic to a biosystem if discharged in high concentrations. Slightly soluble components, such as PCBs or heavy metals, can accumulate in a biological system through adsorption and bioconcentration and may, thereby, reach levels inhibitory to the process.

Microbial populations may be specifically adapted to certain chemicals to successfully achieve oxidation. Examples of materials that are biodegradable under acclimated conditions are cyanide, phenol, formaldehyde, acrylonitrile, and hydroquinone. However, some cyanide compounds, such as the metallic-cyanide complexes, are highly resistant to degradation, even under strongly acclimated conditions.

The activated-sludge process is a continuous system in which aerobic bacteria are mixed with waste water and then physically separated by gravity clarification or by air flotation. As shown in Figure 10–15, the concentrated sludge is recycled to the reactor to mix with the incoming waste. O_2 is provided in a variety of ways,

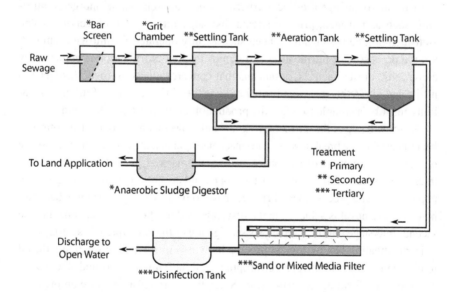

FIGURE 10–15. Schematic of typical modern waste water treatment plant using aeration tank digestion.

for example, by diffused aeration, surface aeration, or static mixers, and may be introduced either as air, pure O_2, or O_2-enriched air. The waste product from the activated-sludge process is excess sludge which must ultimately be disposed of, generally after partial reduction by anaerobic digestion.

As the concentration of organics in the remaining waste waters decreases, the rate of biological removal also decreases, since the remaining organics are progressively more difficult to remove. In the treatment of industrial wastes, a significant fraction of the organics may prove nonbiodegradable. Thus, although the BOD removal may be excellent, the removal of COD may be quite low. The activated-sludge process is broadly applicable. It is used in the treatment of soluble organic wastes from many industries, including food processing, meat packing, pulp and paper, oil refining, leather tanning, textiles, organic chemical, and petrochemical.

The continuous addition of powdered activated carbon to aeration tanks of suspended-bacterial systems can improve the operation and performance of the activated sludge. The activated carbon gives the system the capability of adsorbing nonbiodegradable organic matter present in the waste waters, thereby providing a degree of tertiary treatment and resistance to shock-loading (i.e., sudden input of a large amount of a chemical). The carbon is regenerated by pyrolysis.

A major modification to the conventional air-aerated activated sludge process has been the use of high purity O_2. Some advantages of this compared to the conventional system are lower land requirements; the ability to supply O_2 at high rates, making it amenable to high-strength wastes; and increased self-neutralization of highly alkaline wastes.

Increasingly stringent effluent criteria require the removal of inorganic nutrients, such as nitrogen, prior to stream discharge into a natural receiving water. Activated-sludge systems have been successfully operated for nitrogen removal, via the processes of nitrification and denitrification. Nitrification is achieved by autotrophic nitrite bacteria, organisms that convert reduced nitrogen (as NH_3) to nitrite and nitrate bacteria that further oxidize the nitrite to nitrate. Sufficient alkalinity must be available to offset the production of nitrous and nitric acids.

The growth rate of the nitrifiers is significantly lower than that of the heterotrophic bacteria used in the breakdown of carbonaceous organic matter. Thus, in the presence of significant amounts of carbonaceous organic material, the nitrifiers are unable to compete successfully with the heterotrophs and cannot accumulate as significant populations. The process may, therefore, have to be a two-stage system, where carbonaceous removal is achieved in the first stage and nitrification is performed in the second. Interstage clarification allows segregation of the two types of bacteria.

Denitrification is a biological process by which nitrite and nitrate are reduced to nitrogen gas. Facultative, heterotrophic bacteria found in activated sludges are capable of performing denitrification. Since these organisms require an organic-carbon source for cell growth, and because this process generally follows secondary treatment, some form of organic compound must be present or added.

Approximately 4.5 lb of COD, typically as methanol per pound of nitrate, are required to reduce the nitrate to N.

The predominant fixed-film biological process is the trickling filter. This is a packed bed of some medium, such as stones, covered with a biological slime, over which waste water is passed. Oxygen and organic matter diffuse into the slime film, where degradation reactions occur. End products such as CO_2, NO_3^-, and others counterdiffuse back out of the film and appear in the filter effluent. Trickling filters have, however, found only limited application for industrial waste waters. Removal rates for soluble industrial wastes are typically low, making filters unattractive for high BOD-removal efficiency. The process is, however, capable of accepting highly variable loading and thus may be used as a roughing filter to provide partial treatment.

Another type of fixed-film biological reactor is the rotating disk process. Plastic disks are mounted on a shaft and placed in a tank conforming to the general shape of the disks. The disks are slowly rotated while approximately half is immersed in the waste water. Rotation brings the attached bacteria culture into contact with the waste water for removal of organic matter. Rotation also provides a means of aeration, by exposing a thin film of waste water on the disk surface to the air.

Biological oxidation may be achieved in ponds, lagoons, and basins or in soil after land application of waste water. The land requirements for O_2-demand removal are greater in these systems than those required for fixed-film processes and are much greater than those for activated-sludge processes. In aerated stabilization basins, air is pumped into the water or the water is sprayed through the air to increase the O_2-demand removal capacity. In ordinary stabilization and oxidation ponds and in land irrigation, natural aeration rates are the limiting factor in determining the capacity for O_2-demand removal.

Chemical treatment

Chemical treatment involves use of oxidants, and those widely used today are chlorine, ozone, and hydrogen peroxide. Their historical use, particularly for chlorine and ozone, has been for disinfection of water and waste waters. They are, however, receiving increased consideration for removing from waste water organic materials that are resistant to biological or other treatment processes.

Ozone is found to be effective in many applications, such as for color removal, disinfection, taste and odor removal, iron and manganese removal, and the oxidation of many complex organics, including lindane, aldrin, surfactants, cyanides, phenols, and organometal complexes. With the latter, the metal ion is released and can be removed by precipitation.

Chlorine, and its more readily storable form, hypochlorite ion, has long been used to purify water, destroy organisms in waste water and swimming pools, and oxidize chemicals in aqueous solutions. The destruction of cyanide and phenols by chlorine oxidation is well known in waste treatment technology. The continued use of chlorine for these applications is, however, uncertain, because of concern about the toxicity and carcinogenicity of chlorine oxidation products.

Advanced waste-water treatment processes

Most conventional waste-water treatment facilities use primary and secondary treatment stages. Table 10–7 shows typical removal efficiencies for BOD, suspended solids, and dissolved solids by these processes; dissolved solids, which may include many toxic chemicals, are not effectively removed by this treatment. Some emerging technologies for bioremediation are described in Table 10–8.

Various physical and chemical processes are used in tertiary or advanced treatment to bring the water to a higher quality level than that achievable by conventional primary and secondary treatment. The methods include ammonia stripping, electrodialysis, activated carbon adsorption, reverse osmosis, distillation, and ion exchange.

Ammonia can be removed from waste water by the technique of air stripping, which involves pH control of the water. Ammonium ions in the water exist in equilibrium with ammonia and hydrogen ions. As the pH increases, usually by the addition of lime, the equilibrium shifts to the right, and above a pH of 9 ammonia may be liberated as a gas by agitating the waste water in the presence of air.

Electrodialysis uses an induced electric current to separate the cationic and anionic components of a solution. It relies on membranes, which are at right angles to the line of electric current flow, that permit ions to pass from a dilute solution on one side to a concentrated solution on the other.

Activated carbon adsorption is used for the removal of soluble organic matter not removed by conventional methods. Water passes through granular activated carbon, where organic molecules attach to the carbon surface. When the carbon reaches its adsorptive capacity, it is regenerated by pyrolysis. Large particles are removed by filtration or coagulation–sedimentation prior to carbon adsorption.

TABLE 10–7 Approximate Removal Efficiencies of Waste Water Treatment Stages

	PERCENTAGE REDUCTION FROM ORIGINAL WASTE WATER		
CONTAMINANT CATEGORY	PRIMARY TREATMENT	SECONDARY TREATMENT	TERTIARY TREATMENT
BOD	30	90	99.8
Suspended solids	60	90	100
Dissolved inorganic solids	0	5	99.5

BOD, biochemical oxygen demand.

TABLE 10-8. Current Feasibility of Bioremediation

CHEMICAL CLASS	FREQUENCY OF OCCURRENCE	STATUS OF BIOREMEDIATION	EVIDENCE OF FUTURE SUCCESS	LIMITATIONS
Hydrocarbons and derivatives				
Gasoline, fuel oil	Very frequent	Established		Forms nonaqueous-phase liquid
Polycyclic aromatic hydrocarbons	Common	Emerging	Aerobically biodegradable under a narrow range of conditions	Sorbs strongly to subsurface solids
Creosote	Infrequent	Emerging	Readily biodegradable under aerobic conditions	Sorbs strongly to subsurface solids; forms nonaqueous-phase liquid
Alcohols, ketones, esters	Common	Established		
Ethers	Common	Emerging	Biodegradable under a narrow range of conditions using aerobic or nitrate-reducing microbes	
Halogenated aliphatics				
Highly chlorinated	Very frequent	Emerging	Co-metabolized by anaerobic microbes; co-metabolized by aerobes in special cases	Forms nonaqueous-phase liquid
Less chlorinated	Very frequent	Emerging	Aerobically biodegradable under a narrow range of conditions; co-metabolized by anaerobic microbes	Forms nonaqueous-phase liquid

(continued)

TABLE 10–8. (*Continued*)

CHEMICAL CLASS	FREQUENCY OF OCCURRENCE	STATUS OF BIOREMEDIATION	EVIDENCE OF FUTURE SUCCESS	LIMITATIONS
Halogenated aromatics				
Highly chlorinated	Common	Emerging	Aerobically biodegradable under a narrow range of conditions; co-metabolized by anaerobic microbes	Sorbs strongly to subsurface solids; forms nonaqueous phase-solid or liquid
Less chlorinated	Common	Emerging	Readily biodegradable under aerobic conditions	Forms nonaqueous phase-solid or liquid
Polychlorinated biphenyls				
Highly chlorinated	Infrequent	Emerging	Cometabolized by anaerobic microbes	Sorbs strongly to subsurface solids
Less chlorinated	Infrequent	Emerging	Aerobically biodegradable under a narrow range of conditions	Sorbs strongly to subsurface solids
Nitroaromatics	Common	Emerging	Aerobically biodegradable; converted to innocuous volatile organic acids under anaerobic conditions	
Metals (Cr, Cu, Ni, Pb, Hg, Cd, Zn, etc.)	Common	Possible	Solubility and reactivity can be changed by a variety of microbial processes	Availability highly variable and controlled by solution- and solid-phase chemistry

Reverse osmosis forces water at high pressures through cellulose acetate membranes against the natural osmotic pressure. The mechanisms responsible for the action of the membranes in reverse osmosis include sieving, surface tension, and hydrogen bonding. The technique removes dissolved solids.

Distillation is a vapor–liquid transfer operation for removal of dissolved solids, in which vapor is driven from waste water by heat. Because of its high cost, distillation has limited application in waste-water reclamation.

Ion exchange is a very effective technique for the removal of ions from waste water. Various natural materials and synthetic ion-exchange resins may be used. Some common applications are for Cl^-, NH_4^+, PO_4^{-3} and metal ion removal.

Control of Chemical Contamination from Diffuse Sources

Adequate control of discharges into surface waters from sewage treatment plants and industrial operations is technically feasible, even if not always achieved. It is more difficult to control the discharge of chemical contaminants associated with diffuse sources, such as the drainage from agricultural land. Such drainage contains a variety of pesticides and nutrients, which can adversely affect aquatic organisms and stream productivity. Some agricultural operations use oxidation lagoons to reduce the O_2 demand of animal wastes. However, the best prospects for reducing fertilizer and pesticide burdens in runoff water appear to lie in educating farm operators about the problem, with the hope that they will apply these chemicals in quantities and by procedures that will minimize adverse environmental effects. Some of the problems have been, and will continue to be, resolved by the removal of the more toxic pesticides from the market.

Control of Drinking Water Quality

The provision of bacteriologically pure drinking water is one of the more notable public health success stories in this country. Disease transmission via public drinking water supplies is, fortunately, a rare event. There are two ways to accomplish the delivery of high-quality drinking water. One is to have an adequate volume supply of clean natural water. The other is to purify and disinfect available water to the point that it is safe to use. A typical modern treatment plant system for the supply of drinking water is illustrated schematically in Figure 10–16. Another key factor is continual monitoring of the purity of the water, not only at central distributing points but also at various points within the distribution system where it is used, to make sure that clean water is delivered to the consumer.

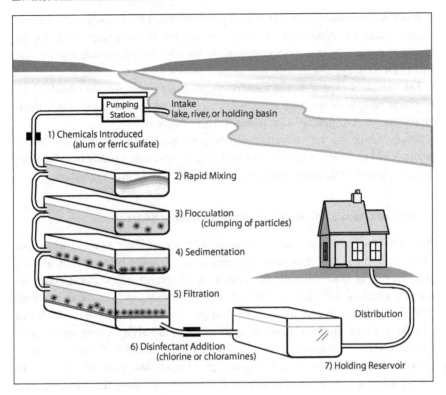

FIGURE 10–16. Typical water treatment plant system.

Selection of Clean Natural Water Sources

Natural waters, with little or no treatment other than light disinfection with chlorine, are used as drinking-water supplies by a large proportion of the people in this country. Some cities, most notably New York and Boston, maintain protected watershed areas and reservoirs that collect and distribute rainwater. Many other cities and smaller towns use groundwater from deep wells for drinking purposes. Such groundwater may be "hard" but is usually of very good quality in terms of bacterial content. Water from shallow wells in urban areas can readily by contaminated and must be monitored carefully. Little attention was paid to the chemical purity of drinking water until the mid-1900s. The extent of the problem arising from the increasing content of organic compounds and trace elements in drinking water has not yet been adequately defined.

Purification of Contaminated Groundwater

When groundwater is used as a potable water supply, it must either not be contaminated by surface sources of toxic chemicals or it must be pretreated as are surface

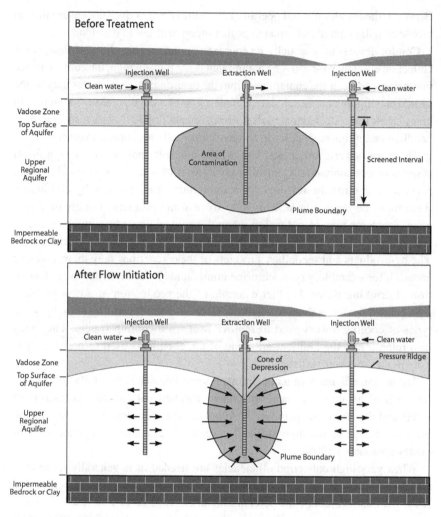

FIGURE 10–17. Treatment techniques for contaminated groundwater.

water supplies. Techniques are being developed to clean up locally contaminated groundwater supplies. An example is shown in Figure 10–17.

IMPACT OF CONTAMINANT CONTROLS
ON ENVIRONMENTAL QUALITY

The intended effect of controls, and generally their main impact, is to reduce or eliminate releases of toxic or otherwise harmful chemicals to the workroom air or general environment. To the extent that controls are successful, they are beneficial.

However, the installation and operation of control devices does not ensure that all problems will be totally eliminated or that others will fail to develop.

Control devices may actually be sources of contaminants. For example, if the collection efficiency for a given contaminant is not high enough, the control device may release some of the contaminant into the environment. A collector may be the source of a highly concentrated toxic material or of a large volume of material. In many cases, the collected materials pose significant problems with respect to their handling and ultimate disposal in an environmentally acceptable manner.

There are several other ways in which controls may serve as significant sources of contaminants. Controls may consist of a network of conduits that prevent or control the releases from numerous small sources by combining them into one large source. This may solve some problems but create larger ones. Problems may also develop when the control procedure involves chemical reactions for the purpose of contaminant degradation or neutralization. The by-products and secondary products of these reactions may themselves be harmful, for example, organochlorine compounds resulting from the chlorination of drinking water. Another example of the production of a new problem by a control device is the release of excessive moisture, forming fogs, by some types of cooling towers used to remove heat from cooling waters. They may also release into the ambient air some of the chemicals used to prevent algae growth on the tower walls.

The design efficiency of a contaminant-control device is generally selected on the basis of its ability to meet an effluent standard with an assured margin of safety and the cost of its purchase, instillation, and operation. These two considerations become increasingly difficult to reconcile as the desired collection efficiency approaches 100%.

When very high collection efficiencies are needed, it is generally more constructive to consider penetration rather than efficiency; the percent penetration (P) is the complement of the percent efficiency (E); that is, $P = 100 - E$. There may not seem to be a great difference between efficiencies of 90%, 95%, 98%, 99%, and 99.5%, but their significance becomes clearer when considered as a penetration of 10%, 5%, 2%, 1%, and 0.5%, respectively. Each becomes different from the other by a factor of approximately 2, and the increased cost of achieving each increment may be quite large.

The strengths of collection devices as contaminant sources vary directly as their penetration efficiencies. Thus, for example, a new power-plant electrostatic precipitator with a 99% collection efficiency that replaces one with a 90% efficiency will discharge 10 times less fly ash. While this may have a large effect on airborne total suspended particulate and some trace-metal concentrations, it will have relatively little impact on the ash-disposal problem. There would be only a 10% increase in collected fly ash and less than that for the combined fly-ash and bottom-ash mass.

In collectors that act as simple traps, where contaminants are removed from the effluent stream by physical processes such as filtration, electrostatic precipitation, scrubbing, adsorption, and condensation, the collected materials are not changed chemically, although in wet collectors they may be dissolved or suspended in water. On the other hand, in collectors utilizing chemical reactions to capture, oxidize, or neutralize the waste, the residual reaction products will differ in composition from the original materials. The major reaction products will usually be innocuous or relatively easy to handle. If this was not the case, the process would not have been selected in the first place. Problems may arise, however, from the presence of trace contaminants in the waste.

The problem of the release of trace contaminants may be exemplified by considering incineration, which is an effective means of reducing the weight and bulk of household refuse and waste paper and, to an increasing degree, may be favored as a means of resource recovery in terms of its heating value. However, mixed refuse may contain, for example, chemically treated papers; halogenated plastics; discarded batteries containing Pb, Hg, Ni, and Cd; and a host of other chemicals. These constituents may be vaporized and injected into the atmosphere along with the waste gases.

A classic example of the formation of a secondary contaminant in a control device involves the adoption of catalytic converters in motor vehicle exhaust systems. As discussed in chapter 3, no consideration was initially given to the oxidation of the SO_2 to SO_3 in the exhaust and the hydrolysis of the latter in the exhaust stream to sulfuric acid mist.

Disposal of Collected Materials and Process Sludges

A first consideration in disposing of waste materials is to reduce the volume of materials needing disposal. The first consideration should always be waste minimization through process changes that reduce waste generation. A second consideration is the option for recycling the waste materials to recover the economic values for productive usages.

The technology of resource recovery has advanced greatly in recent years. This is illustrated for materials recovery in Figure 10–18, and for energy recovery in Figure 10–19.

The ultimate and safe disposal of liquid and solid wastes and treatment plant residues is one of the most important current concerns of environmental protection and public health authorities. There are no easy, foolproof, or inexpensive solutions to most of the current problems.

Control devices such as incinerators, which destroy chemical contaminants, will produce relatively little residue material that requires disposal. On the other

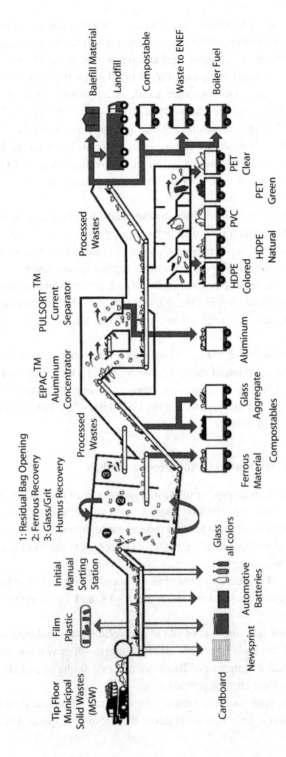

FIGURE 10-18. Schematic diagram of advanced technology for separation and recovery of materials in municipal solid wastes.

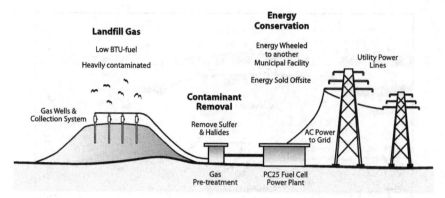

Landfill Gas
Low BTU-fuel
Heavily contaminated

Energy Conservation
Energy Wheeled to another Municipal Facility
Energy Sold Offsite

Utility Power Lines

Gas Wells & Collection System

Contaminant Removal
Remove Sulfer & Halides

AC Power to Grid

Gas Pre-treatment

PC25 Fuel Cell Power Plant

FIGURE 10–19. Schematic diagram of recovery and use of landfill gases. (*Source*: Adapted from EPA 600-R-97-008.)

hand, most control devices that extract chemicals from air or liquid streams, or react with them to form secondary chemicals, create difficult disposal problems. The wastes may be dry solids, concentrated suspensions of solids in water (slurries and sludges), or concentrated organic or aqueous solutions. These wastes may require further treatment or degradation before they can be disposed of in an environmentally acceptable manner.

Incineration

For combustible wastes, incineration has the advantages of relatively low cost and possibilities for heat recovery. The residual ash occupies less than 10% of the volume of the original waste and is usually a relatively innocuous, soil-like material. However, incineration can create smoke and odor nuisances. Even with nominally complete combustion, there will be unburned and noncombustible volatile contaminants in the exhaust, as previously discussed. Exhaust gas cleaners, when installed, reduce the cost advantages of incinerators and may not collect some of the materials of interest, for example, mercury vapor. Contaminants that escape into the atmosphere will eventually be deposited on land or surface waters, becoming available for uptake into the biosphere.

Land Disposal

Most solid refuse is disposed of in sanitary landfills. The basic construction of a modern landfill is illustrated in Figure 10–20. Each layer of refuse is compacted and covered with a layer of earth or clean fill. The earth cover absorbs odorous gases of decomposition and acts as a barrier to pests and vermin. Filled land

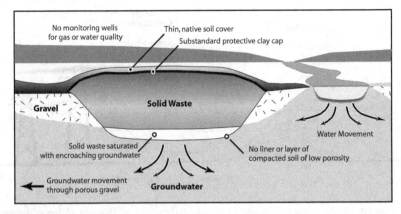

Modern Sanitary Landfall

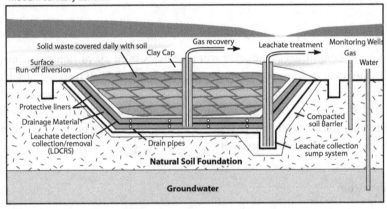

FIGURE 10–20. Schematic diagram of conventional (top) and more modern sanitary landfills. A major problem in the selection of suitable areas for landfill, and their successful operation, is the isolation from groundwater of the leachate solutions that form in the wastes. It is generally necessary to line the bottom of the landfill with an impervious layer of clay to control the drainage patterns, and to pump out the leachate for subsequent biological treatment.

gradually settles as the waste decomposes. Thus, these areas should not be used for residential or commercial purposes without special precautions, not only because of the settlement problem but because explosive concentrations of methane gas can build up in basement areas. Landfills in urban areas have usually been used for recreational purposes, such as parks and golf courses.

A major problem in the selection of suitable areas for landfill, and their successful operation, is the isolation from groundwater of the leachate solutions that form in the wastes. It is generally necessary to line the bottom of the landfill with an impervious layer of clay to control the drainage patterns and to pump out the leachate for subsequent biological treatment.

Underground Disposal

Some highly toxic and radioactive solid and liquid wastes are stored in natural underground caverns or in chambers excavated in selected geological strata such as salt domes. The sites are selected for their inaccessibility and isolation from groundwater and for their supposed geologic stability.

Large volumes of liquid wastes are disposed of by deep-well injection into underground strata. Wells are drilled into suitable permeable strata that are isolated from groundwater sources by impermeable strata and the well casing. A schematic diagram of a waste-liquid injection well is shown in Figure 10–21. Strata suitable for deep-well disposal are located in many parts of the US, primarily in the western parts of the country.

Problems occasionally develop due to the instability of the underground rock strata. Liquids pumped into certain strata at high pressures can act as lubricants, causing slippage of contact surfaces; this effect is manifested as earthquakes at the surface. The close correlation between the times of active pumping of wastes by

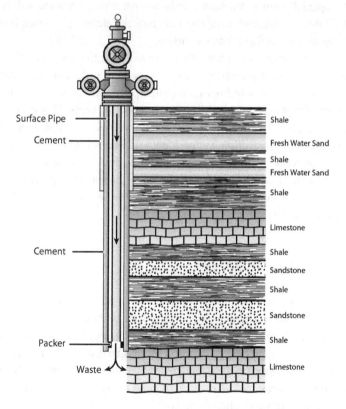

FIGURE 10–21. Deep-well disposal system. The well casing must be cemented in place to protect ground water.

the US Army Rocky Mountain Arsenal near Denver and the occurrence of earth-quakes in the Denver region in the mid-1960s led to discontinuance of deep-well disposal there.

Disposal in the Oceans

There are several different kinds of ocean disposal. One is the disposal of waste liquid or slurries into surface waters relatively close to shore. Huge volumes of sewage sludge and industrial acid wastes have been disposed of in this manner, with the implicit rationalization that the wastes were degraded and/or neutralized within the ocean and did not accumulate excessively or do permanent damage to marine life.

Near-shoreline marine disposal is also done using pipelines that extend for distances up to several miles out to sea. In this case, the discharge is at the lower, rather than the upper, surface of the water. Such systems have been used for disposal of raw sewage by coastal cities and may appear to be adequate initially. However, they are usually found wanting as population and sewage volume increase, especially when on-shore winds deposit floating sewage solids on the beaches. Problems may also arise when the pipe develops leaks, since inspection and maintenance are difficult and expensive.

Deep-ocean dumping is theoretically a safe and effective means of disposal for dangerous wastes. As discussed in chapter 4, vertical diffusivity in ocean water is extremely low and should provide an effective barrier against transfer to the surface water for periods of hundreds to thousands of years. However, diffusivity is the controlling factor only for dissolved materials; for PM, their effective density would be the controlling factor. Materials with low density, or particles of high-density materials with attached gas bubbles, could rise rapidly through the water. In addition, if the materials are ingested by marine organisms that migrate vertically or are consumed by predators who do, these chemicals can be readily transported toward the surface.

RISK REDUCTION OPTIONS

This chapter has emphasized the technical aspects of risk reduction, which is a major component of risk management. However, it should be remembered that risk management decisions will always also be influenced by other valid considerations. In practice, there are major information gap differences in risk management priorities within the political and governmental agency cultures that make it quite difficult to allocate political and economic resources when making choices among the various risk reduction options.

To illustrate these issues, consider ambient air pollution. When attempts are made to determine the extent of any health risk existing among the members of the population of concern resulting from the inhalation of airborne chemicals, information needed includes the distribution of the concentration of the agent in the air and, for airborne PM, the distribution of particle sizes and the unit risk factor (i.e., the number of cases and/or the extent of the adverse effects associated with a unit of exposure). For more sophisticated analyses, more information is also needed about the population of concern, such as the distribution of ages, preexisting diseases, predisposing factors for illness, such as cigarette smoking, dietary deficiencies or excesses, and so on.

When basic information on ambient levels and unit risks is available, it is relatively straightforward to compute, tabulate, and compare the risks associated with the different chemicals in community air. However, such direct comparisons can, in practice, only be made with any quantitative reality for a handful of chemicals, e.g., the criteria pollutants whose ambient air levels are routinely monitored and for which directly measured human exposure–response relationships have been developed. For hundreds of other airborne chemicals, known collectively as hazardous air pollutants (HAPs) or air toxics, there are neither extensive ambient air concentration data nor unit risk factors that do not intentionally err on the side of safety. This disparity has resulted from the different control philosophies built into the Clean Air Act and maintained by the EPA as a part of its regulatory strategy. The rationale for the distinction is that criteria pollutants come from numerous and widespread sources, have relatively uniform concentrations across an airshed, and require statewide and/or regional air inventories and control strategies for source categories (motor vehicles, space heating, power production, etc.) focused on the attainment of air quality standards (concentration limits) whose attainment provides protection to the public health with an adequate margin of safety. There is also a long history of routine, mostly daily measurements of criteria pollutant concentrations throughout the country.

By contrast, HAPs sources are far fewer in number and are considered to be definable point sources at fixed locations. Downwind concentrations are highly variable and generally drop rapidly with distance from the source due to dilution into cleaner, background air. The national emission standards for hazardous air pollutants are based on technologically based source controls and are intended to limit facility fence-line air concentrations to those that would not cause an adverse health effect to the most exposed individuals living at the fence-line. Also, until quite recently, there has been no program for routine measurements of air toxics in adjacent communities.

Most of the unit risk factors for air toxics are based on cancer as the health effect of primary concern. In these studies, and in studies to assess noncancer effects, the data are most often derived from controlled exposures in laboratory animals at maximally tolerated levels of exposure. The translation of the results of these studies to unit risk factors relevant to humans exposed at much lower levels in the environment is inherently uncertain and is approached conservatively, following the model pioneered for food and drug safety beginning in the 1930s by the Food and Drug Administration. The resulting unit risk factors are generally based on an assumption of no threshold and a linear extrapolation to zero risk at zero dose. They are generally described in terms of being 95% upper-bound confidence limits, but this descriptor is undoubtedly conservative in itself. When these conservative unit risk factors are used for the prediction of the consequences of human exposures, they are multiplied by estimates of predicted ambient air concentrations that are themselves, in the almost universal absence of air concentration measurements, almost certainly upper-bound estimates from pollutant dispersion models that apply to the most highly exposed individuals in the community.

The resulting estimates of health risk are therefore highly conservative upper-bound levels. Thus, they are inherently incompatible with population impacts estimated for the more widely dispersed criteria pollutants. The margins of safety for criteria pollutants are generally less than a factor of two, rather than the multiple orders of magnitude of safety factors built into the risk assessments for air toxics.

Comparative risk analysis, as currently practiced, has other inherent limitations as well. Even when there are available reliable estimates of the exposure-related numbers of cases of premature mortality; hospital admissions; other uses of medical, clinical, and pharmaceutical drug resources; lost time from work or school; reduced physiological and functional capacities; and daunting societal equity and valuation, challenges are faced in intercomparing numbers of incident cases of quite variable clinical severity and psychological impacts. For carcinogenic agents, it has become customary to expect regulations to be effective in limiting the risks of lifetime exposures to no more than one in 10,000 and often to less than one in 1 million. For other diseases that can also reduce lifespan, such as chronic obstructive pulmonary and cardiovascular disease, which also are exacerbated by air pollutant exposures, a much higher risk level has been considered acceptable by regulators and the public.

In summary, current abilities to determine residual risks of air toxics and/or to compare risks quantitatively are quite limited by key gaps in knowledge and by reliance on unvalidated predictive models for exposure and for dose–response. A major part of the problem is the existence of two very different cultures of risk assessment, namely for carcinogens and for other toxicants. Carcinogen risk assessments seldom have been based on relevant data on either low-dose exposure on human exposure–response data at concentrations anywhere near ambient levels. They require high-dose to low-dose extrapolations and generally animal to human extrapolations as well, often using unvalidated predictive models. In the face of such a high degree of uncertainty in the output of the models, conservative assumptions are used to ensure that potency and exposures are not underestimated. Thus, yields of risk estimates are often higher than the real risks. Such risk estimates cannot be fairly compared to the risks associated with criteria air pollutants, which are determined largely from the product of measured air pollutant concentrations and measured responses among humans exposed to either ambient air or, for short-term responses, to controlled exposures in chambers. Fair comparisons can only be done within the separate categories of pollutants.

Comparative risk assessment is an idea whose time is coming, and with appropriate research resources to harness the new technical approaches and sophisticated research tools now emerging to fill in key knowledge gaps, comparative risk assessment will be made more useful and feasible.

Options for reducing environmental risks to human health may include measures for intervening at any point in the sequence of steps typically involved in the process by which a potentially hazardous agent is produced, is released, transported through the environment, reaches a susceptible individual, is taken up by the individual, and subsequently gives rise to a reaction adversely affecting the health of the individual and/or his or her offspring (Fig. 10–22). Some options are not possible without action at the community level, whereas others lie within the power of the individual acting alone. All options, however, depend to varying degrees on understanding each of the risks in question and on having the skills needed to reduce them. Research and education in the relevant aspects of environmental health and safety are, therefore, essential for arriving at sound policies for risk reduction at the individual and the community levels.

At the community level, the options for risk reduction encompass a broad range of activities, including support of research for identifying potentially hazardous agents, elucidating their toxicity and modes of action, and defining their relevant dose–effect relationships; systematic monitoring of the extent to which individuals

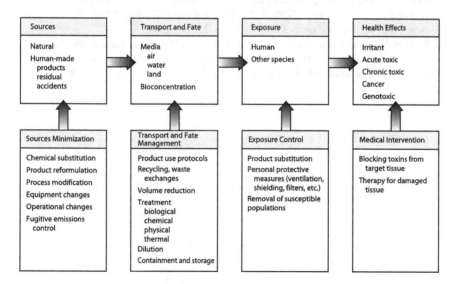

FIGURE 10–22. Summary of options for reducing environmental risks to human health.

or populations may be exposed to harmful agents via air, water, soil, food, or other media and proper assessment of the magnitude of any risks that may result from such exposures; identification of individuals or groups at unusually increased risk because of heightened susceptibility and/or level of exposure; formulation and enforcement of standards and regulations for limiting the exposure of individuals or populations to potentially harmful agents, along with engineering measures for controlling the production and/or release of potentially toxic agents planning ahead to cope with emergencies that may result from the accidental release of hazardous agents; maintaining in readiness the organizational capability needed to cope with environmental emergencies; mounting programs of public and professional education in environmental risk reduction (including information clearinghouses, workshops, telephone hot lines, Internet web pages, etc.).[1,2]

At the individual level, the options for risk reduction include staying abreast of relevant information received via the media and/or communications from health authorities, environmental protection agencies, and other sources; modifying one's own behavior, diet, and lifestyle to minimize risks to oneself and to others; carefully observing any special precautions that may be called for to protect oneself and one's fellow workers against hazardous agents in the workplace; and joining with others in efforts to promote collective awareness of environmental risks and to reduce such risks.[1]

REFERENCES

1. Griffith, R. and Saunders, P. Reducing environmental risk. In: R. Detels, W. Holland, J. McEwen, and G.S. Omenn (eds.), *Oxford Textbook of Public Health.* New York: Oxford University Press, 1997, pp. 1601–1620.
2. Omenn, G.S. and Faustman, E.M. Risk assessment, risk communication, and risk management. In: R. Detels, W. Holland, J. McEwen, and G.S. Omenn (eds.), *Oxford Textbook of Public Health.* New York: Oxford University Press, 1997, pp. 969–986.

11

Contaminant Criteria and Exposure Limits

Excessive levels of chemical contaminants can produce adverse health effects, as well as degradation of environmental quality. Since the extent and frequency of the effects vary with both the magnitude and duration of exposure, it follows that the effects can be eliminated, or at least reduced in frequency and severity, by reducing the exposure.

Eliminating exposures entirely can sometimes be accomplished by banning the use or consumption of a material, as was done by the US Food and Drug Administration (FDA) ban, in 1970, of the artificial sweetener cyclamate. Since cyclamate is not found in nature, human exposures gradually fell toward zero as inventories were depleted. The cyclamate ban was readily accepted by the public, in part because of the availability of an alternative artificial sweetener that was considered to be "safe," that is, saccharin. When a ban on saccharin was subsequently proposed by the FDA in 1977, there was a much greater reluctance on the part of the public to accept this action, and implementation was forestalled. On the other hand, the alternative to a ban on the use of a toxic material is a permissible or acceptable product content or an acceptable level of exposure. As examples, there are permissible levels, or "tolerances," for the carcinogen aflatoxin, which can be found in peanuts and peanut products; for coliform bacteria in dairy products; for rodent hair and feces in bakery products; and for mercury (Hg) in fish.

In practical terms, a contamination level is deemed to be acceptable as long as it produces no detectable adverse effects, and what are considered acceptable levels may change periodically to reflect a wider range of potential effects, as well as shifting societal values on what actually are acceptable effects.

Contaminant criteria and exposure limits are developed to ensure protection of public health and/or to prevent environmental degradation and are generally termed guidelines or standards. In broad terms, "guidelines" are nonbinding recommendations prepared by knowledgeable professionals to assist other professionals and public health authorities in evaluating the nature and extent of health risks associated with chemical exposures. They are an essential part of the risk assessment process (as discussed in chapter 8). By contrast, concentration limits for environmental media or emission limits for anthropogenic sources that generally have the force of law behind their enforcement are commonly known as "standards." Standards are generally established and enforced by governmental regulatory agencies. There are also, however, what are known as "consensus standards," such as those established by the International Standards Organization and those adopted by independent consensus organizations, such as the American National Standards Institute and/or affiliated national standards organizations, that have the force of law behind them only when they are also adopted by regulatory authorities. There are also published "standards" that are recommendations of professional societies. For example, the American Society of Heating, Refrigeration, and Air Conditioning Engineers has published guidelines for indoor air quality that it has called standards. Some of these are legally binding but only in those parts of the United States (US) that have included them in local codes.

In terms of recommended exposure limits on the international level, the lead agency is the World Health Organization (WHO), which has established guidelines. The purpose of these is to provide a basis for protecting public health and welfare from adverse effects of pollution and for eliminating, or reducing to a minimum, those contaminants that are known or likely to be hazardous to human health and well-being. While guidelines should provide information useful for eventual standard setting, their use is not restricted to this. In moving from guidelines to standards, prevailing exposure levels and environmental, social, economic, and cultural conditions can be taken into account. For example, WHO explicitly acknowledges that, in certain circumstances, there may be valid societal reasons to pursue policies that will result in pollutant concentrations above or below its guideline recommendations.

As noted, a true "standard" is generally a description of a level of pollution that is adopted by a regulatory authority as being enforceable. At its simplest, an environmental quality standard should be defined in terms of one or more concentrations and their associated averaging times. In addition, information on the form of exposure and monitoring, which are relevant in assessing compliance with the standard, and on methods of data analysis, quality assurance, and quality control requirements should also be parts of a standard.

In some countries, a standard is further qualified by defining an acceptable level of attainment or compliance. Levels of attainment may be defined in terms of the fundamental units of definition of the standard. Often, percentiles have been used; for example, if the unit defined by the standard for short-term exposure is the day, then a requirement for 99% compliance may allow three days' exceedance of the standard within a year.

The process by which contamination criteria and exposure limits are established and subsequently modified is inherently difficult, slow, and often contentious. There are generally conflicting forces at play, some with major economic consequences, and public health and environmental quality concerns. While often available, exposure–response relationships for biological effects are sometimes inadequate, and the standards-setting process, once begun, can seldom be delayed sufficiently to await the availability of more or better data. Laboratory-based studies of informative biological effects generally are expensive and time-consuming. Similarly, large-scale surveys of environmental quality or public-health parameters are frequently very expensive, even when the measurement technology is adequate. Often, successful field evaluations require the prior development of better measurement techniques, and such developments often take many years. Criteria and limits are, therefore, often heavily dependent upon "informed judgment"; they are, in the end, educated guesses by well-informed professionals who have considered and weighed the available data, as well as views of interested advocacy groups.

While tolerance limits for contaminants in food have been established in the US by the federal government for more than 75 years, it is only within the past 50 years that it has become heavily involved in setting standards in other aspects of the environment. Occupational exposure limits were initially proposed by professional society committees and/or by consensus-type voluntary standards-setting organizations. These recommendations were often incorporated into code limits by local governments. Air and water contaminant limits were also established by state and municipal governments. As a result, there were often different "standards" in different jurisdictions and relatively little real incentive to enforce those that existed. Standards-setting in the US is now largely done by the federal government and/or certain states, which have sometimes established more stringent standards than has the US government for specific contaminants.

BASES FOR ESTABLISHING CONTAMINANT CRITERIA AND EXPOSURE LIMITS

There are many possible bases for the establishment of contamination criteria and the setting of exposure limits. These include:

1. Epidemiological studies of populations of workers exposed in their occupations, or of community-level populations exposed through contamination of

food, drinking water, or community air. Such studies can provide statistical associations between contaminant levels and reported effects.

2. Toxicologic studies, that is, controlled studies of groups of animals intention-ally exposed in laboratory experiments where it is possible to define the doses, their frequency of application, and their route of administration. Such stud-ies can provide more complete information on metabolic pathways, storage depots, and the types and degrees of biological damage that the agents pro-duce, including information on whole-body effects.

3. Clinical studies of humans undergoing short-term exposures under controlled conditions.

4. Extrapolation of available epidemiological and toxicological data on other related materials to the material in question, based on their similarities in chemical structure and metabolism and perhaps their effects in simplified bio-logical test systems, such as bacteria and cell tissue cultures.

Existing contaminant limits were not all designed to protect human health. Lower limits may sometimes be needed to prevent health effects in domestic and wild animals and to limit welfare effects, such as economic disruptions and ecological impacts. For example, some materials, such as ethylene gas, which are essentially nontoxic to mammalian species, can produce severe damage to rooted plants and trees. Controls on the levels of aerosol mass have been based on the soiling properties of particles and the attendant costs of cleaning and earlier replacement of clothing, building materials, and so on. On a broader spatial scale, the effects of submicrometer-sized particles on atmospheric visibility, and the effects of aerosols and carbon dioxide on climate, may provide a basis for control-ling their airborne concentrations.

Epidemiology

The primary advantage of epidemiological data for establishing safe limits for human exposure is that they are not complicated by the uncertainties of inter-species variations, a limitation that is inherent in the interpretation of animal response data, as previously discussed. On the other hand, the number of materi-als whose standards can be set on the basis of measured adverse health effects in human populations is, fortunately, quite limited. Intentional exposures of humans to known or potentially toxic materials are severely constrained by ethical and legal considerations; therefore, controlled studies in animals are generally needed in place of, or to supplement, available human data.

The application of available epidemiological data in the standards-setting pro-cess is also often limited by either poor quality or other inherent failings of many of the studies. This is less a reflection on the epidemiologists than on the quantity and quality of the information that is generally available to them. As discussed in chapter 6, the populations at risk may be hard to define in terms of their number; ages; ethnic and educational backgrounds; prior smoking habits; occupational and residential histories; past and current consumption of alcohol or drugs; past and current dietary practices; or the number, extent, and severity of their prior diseases and medical treatments. These factors can affect the reported incidence of many diseases of nonspecific etiology and may affect the recognition of a disease frequency being influenced by the environmental factor in question. One of the most important and difficult tasks in designing an epidemiologic study is the selection of a suitable control population. Ideally, it should share all of the pertinent characteristics of the study population with the single exception of the chemical exposure of interest.

Another major problem is the characterization of the health endpoint and its adversity. While death is a generally reliable and easily definable statistic, its cause(s) may be more difficult to ascertain. At a minimum, it is necessary to locate the place of death and have access to the death certificate to determine the primary and secondary causes. However, the reporting physician may not have entered the secondary causes, and these may be the most informative for the study in question. For example, a sandblaster with advanced silicosis may die of a heart attack; but it is important to know that the lung disease was present, since it likely was contributory to the ultimate and reported cause of death. The influence of other contributory factors, such as cigarette smoking, on mortality are more difficult to obtain than are data on the death certificate, and few investigations have the resources to do so.

Morbidity statistics have additional complications. There is often a lack of consistency in the criteria for defining cases of disease and/or dysfunction, and data collected for other purposes are generally not considered to be useful. While physicians must report deaths of patients in their care, they are under no obligation to report cases of most nonfatal diseases. Furthermore, they tend to differ on the severity of symptoms needed to classify an individual as "diseased." In addition to the individual variability in the reporting of disease in a given geographic area, there are also regional and national differences in reporting criteria.

The utility of both mortality and morbidity studies is often limited by the absence or poor quality of the data on the exposure of the population of interest and/or of the control population to the agent under study. In some cases, no environmental measurement data are available at all. In others, measurement data are available for the recent past but not for exposures in earlier times when exposures may have been much greater. In any case, the environmental parameters measured may not be the most appropriate ones. In many community pollution problems, the evidence for health effects may be an increase in the incidence of a chronic nonspecific disease and the specific causative agent, if any, may not be known. Finally, even if community-level measurements are made on the appropriate chemical species and are accurate, they still may not be indicative of the extent and range of the exposures within the overall community of interest. For example, the relation between the concentration of an air contaminant as measured on the roof of a building and the actual concentration in the breathing zones of various individuals in the community can be uncertain and highly variable. There are spatial variations within the outdoor environment, and there may be large differences between indoor and outdoor concentrations because of differences in air exchange rates, particle sizes, chemical reactions between infiltrating pollutants, indoor surfaces, and other airborne chemicals.

Epidemiologic studies on community air and water contaminants are generally not definitive with respect to actual cause and effect and only rarely establish reliable exposure–response relationships. This should not be surprising, considering the extraordinary difficulty of characterizing population exposures and the fact that many diseases associated with exposure to chemical contaminants can also be caused or exacerbated by smoking and other personal risk factors.

Epidemiologic studies of working populations can more frequently establish cause and effect and exposure–response, because industrial exposures to chemical agents may be much higher than community exposures. They are, therefore, more likely to produce clear-cut symptoms and/or lesions, including many not commonly seen in the general population. Also, the populations exposed are more easily defined, and their exposure levels and intervals can be more easily determined. Furthermore, supplementary indications of exposure, and of biological effects, can frequently be obtained by physiological testing and collection and analysis of samples of, for example, blood, urine, hair, and exhaled air. Information may also be available on other factors that affect their response, for example, smoking histories, preexisting diseases, residential locations, ethnic backgrounds, and so on.

The major problems with epidemiologic data on occupational groups are generally the relatively small population sizes at risk for any given exposure and the reluctance of most employers to permit publication of data that might increase their legal and financial liabilities. The problem of limited population size can sometimes be overcome by establishing industry-wide studies. In any case, the human health effects data that are available on occupationally exposed populations are of considerable value in establishing exposure limits for workers. Fortunately, there are not very many materials that have adversely affected enough workers, and, therefore, most occupational exposure limits have been based on studies in experimental animals.

Human health effects data from occupational exposures are, however, of limited value in establishing safe limits for general community exposures. Community exposure limits are almost always much lower than occupational exposure levels for a variety of reasons. The most important of these are discussed next.

Intermittent versus continuous exposure

Employee groups are exposed during working hours and generally have at least 16 hours per day of essentially negligible exposure outside of their workday exposures. During this time, some of the accumulated dose can be eliminated and other fractions can be neutralized or immobilized. Community exposures to some contaminants can, on the other hand, be essentially continuous.

Selected versus total population

Working populations are inherently more resistant to environmental stresses since they do not include most of the more vulnerable segments of the overall population, such as young children, the aged and infirm, and those people with chronic diseases or disabilities too severe to permit them to engage in full-time work.

Voluntary versus involuntary exposure

While working populations are entitled, by right and by law, to safe and healthy working environments, there will always be some risk of accidents and excessive chemical exposures due to unknown or unanticipated events and toxicities of materials. In accepting employment, each employee implicitly accepts some risks. The people in the surrounding community, on the other hand, seldom obtain direct benefits from contaminants emitted into their air and water and therefore are much less willing to accept any significant risk. At most, the risk should be proportionate to the indirect benefits of the activities generating the contaminant releases, for example, a healthy local economy and minimal rises in the cost of living.

To these one may add an additional practical reason for differences between occupational and community exposures for the same chemical: the probability of a detectable incidence of measurable effects. As indicated earlier, industrial populations exposed to a particular chemical under defined conditions are generally relatively small and have seldom exceeded a few thousand. Thus, considering population variations, conditions that elevate the number of cases of, for example, emphysema, lung cancer, liver cirrhosis, or heart disease by one in 1,000 will seldom, if ever, be detected. On the other hand, an increase in the incidence of a significant disease by one in 1,000 in the total US population would result in more than about 300,000 excess cases and would be considered a major public-health crisis.

Toxicology

A major advantage of toxicological data for establishing safe exposure limits is that the experimental design and protocols are set by the investigator and the number of uncontrolled variables can be minimized. Another is that the lifespans of most test species are relatively short and life-term studies, such as required for assessing carcinogenic potential of a chemical, can be conducted within a reasonable time frame. Other advantages are that statistical uncertainties can be reduced by using large numbers of animals in each test group, and inherent interindividual variability can be reduced by using inbred strains of test animals of one gender and of essentially the same age. The effects of extraneous environmental factors can be limited by providing a diet compatible with the study objective and one constant over time, maintaining a uniform temperature and humidity in the housing and dosing facilities, and having a clean animal facility that is free of infectious agents and other contaminants. The exposures can be made uniform for each test group with respect to the manner and amount of administered dose to each individual animal and with respect to the number of doses and the intervals, if any, between them, and the effects of the exposure protocols themselves can be compensated for by giving the control group "sham exposures," in which the animals are handled under the same conditions as the experimental group with the exception of exposure to the chemical of interest following which these animals undergo the same effects assays as the exposure group.

To the extent that an animal test system is idealized, it can become somewhat unrealistic as a model for human exposure. Many of the effects that environmental agents produce in humans are seen primarily in people who have preexisting diseases and/or physiological dysfunctions. While there are animal models available that mimic some human diseases, the extent of the correspondence is often debatable. In any case, purebred strains can have very different responses or sensitivities to some chemicals than do other strains in the same species, and there can be substantial further variations among the common animal test species. There are well-established reasons for much of this interspecies variability. Some materials are not very toxic themselves but undergo biotransformation into more toxic chemicals, as noted in chapter 6. The enzymes that catalyze these transformations, and hence the products of metabolism, may vary substantially among species. Thus, even with animal test data that are consistent and reliable, the degree of confidence with which they can be extrapolated to indicate the potential effects in humans who have a broad range of genetic variations is often limited. Therefore, it is generally desirable to have data on several different species and to have some other information to indicate which species most closely resemble humans in terms of metabolism of the chemicals in question.

While the absence of observed effects does not necessarily mean that there was no toxicity, and animal tests can never provide absolute certainty that a chemical is safe for its intended uses, animal testing does provide an essential screening step for any chemical for which either groups of workers or the general public can be expected to receive significant exposure.

Chemical contaminants in the environment can exist in many possible molecular and ionic forms, and many form isomers with different geometric configurations; these differences can have major effects on the expression of toxicity. It is obviously impossible to perform toxicity tests on all of the forms of a chemical in the environment. Thus, frequent use is made of the available toxicity data, in conjunction with available information on the effects of physical form and structure as modifiers of biological effects, to develop estimates of acceptable levels of those forms not tested, as discussed in chapter 6. Thus, many concentration limits for air, water, and food may appear to be specific but, in fact, are actually generic. This is especially true for the metals. The toxicity may vary considerably among the various compounds, but the only distinction made, if any, will be between "soluble" and "insoluble" forms, or between organic and inorganic compounds of metals. In such cases, experimental data on one or several compounds have, in effect, been extrapolated to cover related materials.

SETTING CONTAMINANT AND EMISSION LIMITS

The ultimate goal of chemical contamination criteria and exposure limits is to protect people and other environmental receptors from exposures to chemicals that can produce adverse effects. Exposures of concern can take place via one dominant pathway, such as inhalation, ingestion, or surface deposition, or by a combination of pathways, and one limitation of current environmental quality criteria is that they are generally based on protecting excessive exposures via only a single pathway.

Concentration limits established by the US Environmental Protection Agency (EPA) to control atmosphere contamination that are directly related to human health are known as "primary standards." Those that are based upon recognized adverse effects on animals, vegetation, or building materials; economic losses; or evidence of aesthetic degradation of surface air or water are considered to be effects upon "public welfare" and are known as "secondary standards."

In some cases, the evidence used to establish secondary standards is better defined and more susceptible to realistic cost-accounting than that used to establish standards based on human-health effects. It is not as difficult or time-consuming to estimate the impact of air contaminants on the soiling of clothing, home furnishings, and building materials, on the reduction in their useful lives, and on their maintenance and replacement costs. Similarly, direct economic impacts due to damage to crops and livestock can readily be calculated. On the other hand, there are some effects whose costs are more intangible. These include, for example, the limitation of tree and ornamental plant species that can be grown in contaminated airsheds and the limitation or alteration of aquatic species that live in contaminated surface waters. An example of intangible costs that, however poorly defined, are deemed unacceptable by many are effects on the survival of natural species. The successful pressures to ban general usage of DDT in the US were based largely on its effects on the viability of the eggs of certain wild birds. The potential of DDT and its metabolites to produce human-health effects were, and still are, quite speculative.

Excessive exposures can be controlled by enforcing concentration limits in environmental media (i.e., air, water, soil or food) by enforcing emission limits into ambient air and surface waters and onto land or by a combination of both. Concentration limits are generally most appropriate for monitoring exposures to chemicals arising from multiple and widespread sources, often including natural background sources, while emission limits are generally most appropriate for identifiable point sources that can produce important local impacts downwind or downstream.

When concentration limits are exceeded, control agencies generally need to conduct an emissions inventory in order to determine which specific sources or source categories are causing the elevated exposures and then to impose emission restrictions on those sources or source categories that are responsible for the excess exposures and their effects. When one source is dominant, for example, fluoride contamination downwind of a fertilizer manufacturing plant, that source may be held accountable for whatever problems result from the contamination. Typically, the ambient contaminant level is due to many sources, and excessive levels can only be effectively reduced by an across-the-board reduction in emissions by all sources or by eliminating, or greatly reducing, the strengths of a limited number of major sources.

The attainment of reduced exposure levels through effective control of emissions requires the establishment of rational and enforceable emission limits. The first step is generally an inventory of sources and their strengths. It is also important to know the location of the discharge point and the temporal pattern of emissions, since the latter affects contaminant dispersion. Other important factors are the technological and economic feasibility of the application of controls. Finally, it is important that there be some positive incentives for the installation and maintenance of controls and/or penalties for not installing them or maintaining their effectiveness. For example, the effectiveness of the manufacturer-supplied motor-vehicle emission controls has often been considerably less than their potential because of improper maintenance and some deliberate disconnections and alterations of the system components. In some states, there are no requirements for mandatory inspection or maintenance of these controls. Other types of emission controls have been more effective. These include those for lead (Pb), achieved through the phased elimination of the Pb content of gasoline, and those for sulfur dioxide (SO_2), achieved through reductions in the sulfur (S) content of fossil fuels used in boilers and motor vehicles. In practice, it is much easier to obtain compliance of standards from a relatively small number of suppliers than from a relatively large number of consumers.

Exposure levels have considerable temporal variability. Some effects are determined primarily by cumulative exposure, while others are influenced more by peak levels of exposure. As a result, several different kinds of exposure limits are needed, some based on overall exposure and others based on rate of exposure. Another basic factor affecting the choice of a numerical standard is the degree of protection that is desired. As discussed previously, much lower concentration limits are generally set for general population exposures than for occupational exposures. Within the occupational environment, it may be desirable to have relatively high emergency limits as well as those for routine exposures. Such emergency limits are usually based on brief single exposures at levels that may be expected to produce clearly demonstrable, but largely reversible, adverse effects. The rationale for accepting such effects is that it permits actions that can prevent more serious consequences. These actions include rescue of injured or unconscious workers, fire-fighting, and gaining access to equipment or controls in order to halt release of toxic materials.

Occupational Guidelines and Standards

In the first half of the twentieth century the greatest need for exposure limits for chemical contaminants was in occupational health protection, where clear-cut cases of chemical intoxication and chronic diseases were quite common. In 1940, the American Standards Association, a consensus organization that later changed its name to American National Standards Institute, formulated allowable air concentrations of carbon monoxide (CO), Pb, and various solvent vapors. Its standards currently are known as maximum acceptable concentrations (MACs) and represent upper bounds on the excursions of air concentrations at industrial operations.

An alternate source of occupational exposure limits is the American Conference of Governmental Industrial Hygienists (ACGIH), which has published threshold limit values (TLVs) for airborne substances since 1947. The first list contained approximately 140 listings, while the 2016 list has more than 800.[1] The lists also contain further guidance on occupational exposures to carcinogens, substances of variable composition, mixtures, nuisance particles, and asphyxiants. Special consideration was given to some TLVs by the added notation "skin." These substances are those for which there is a substantial potential contribution to the overall exposure by the percutaneous route, including mucous membranes and the eye, either via air or, more particularly, by direct contact. This designation was intended to suggest appropriate measures for the prevention of cutaneous absorption so that the threshold air concentration limit is not invalidated.

Originally, all TLVs were defined as time-weighted average (TWA) concentrations over the course of the workday, representing conditions under which it was believed that nearly all workers may be repeatedly exposed daily without any adverse effect. Later, some listings were given a "C," or ceiling, a notation that, in effect, makes them equivalent to a MAC (i.e., an absolute upper limiting value below which all concentrations should fluctuate). Ceiling values apply to chemicals having an immediate response, but for other compounds where the response develops relatively slowly the TWA concentrations are more appropriate. The ACGIH has also provided guidelines for modified TLVs for work schedules that depart significantly from the standardized eight hours per day, five days per week schedule.[2-5]

In 1976, the ACGIH introduced the concept of a short-term exposure limit (STEL) to supplement the TWA TLVs. The STEL values were defined as maximal concentrations to which workers can be exposed for a period up to 15 minutes continuously without suffering from irritation, chronic or irreversible tissue damage, or narcosis of sufficient degree to increase accident proneness, impair self-rescue, or materially reduce work efficiency. The proviso is that no more than four excursions per day are permitted, specifying at least 60 minutes between excursion periods and that the daily TLV TWA is not exceeded. The STEL value is not intended to be used as an engineering design criterion or considered as an emergency exposure level.

While enforcement of airborne concentration limits is generally the best approach to control of occupational overexposures, there are many situations in which it is difficult to characterize airborne exposure and in which routes other than inhalation are important to the overall exposure. This is frequently the case, for example, for maintenance workers, whose irregular schedules and work locations may be difficult to characterize or track. Among the other means that may be useful in determining the extent of exposure are biological exposure indices (BEIs), also published by the ACGIH. These values represent limiting amounts of substances to which workers may be exposed without hazard to their health or well-being. The biologic measurements on which the BEIs are based can furnish two kinds of information useful in the control of worker exposure, namely a measure of the individual worker's overall exposure or a measure of the worker's response.

Measurements of response can furnish an estimate of the health status of the worker and can involve changes in the amount of some critical biochemical constituent, such as activity of a critical enzyme, or changes in some physiologic function. Measurement of exposure may be made by determining the amount of a substance to which the worker was exposed via analysis of blood, urine, hair, nails, and other body tissues and fluids; determination of the amount of the metabolite(s) of the substance in tissues and fluids; or determination of the amount of the substance in exhaled breath. BEIs may be used as an adjunct to the TLVs or in place of them. The BEIs, and their associated procedures for determining compliance, should thus be regarded as an effective means of providing health surveillance of workers.

TLVs and MACs are stated as numbers with one or, at most, one and one-half significant figures, since the epidemiological and/or toxicological database is generally inadequate, especially in view of inherent human intersubject variability, to establish more precise limits. Thus, they were not intended to provide fine lines separating safety and hazard or to provide a basis for regulatory control. However, by the late 1960s, most states had established codes for permissible occupational exposure levels based on the TLVs, primarily because numerical standards were needed for code enforcement and no other reasonably comprehensive and authoritative recommendations were available. However, states still varied considerably in permissible limits because their codes were based on the limits recommended in the years preceding the adoption of the code and the frequency of periodic updating varied considerably among the states. Since the TLV list was continually updated and expanded, many states were left with outdated exposure limits. Some of the confusion of varying state exposure limits was eliminated by the Occupational Safety and Health Act of 1970, which directed the federal government to promulgate uniform standards. The process began with adoption of interim standards, and 22 MACs were used as such. Where TLVs existed but not MACs, as was the case for approximately 280 other materials, the 1968 TLVs were adopted as interim standards. The interim standards were to be replaced, in "due" time, by permanent standards.

TLVs have been used as a pattern for standards-setting by many countries throughout the world. Other countries, especially those in the European Union, have adopted *maximale arbeitsplatz-konzentration* (MAKs), values developed in Germany by the Commission for the Investigation of Health Hazards of Chemical Compounds in the Work Area. The TLVs and MAKs differ in some respects but are in general agreement in most respects. However, exposure standards that were developed in the Soviet Union and in the countries that followed its lead were quite different, and frequently much lower, than the corresponding TLVs or MAKs. The Soviet standards, which are all ceiling values, are based primarily on neurophysiological responses in experimental animals and on behavioral responses in humans and are set at a no response level. On the other hand, these standards have not, in practice, been treated as limits to be enforced but rather as goals and design criteria for new facilities. Discussions of occupational health standards used in various countries are available.[6-8]

Air Contamination Guidelines and Standards

The air pollution episode that occurred in Donora, Pennsylvania, in 1948 was found, in retrospect, to have resulted in an estimated 20 deaths and in other health effects among 43% of a total population of 13,800. The team of engineers and physicians dispatched by the US Public Health Service (PHS) to investigate this disaster was drawn from the staff of its Division of Occupational Health. At that time there was no ambient air pollution group in the PHS, indicative of the fact that air pollution was not yet considered a significant health problem at the national level.

Air contamination was still considered to be a local concern and more of a nuisance problem than a health hazard even by the mid-twentieth century. In eastern and midwestern US cities, it was primarily a soot problem, involving soiling and visibility; in Los Angeles, it was causing visibility issues and eye irritation. In both cases, contaminants in the air were causing the effects, but it was not at all clear what the specific offending chemicals were or what the threshold levels were for their effects.

The only community air contamination standard that was in widespread use at the time was an emission standard for black smoke from point sources. This was the Ringelmann chart, named after the French engineer who developed it in 1895. It consisted of a row of squares having 20%, 40%, 60%, and 80% in coverage with black ink (Fig. 11–1).

The Ringelmann chart was simple, inexpensive, easy to use, and unambiguous in interpretation. A smoke inspector would hold it up while looking at a smokestack. All one had to do was decide which square was closest in blackness to the plume from the stack. Ringelmann charts were still used as enforcement tools into the late 1970s.

Another index of air pollution that had widespread use during that time was dustfall. Dustfall is the weight of the particles that fall into an open pot during one month, normalized for the cross-sectional area of the pot entrance. The results were reported in terms of tons/square mile/month and were useful in gauging trends, provided that there were no transient sources, like construction activities, and no significant changes in airflow patterns around the sampling site. Dustfall is an index of very large contaminant particles but is useless as an index of health risk; particles small enough to be inhaled do not fall into the pot.

Since air contamination was considered to be a local problem, standards for contaminant levels that did exist were established for local jurisdictions such as cities, counties, and states, and different jurisdictions regulated different air

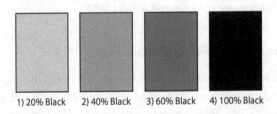

1) 20% Black   2) 40% Black   3) 60% Black   4) 100% Black

FIGURE 11–1. Modified Ringelmann's scale for grading smoke density.

contaminants and often had different standards for the same contaminants. The most frequently regulated contaminants in this regard were black smoke, CO, and SO_2.

With the passage of the Air Quality Act of 1967, air pollution control in the US became a federal responsibility to be shared with states, which developed control programs meeting federal standards. Since contaminant problems were not confined to single political jurisdictions, provision was made to establish air quality control regions based, in part, on airshed configurations. The governors of states with designated air quality control regions within their boundaries had 90 days to notify the federal government that they intended to adopt air quality standards for the contaminants and then had 180 days to hold public hearings and adopt standards and another 180 days to adopt plans for enforcement. The standards adopted by the states were subject to review and approval, and in the event no local standard was approved, the federal government had the right to impose one.

One provision of the Air Quality Act of 1967 mandated development of "criteria and control documents" for certain pollutants. The first of these documents were issued in 1969 and 1970 for particulate matter (PM), sulfur oxides (SO_x), CO, photochemical oxidants, hydrocarbons, and nitrogen oxides (NO_x) These chemicals were designated as "criteria" air pollutants and were the basis for the newly established US EPA in setting, in 1971, the initial suite of National Ambient Air Quality Standards (NAAQS). Since then, Pb was designated as a criteria air pollutant, and hydrocarbons were removed from this category. Table 11–1 lists the current NAAQS pollutants. For certain other air contaminants called toxic air pollutants, or air toxics (primarily substances that were carcinogenic), the establishment of standards has been difficult. A primary question is whether there are threshold concentrations below which they have no biological effects and, if not, how much risk to the public can be considered acceptable.

In addition to primary responsibility for the enforcement of federal air pollution legislation, the EPA has responsibility for controlling existing mobile or stationary sources of contaminants to bring air quality to levels defined by the NAAQS and to set national emission standards for new or existing hazardous air pollutants for which ambient air quality standards are not applicable, for example, asbestos, beryllium (Be), and Hg, and to set nationwide performance standards for new or modified stationary air contaminant sources so as to prevent the general occurrence of new air contamination problems by requiring the installation of the best controls during initial construction, when the installation of such controls is the least expensive. The new standards were not, however, to be applied to existing sources.

The performance standard was defined as "a standard for emissions of air pollutants which reflects the degree of emission limitation achievable through the application of the best system of emission reduction which (taking into account the cost of achieving such reduction) . . . has been adequately demonstrated." A new source

TABLE 11–1. National Ambient Air Quality Standards in Effect as of December 2016

POLLUTANT	PRIMARY (HEALTH RELATED)		SECONDARY (WELFARE RELATED)	
	TYPE OF AVERAGE	STANDARD LEVEL CONCENTRATION[a]	TYPE OF AVERAGE	STANDARD LEVEL CONCENTRATION
CO	8-hour[b]	9 ppm (10 mg/m^3)	No secondary standard	—
	1-hour[b]	35 ppm (40 mg/m^3)	No secondary standard	—
Pb	Maximum quarterly average	0.5 μg/m^3	Same as primary standard	Same as primary standard
NO$_2$	Annual arithmetic mean	0.053 ppm (100 μg/m^3)	Same as primary standard	Same as primary standard
O$_3$	8-hour[d]	0.070 ppm	Same as primary standard	Same as primary standard
PM$_{10}$	24-hour[e]	150 μg/m^3	Same as primary standard	Same as primary standard
PM$_{2.5}$	Annual arithmetic mean[f]	12.0 μg/m^3	Same as primary standard	15.0 μg/m^3
	24-hour[g]	35 μg/m^3	Same as primary standard	Same as primary standard
SO$_2$	1-hour[b]	0.075 ppm	3-hour	0.5 ppm

[a]Parenthetical value is an approximately equivalent concentration.
[b]Not to be exceeded more than once per year.
[c]Not to be exceeded more than once per year on average.
[d]3-year average of annual fourth highest concentration.
[e]The preexisting form is exceedance-based. The revised form is the 99th percentile.
[f]Spatially averaged over designated monitors.
[g]The form is the 98th percentile.
Source: Section 40 Code of the Federal Register, Part 50.

was defined as "any stationary source, the construction or modification of which is commenced after the publication of proposed regulations for that source type." Modification is defined as "any physical change in the method of operation of a stationary source which increases the amount of any air contaminant emitted by the source or which results in the emission of any air contaminant not previously emitted." Examples of the source categories for which the EPA has promulgated standards of performance are fossil fuel fired steam generators, municipal incinerators, Portland cement plants, nitric acid plants, and sulfuric acid plants.

Source controls for CO, NO_x, and hydrocarbons were initially focused on the internal combustion engine and current emission standards, expressed as g/mi, exist for NOx, CO, formaldehyde, PM, and nonmethane organic gases. The values have been reduced over the years since their promulgation due to various engineering improvements in engines, including evaporative controls, positive crankcase ventilation, exhaust-gas recirculation, and the addition of oxidation catalysts on the exhaust line. In 1999, the EPA mandated that the S content of motor vehicle fuels be drastically reduced.

Amendments to the Clean Air Act (CAA) in 1977 clarified several issues of concern not specifically addressed in earlier legislation. One was the prevention of significant deterioration in areas that were cleaner than the NAAQS. Areas with the purest air were designated as Class I; this designation was mandatory for national wilderness sites. Areas where the air was not as pure as the Class I regions but was cleaner than national standards were classified as Class II. Allowable contaminant levels were highest in Class III areas. A state may reclassify any area other than a mandatory Class I area by following a procedure set out in the CAA amendments.

The CAA also provided for limited allowable increments, which were the permissible increases in contaminant levels in any Class I, II, or III area. The smallest increments are allowed in Class I, the next largest in Class II, and the largest increments are allowed in Class III areas. However, while a variance above the established Class I increment can be granted by a state governor—up to 8% above the allowable increment for low-terrain areas and 15% for high-terrain areas—the President of the US is the arbitrator regarding approval of a variance in cases where there is a disagreement between the state and federal land managers.

Another area of concern addressed by the 1977 CAA amendments was the leeway granted to the EPA when air quality goals were not attained. The amendments endorsed the EPA's "offset" policy for new or modified major sources of air contaminants in areas that do not meet air quality standards. The offset policy allows new development if the net effect is an improvement in overall air quality due to decreases from other sources. However, there was a provision for waivers of offset requirements where the state has an adequate program for incremental reductions in emissions that would assure attainment of the standards by the deadlines, namely 1982 for contaminants other than those that were automobile related and 1987 for those that were not auto related.

The EPA is required to apply different levels of stringency to airborne emissions in areas that meet the existing NAAQS (attainment areas) versus those that do not (nonattainment areas). Furthermore, to ensure that pollution would not increase in attainment areas, new stationary sources in such areas are required to use the best available control technology. Furthermore, the EPA is required to take into account the costs of compliance. Existing sources in nonattainment areas are required to use "reasonably available control technology," which represents a lesser level of control that can be achieved at a lower cost.

A later amendment to the CAA, in 1990. required the separation of emission standards into several classes. These included risk-based standards designed to protect public health, technology-based standards requiring application of various levels of control technology, and technology-forcing standards designed to ensure that industry develop and apply the very best control technology. In many respects, these changes amplified the requirements mandated under the 1970 amendments. At the same time, Congress mandated the regulation of 189 (now 187) air toxics.

The 1990 CAA amendments also mandated further reductions in motor vehicle emissions, as well as a 50% reduction in power plant emissions of SO_2 and NO_x; called for the establishment of a new permit system consolidating all applicable emission control requirements; and mandated a production phaseout by the year 2000 of the five most destructive ozone-depleting chemicals.

While the traditional approach to mobile sources has been to require that pollution controls be incorporated by the manufacturer into the production process, the primary approach for limiting airborne emissions from stationary sources, for example major manufacturing facilities, has been to apply a combination of controls and to enforce the requirements through the granting of operating permits. To assure successful control, each state must develop an implementation plan describing how the federally specified standards will be met.

Among the new approaches incorporated into the 1990 CAA amendments was a provision that permits the buying and selling of air pollution emission allowances. The goal was to encourage those industries that can remove pollutants at minimal cost to sell their polluting allowances to industries whose costs are higher. Setting a limit on the total amount of pollution that can be released enables companies to trade their emission allowances at market prices. The market for SO_2 allowances has already worked very well. As discussed in chapter 8, reduced emissions have been achieved at a much lower overall cost than using any other approach. Economic considerations also led Congress to mandate cost–benefit analyses under the 1990 CAA amendments. A retrospective cost–benefit analysis for the CAA for 1970 to 1990 was completed; findings of the analysis are discussed in chapter 12.

Water Quality Standards and Guidelines

Contaminant criteria for water have been more narrowly focused than those for air, which is reasonable considering the basic differences in contaminant dispersion within the two media. The atmosphere is one large and continuous mantle, whose motion distributes contaminants released into it in all directions. Surface water, on the other hand, flows only downhill and generally within confined channels. Thus, water supplies used for drinking water or other specific purposes can be selected on the basis of their freedom from excessive contamination or can be purified to specified quality criteria prior to delivery for their desired use.

Historically, there have been three types of standards established by public agencies for the maintenance of environmental water quality: drinking-water standards, surface-water standards, and effluent-quality regulations. There are also specific industry standards that define the quality factors required of water supplies for cooling and process operations. In the US, standards for toxic chemicals in drinking water in the states are based upon recommendations of the federal government. The initial US standards promulgated in 1975, which also included limits for a variety of chlorinated hydrocarbon pesticides, were set under the Safe Drinking Water Act (SDWA) of 1974. This act was also designed to protect underground sources of drinking water by prohibiting wastewater disposal in areas that rely upon aquifers as the principal drinking-water source.

Under the SDWA, the EPA established national drinking-water quality standards, including the specification of maximum contaminant levels (MCLs) for specific substances in water. The MCLs were set at a level to prevent known or anticipated adverse health effects. Some limits are risk based, and others are technology based. The EPA may either establish a maximum contaminant level for a specific substance or prescribe a technique for its control. In the former case, the limiting technology may be the ability to detect the contaminant in the water. When the technology for monitoring the level of a contaminant at the required sensitivity is readily available, establishment of an MCL is generally the preferred approach. Since the feasibility of achieving a specified MCL, or of implementing a given treatment, will change with advancing technology, the act requires the EPA to revise and update the regulations on a continuing basis.

Also reflected in the 1977 SDWA amendments were concerns about the potential health effects of by-products that occur in drinking water as a result of the use of disinfectants, such as chlorine. The EPA was directed to study the reactions of chlorine with humic acid, a commonly occurring naturally in surface waters, and to evaluate the potential health effects, including any possible carcinogenic nature, of the new chemical products that result. The SDWA was further amended in 1986 and 1996 to specify additional contaminants to be regulated and acceptable treatment techniques for each contaminant. The amendments required disinfection of all drinking-water supplies, prohibited the use of Pb products in drinking-water conveyances, and emphasized the need for protection of groundwater sources. A major stimulus for the 1996 amendments was the recognition that many water suppliers were not analyzing for some of the emerging biological contaminants, such as *Cryptosporidium*. The new amendments required that such analyses be performed and that consumers be provided with data on the concentrations of such contaminants in their water supplies.

As noted, water quality criteria define acceptable levels of contaminants for the maintenance of specified water uses. While maximum levels are set by the federal government for some chemicals in water that is to be used for irrigation, standards for surface-water quality have traditionally been established by state governments and generally on the principle of assigned best usage. Water quality standards consist of beneficial uses, numeric and narrative criteria for supporting each use, and an antidegradation statement; designated beneficial uses are the desirable uses that water quality should support, such as drinking-water supply, primary contact recreation, such as swimming, and aquatic life support. Each designated use has a unique set of water quality requirements or criteria that must be met for the use to be realized, and states may designate an individual waterbody for multiple beneficial uses. Numeric water quality criteria establish the minimum physical, chemical, and biological parameters required to support a beneficial use. Physical and chemical numeric criteria may set maximum concentrations of pollutants, acceptable ranges of physical parameters, and minimum concentrations of desirable parameters, such as dissolved oxygen. Numeric biological criteria describe the expected attainable community attributes and establish values based on measures such as species richness, presence or absence of indicator taxa, and distribution of classes of organisms.

Narrative water quality criteria define, rather than quantify, conditions and attainable goals that must be maintained to support a designated use. Narrative biological criteria establish a positive statement about aquatic community characteristics expected to occur within a waterbody; for example, "Ambient water quality shall be sufficient to support life stages of all indigenous aquatic species." Narrative criteria may also describe conditions that are desired in a waterbody, such as, "Waters must be free of substances that are toxic to humans, aquatic life, and wildlife." Finally, antidegradation statements protect existing designated uses and prevent high-quality waterbodies from deteriorating below the water quality necessary to maintain existing or anticipated designated beneficial uses.

The Clean Water Act (CWA) provides primary authority to states to set their own standards but requires that all state beneficial uses and their criteria comply with the "fishable and swimmable" goals of the act. At a minimum, state beneficial uses must support aquatic life and recreational use. Where possible, states must identify the pollutants or processes that degrade water quality and indicators that document impacts of water quality degradation. Pollutants include sediment, nutrients, and chemical contaminants. Processes that degrade waters include habitat modification, such as destruction of streamside vegetation, and hydrologic modification, such as flow reduction. Indicators of water quality degradation include physical, chemical, and biological parameters. Examples of biological parameters include species diversity and abundance. Examples of physical and chemical parameters include pH, turbidity, and temperature.

The EPA is also charged with developing nationally consistent guidelines limiting pollutants in discharges from industrial facilities and municipal sewage treatment plants. These guidelines are then used in permits issued to dischargers under the National Pollutant Discharge Elimination System (NPDES) program. Additional controls may be required if receiving waters are still affected by water quality problems after permit limits are met. Any link between water quality standards and point/non-point source (NPS) source pollution control actions, such as permits or best management practices, is provided by the development of total maximum daily loads.

All industrial and municipal facilities that discharge wastewater must have an NPDES permit and are responsible for monitoring and reporting levels of pollutants in their discharges. The EPA issues these permits or can delegate that permitting authority to qualifying states. The states and the EPA inspect facilities to determine if their discharges comply with permit limits. If dischargers are not in compliance, enforcement action is taken. In 1990, the EPA promulgated permit application requirements for municipal sewers that carry stormwater separately from other wastes and serve populations of 100,000 or more and for stormwater discharges associated with some industrial activities.

Under the National Combined Sewer Overflow Control Strategy of 1989, states develop and implement measures to reduce pollution discharges from combined storm and sanitary sewers. The EPA works with the states to implement the national strategy. The CWA also established pollution control and prevention programs for specific waterbody categories, such as the Clean Lakes Program. Other statutes that also guide the development of water quality protection programs include: (a) the Safe Drinking Water Act, under which states establish standards for drinking water quality, monitor wells and local water supply systems, implement drinking-water protection programs, and implement underground injection control programs; (b) the Resource Conservation and Recovery Act, which establishes state and federal programs for groundwater and surface water protection and cleanup and emphasizes prevention of releases through management standards in addition to other waste management activities; (c) the Comprehensive Environmental Response, Compensation, and Liability Act (Superfund Program), which provides the EPA with the authority to clean up contaminated waters during remediation at contaminated sites; and (d) the Pollution Prevention Act of 1990, which requires the EPA to promote pollutant source reduction rather than focus on controlling pollutants after they enter the environment.

The maintenance of water quality depends on the control of noxious discharges. Prior to 1971, the control of contaminant effluents was entirely up to the states. However, in that year with the enactment of the National Environmental Policy Act, the permits required for discharges into navigable waterways under the 1899

Refuse Act had to be accompanied by an environmental impact statement, and the Army Corp of Engineers had to assume responsibility for evaluating and regulating the impact of toxic effluents on streams.

The confusion that resulted was partially ended with the passage of the Federal Water Pollution Control Act of 1972. A major feature of this act was its emphasis on effluent limitations. The regulations provided for a NPDES, which required that point sources that discharge contaminants into waterways obtain discharge permits and follow Effluent Guidelines and Standards, Pretreatment Standards, oil and hazardous substances rules, ocean dumping rules, and toxic-pollutant standards. The act mandated that Effluent Guidelines and Standards be established for industrial plants. These standards are end-of-pipe limitations expressed as either pound of contaminant per 1,000 pounds of product or as milligram of contaminant per liter of effluent.

Regulations apply to "liquid effluents" discharged from "point sources" into "navigable waters." The broad definition applied to these terms makes the Effluent Guidelines and Standards virtually all-inclusive. Liquid effluents include every type of water, from process wastes to uncontaminated stormwater. A point source is a discharge through any type of conveyance (e.g., pipe, channel, etc.). A navigable water is any water other than groundwater. Effluent Guidelines and Standards are technology-based effluent limits that are developed following appraisal of the treatment technologies available to a particular industrial category and the economic consideration associated with the installation of such technology for the particular industry. Current Effluent Guidelines are applied for BOD, COD, total suspended solids, pH, and some other wastewater constituents.

The Effluent Guidelines and Standards and the Pretreatment Standards are closely related. Pretreatment Standards apply to discharges into municipal sewers rather than directly into waterways. This sewer discharge could cause the contaminants to interfere with the operation of the waste treatment facility. These standards eliminate the economic advantages of plants changing their discharge from a navigable water to a municipal system.

Some chemicals are so highly toxic and impose such severe risks to humans or aquatic life that they warrant a special regulatory mechanism to control their presence in waterways. Under the Clean Water Act, the EPA may set up special effluent standards for these toxic contaminants. In 1977, final regulations were prepared by the EPA to control direct discharges of DDT, aldrin/dieldrin, endrin, toxaphene, benzidine, and PCBs. Furthermore, pursuant to a 1976 settlement with the National Resources Defense Council, Inc., and Citizens for a Better Environment, the EPA was committed to a program to investigate 65 designated contaminants that represent substantial concern related to their potential health or environmental effects; these include both organic and inorganic agents, many of which are recognized or potential carcinogens, mutagens, or teratogens.

Amendments to the CWA enacted in 1987 significantly changed the thrust of enforcement by increasing attention paid to the monitoring and control of toxic constituents in wastewater and to discharges of polluted runoff from city streets, farmland, mining sites, and other "nonpoint" sources. These amendments also changed the NPDES program by requiring much stricter discharge limits and expanding the number of chemical constituents that must be monitored in pollutants that reach waterways. Thus, the CWA emphasizes a permitting system for the release of pollutants into lakes and streams. Polluters can apply for discharge permits that specify the required control technology, as contrasted to specifying limits on the concentrations of specific pollutants in the receiving bodies of water. Technology-based, as opposed to harm-based, standards are emphasized.

Food Contamination

The FDA is responsible for establishing tolerances, or permissible levels of contamination, in food. Tolerance levels have also been established for some carcinogenic contaminants, for example, 20 ppb_w for aflatoxin in peanuts and corn.

In the Food Quality Protection Act of 1996, a major concern was the adequacy of food contaminant limits for the protection of children, on the basis that they may have greater percentage of uptake and dose to target tissues than adults consuming the same food products. The EPA was directed to add an additional safety factor of 10 in risk assessment for pesticide chemicals if it could not demonstrate that such an additional margin of safety was not needed.

REFERENCES

1. ACGIH. *2016 TLVs and BEIs*. Cincinnati, American Conference of Governmental Industrial Hygienists, 2016.
2. Brief, R.S. and Scala, R.A. Occupational exposure limits for novel work schedules. *Am. Ind. Hyg. Assoc. J.* 36:467–469, 1975.
3. Mason, J.W. and Dershin, H. Limits to occupational exposure in chemical environments under novel work schedules. *J. Occup. Med.* 18:603–606, 1976.
4. Calabrese, E.J. Further comments on novel schedule TLVs. *Am. Ind. Hyg. Assoc. J.* 38:443–446, 1977.
5. Hickey, J.L.S. and Reist, P.C. Application of occupational exposure limits to unusual work schedules. *Am. Ind. Hyg. Assoc. J.* 38:613–621, 1977.
6. Hatch, T.F. Permissible levels of exposure to hazardous agents in industry. *J. Occup. Med.* 14:134–137, 1972.
7. AIHA. An international review of procedures for establishing occupational exposure limits. Am. Ind. Hyg. Assoc., Fairfax, VA, 1996.
8. Vincent, J.H. International Occupational Exposure Standards: A Review and Commentary. *Am. Ind. Hyg. Assoc. J.* 59:729, 1998.

12

Our Environmental Future

The environmental trends that have taken place over recent decades are likely to continue unless society makes concerted efforts to alter them. For example, there has been exponential growth in atmospheric concentrations of, for example, carbon dioxide (CO_2) and methane (CH_4), constituents closely associated with climate change, as illustrated in Figure 12–1. There have also been increases in ocean acidification and nitrogen in the coastal zone. These changes in environmental quality markers coincide with markers of environmental impacts, such as tropical forest loss and terrestrial biosphere degradation. Such trends can be expected to continue over the next few decades, even if we begin to substantially limit anthropogenic discharges of pollutants.

Because of large-scale social, economic, technological, and institutional changes already underway, future environmental issues may emerge in diverse areas. For example, the health of biosystems and the sustainable use of natural resources will be stressed by a growing human population, expanding energy use, natural resource consumption, and land development. As the stresses on biosystems intensify, the preservation of biodiversity will become increasingly important for both economic and environmental reasons. As populations grow and urban areas expand, heightened competition for the use of land will put new strains on natural habitats. Failure to maintain healthy terrestrial ecosystems could lead

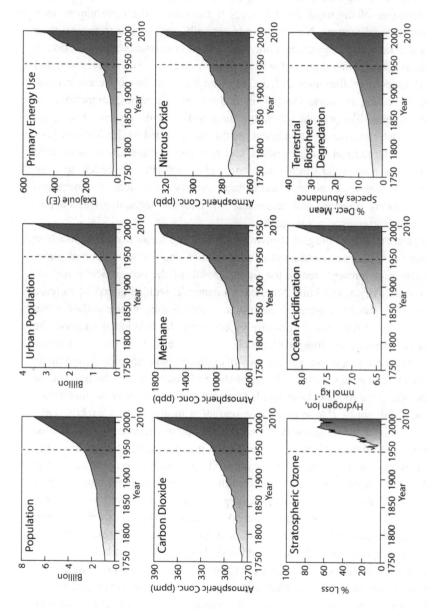

Figure 12–1. Worldwide trends in population, energy consumption, pollutant gases, ocean acidification, and terrestrial biosphere degradation. Adapted from Steffan et al., DOI:10.1126/science.1259855 (2015).

to natural resource damage, irreversible losses of species, and fragmentation of habitats, thus endangering both economic and environmental sustainability and seriously threatening human and ecological well-being.

Perhaps one of the most critical issues is the role of the environment as a strategic national interest. Nations have historically gone to war to protect their access to vital natural resources. Others have used environmental destruction in combat as a major instrument of war. Terrorism, nuclear reactor accidents, and nuclear weapons proliferation all have major implications for public and ecosystem health around the world. Possible natural resource shortages, competition for scarce resources like potable water, and the transborder movement of refugees driven by deteriorating environmental conditions have led to destabilized governments, international disagreements, and regional warfare. Overfishing, acid rain, and raw wastewater discharges along and across national borders are also examples of how environmental and natural resource issues lead to contentious relations among countries and necessitate international negotiations and agreements related to environmental quality.

The future quality of the global environment will be a factor in determining how economic activities are conducted in all countries, including the United States (US). Based on present trends, the future growth of the economies in regions such as Asia, Africa, and Latin America, for example, with an attendant increase in energy use, can be expected to contribute to global atmospheric pollution that historically has been caused primarily by economically developed nations. The loss of biodiversity, resulting from the clearing of rain forests in South America and South Asia for example, and the growth of deserts in in Africa and the Middle East affects everyone on earth. The stripfishing of marine life in the open ocean is diminishing the foodstocks available to global populations over the long term.

Nations will not be able to limit or prevent all or most emerging environmental problems with the same regulatory tools and reactive approaches that they have used in the past. The future will be as challenging as it is uncertain, and new analytical tools, new approaches to decision-making, and new partnerships will be needed. Regulators will need to work more closely with business communities to anticipate any environmental implications of technological innovation, as well as with other agencies, international organizations, and other nations to identify drivers of emerging regional or global problems and then help define possible responses to them. The environmental problems of the future undoubtedly will be facets of large-scale economic, demographic, and technological change. Other organizations, government and nongovernment, will have major responsibilities responding to that change.

Affluence allows some of us to directly control many aspects of our personal environments, such as within our homes and property boundaries. However, as we consider our neighborhoods, villages, cities, states, countries, and global communities, our individual capacity to influence factors affecting our collective environments diminishes, and we become more dependent on political and judicial processes for the protection and/or enhancement of our collective environments, through new or revised legislation, enforcement of environmental regulations, and public information and educational programs.

In the US, the Environmental Protection Agency (EPA), the Department of Health and Human Services, and their state and local counterpart agencies take lead roles in the anticipation, recognition, evaluation, and control of the environmental factors and forces that could or do affect human health. While these agencies can greatly influence our environmental quality and health status, they cannot, and do not, do so alone. Regulatory and control actions taken by other governmental agencies responsible for transportation, energy, agriculture, defense, commerce, and labor can also play significant, and sometimes dominant, roles in shaping our collective environments. Furthermore, two of the most important factors influencing our environmental futures are not directly influenced by regulation in most countries, that is, human population growth and the concentration of people and economic activity within urban areas.

SCANNING THE HORIZON

In the US, the EPA, has been confronted with a series of crises and pressures for rapid responses to environmental issues, such as high concentrations of dioxins in soil at Times Beach, Missouri; a leaking waste dump in Niagara Falls, New York (Love Canal); high concentrations of mercury in fish in the Great Lakes; acid rain in the Adirondack mountains of New York; and *Cryptosporidium* in the public water supply of Milwaukee, Wisconsin. In order to be better prepared for such unanticipated events, the Agency commissioned a study of ways that it could look ahead and be better prepared to anticipate and deal with emerging environmental challenges. The resulting report,[1] released in 1995 and focusing on the time frame between 5 and 30 years ahead, dealt with the "forces of change." It examined forces that elicit change, so-called "drivers," suggesting how changes will affect the future and how the environmental effects of such changes can be altered by current actions. The drivers are interdependent, and the changes can have both positive and negative effects. For example, population growth and higher per capita income create increased demands for energy, natural resources, and manufactured goods. At the same time, higher per capita income, combined with improved education and an expanded range of personal choices, tend to reduce population growth and its pressures, while the availability of cleaner fuels and higher end-use efficiencies could reduce local and global environmental effects. Thus, technological changes can either exacerbate or ameliorate environmental pressures. The drivers of future change are the consequences of personal, community, and national choices and are themselves subject to change. While the report was released in the mid-1990s, the drivers presented are interestingly still very relevant today and are outlined next.

Population Growth and Urbanization

Both the continuing growth in the human population and especially their concentrations in large urban areas pose environmental challenges. With more concentrated populations, environmental problems intensify. Providing safe drinking water, effective wastewater treatment and solid waste disposal systems, and environmentally sustainable transportation systems pose daunting challenges in urban areas.

Economic Expansion and Resource Consumption

Per capita income in many developing countries continues to increase. This development, coupled with population growth, can be expected to result in greater consumption of energy, natural resources, and consumer goods. Although recent US and European experience indicates that energy use does not necessarily grow in direct proportion to economic growth, there is little doubt that energy use will continue to rise dramatically in the developing world.

The choice of energy sources has a profound impact on the environment. With continued reliance by many countries on conventional coal technologies with minimal pollution controls to generate electricity, local, regional, and global environmental impacts can be expected to be substantial. On the other hand, somewhat greater use of an alternative fossil fuel (e.g., natural gas), higher energy efficiency, and greater usage of wind and solar energy has, in recent years, helped to mitigate these impacts in some areas.

Some potentially devastating effects of population growth, economic expansion, and individual behavior on natural resources already are evident in many parts of the world. All major ocean areas presently are being fished at or beyond capacity, and global per capita seafood harvests from natural environments have already declined. Approximately 5% to 10% of the world's living reefs, the rainforests of the oceans, have died because of economic activity along coastlines and in coastal waters.

Technological Development

Throughout history, technological change has been one of the most important factors driving both positive and negative economic and environmental changes. Technology will play an even greater role in the future, as technological development proceeds at a faster pace and has a more pervasive impact on societies and individuals.

In the past, the adverse environmental effects of growing populations and expanding economies have been ameliorated by the development of new technologies for contaminant exposure reduction, such as centralized wastewater treatment systems. Newer technological advances, for example, cleaner fuels, wind and solar sources, more energy-efficient transportation and power distribution systems, and less wasteful manufacturing processes, can be expected to yield substantial environmental benefits.

At the same time, new materials, such as genetically altered organisms, photovoltaic cells, and next-generation batteries, may result in new contaminant exposures and increased risks to human health and ecosystems. Thus, one of the central challenges facing society today is anticipating the likely environmental effects of future technological development

CHANGING ENVIRONMENTAL ATTITUDES AND INSTITUTIONS

Environmental quality is not determined solely by the actions of governments, regulated industries, or nongovernment organizations (NGOs). It is largely a function of the decisions and behavior of individuals, families, businesses, and communities everywhere. Consequently, the extent of environmental awareness and the strength of environmental institutions will be two critical factors driving changes in environmental quality in the future.

Foresight, or futures research and analysis, has, in recent decades, been used by government, private business, and NGOs to anticipate future changes. While most futures studies have focused on the nearer term (i.e., less than five years), some have reached considerably further. For example, the Energy Information Administration within the Department of Energy develops detailed energy use projections as far as 20 years into the future. With a shorter-term focus, the Internal Revenue Service, the Department of Defense, and the intelligence community employ scanning systems and trend analysis as part of institutional planning. The Department of Defense uses "gaming" exercises to anticipate the possible circumstances of future warfare and prepare a range of options in response. In recent years, many regional, state, and local governments have also applied the tools of foresight.

In general, there are three widely used basic techniques to identify possible future conditions. One is a top-down approach, involving the use of "scenarios" that postulate certain circumstances about the future to draw some likely implications. The second technique is a bottom-up approach that draws future implications from early warning signals, that is, those based either on the extrapolation of current data and trends or on the observations of knowledgeable individuals participating in "look-out panels." The third technique, scanning, involves a continual, planned, deliberate, and thorough review of published information and contacts with other "futures watching" organizations. All three approaches individually, and particularly in combination, can provide valuable insights into the possible emergence of environmental problems within the foreseeable future.

In the top-down approach, scenarios are constructed to study the environmental implications of assumed future developments in drivers like energy use, population growth and density, technological advances, waste generation, and demand for natural resources like potable water. These images of possible futures can be systematically evaluated to estimate when and where environmental problems could emerge and to assess different types of policies that could be used to forestall them.

Within a given scenario, assumptions concerning the future can be varied to reflect different rates of change, for example in energy use or population growth. As long as these scenarios display changes in important variables over time within a consistent analytical framework, they can be useful tools for anticipating environmental problems in the future and analyzing the range of possible responses.

In the bottom-up approach, a specialized look-out panel can provide perceptions, observations, and information about important environmental changes. Lookout panels can include laboratory-based scientists, professional field data collectors and analysts, or neighborhood volunteers. Through systematic questioning and feedback, panelists can provide observations about the environment than can serve as early warnings of environmental changes, and they can assess the implications of these changes to human health and ecosystem viability.

In the scanning approach, information related to emerging environmental problems can be gleaned from scholarly journals, newspapers, newsletters, business plans, and science-oriented computer bulletin boards. Such sources of information include literature and academic disciplines well beyond the bounds of traditional environmental science. Scanning also can be part of the foresight activities of lookout panels.

All three approaches are independently useful in identifying early signals that can warn of emerging environmental problems. In addition, the techniques reinforce one another by providing early warnings from different perspectives. Scenario analyses tend to raise top-down issues generated by the assumptions used in the scenarios (e.g., global CO_2 buildup as a result of the energy strategies of large countries like the US, China and India). The lookout panels call attention to specific emerging issues, such as the widespread introduction of new toxic chemicals. Scanning cuts across both approaches. Thus, all three techniques can help identify potential environmental issues that could be subjected to in-depth risk analysis. All three, if used continuously and interactively, could serve as a first line of defense in protecting future environmental quality.

While various governmental agencies have critical roles and responsibilities for influencing and controlling sources and activities that influence environmental changes, there are other, sometimes major, forces at work as well, and many of them have been at work over long periods of time. A view of some of these forces and trends was provided by Ausubel.[2] The following section summarizes some of his perspectives on our environmental future in terms of environmental foresight, with special emphasis on the implications of long-term trends, which we expect will remain relevant as we go forward in time. These can be grouped in terms of energy, land, water, and materials.

Energy

Figure 12–2 summarizes historic changes and trend projections for usage of energy since 1860. It shows fractional, not absolute, usage, which has greatly increased over time. In 1860, wood was still the dominant source of anthropogenic thermal energy. However, wood could not satisfy the energy demands of a growing population, especially in urban and industrial areas. By 1880, it was supplanted by coal, which was a fuel that could be readily extracted from the ground and delivered to consumers at a relatively low cost. However, burning coal efficiently and completely has always been technologically challenging, and incomplete coal combustion has had significant impacts on human health and the environment. Thus, coal was gradually supplanted by other fuels that were easier to transport and burn (i.e., petroleum and natural gas). Coal, which was, for a long time, our most dominant source, was succeeded in that role by oil by 1970 and by natural gas by 1995. Coal usage, in terms of absolute tonnage, did not actually decline between the 1930s and the early part of the twenty-first century, when there was a major shift from coal to natural gas for electric power generation.

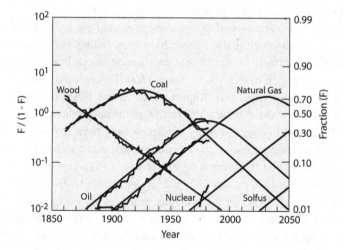

FIGURE 12–2. Global primary energy substitution from 1860 to 1982 and projections for the future, expressed in fractional market shares (F). Smooth lines represent model calculations and jagged lines are historical data. Solfus was the term employed by Ausubel to describe an emerging major new energy technology, for example, solar or fusion.

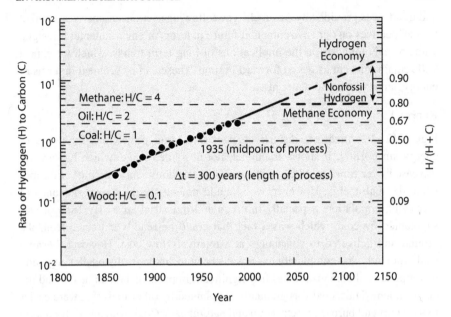

FIGURE 12–3. Ratio of hydrogen (H) to carbon (C) for global primary energy consumption since 1860 and projections for the future, expressed as a ratio of hydrogen to carbon (H/C).

Much of the coal usage has historically been for space heating, railroad locomotives, and steel production, but its remaining usage in developed countries is confined largely to electric power generation in large facilities.

The temporal shift from wood to coal to oil to natural gas has had major implications for environmental quality. There have been reductions in the emissions of hydrocarbon products of incomplete combustion, as well as of sulfur dioxide (SO_2) and nitrogen oxides (NO_x). since 1860 (Figure 12–3). In addition, an increasing H/C ratio reduces the impact of fossil fuel usage on global climate change associated with the secular rise in CO_2 in the atmosphere.

Parts of the gain in overall energy efficiency have been attributable to gains in: the efficiency of combustion, conversion of heat into useful power, delivery of power to users, and the increasing efficiencies of devices using the power. The production of a good or service in the US has, since 1800, required 1% less energy on average than it did the previous year.[2] As shown in Figure 12–4, in the 300 years since the invention of the steam engine, engine efficiencies have grown from approximately 1% to approximately 50%. In terms of illumination, Edison's first electric light in 1879 was 15× more efficient than a paraffin candle. The first fluorescent lamp in 1942 was 30× more efficient than the first incandescent lamp. The progress in illumination efficiency is also shown in Figure 12–4.

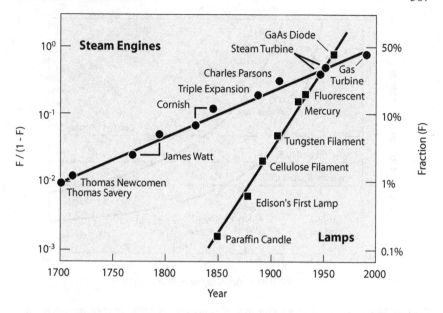

FIGURE 12–4. Improvement in the efficiency of motors and lamps analyzed as a sigmoid (logistic) growth process.

Land

At the end of the twentieth century, population growth in most economically developed countries had virtually ceased, and there were indications of reduced rates of population growth in many less well developed countries. However, the global population may grow by approximately 50% to 100% by 2050, accounted for mostly by growth in the less developed countries. At the same time, the standard of living and food consumption rates can also be expected to rise. During the second half of the twentieth century, the amount of land devoted globally to agriculture remained stable, while the population doubled. The question then arises: Can the food supply meet the demand?

Since 1940, US wheat yields tripled and corn yields quintupled.[2] The potential to increase yields everywhere remains strong, even without invoking such new technologies as the genetic engineering of plants. Future caloric intake per person will likely range between the 3,000 per day for an ample vegetarian diet and 6,000 per day for a diet that includes meat consumption. If global average yields do not rise, people will have to reduce their portions to keep cropland to its current extent. If the global average yield rises approximately 1.5% per year over the next six or seven decades, to the level of today's European wheat, 10 billion people can enjoy a 6,000-calorie diet and still spare close to a quarter of the present 1.4 billion hectares of cropland. Reaching the level of today's average US corn grower would spare for 10 billion people half of today's cropland for nature, an area

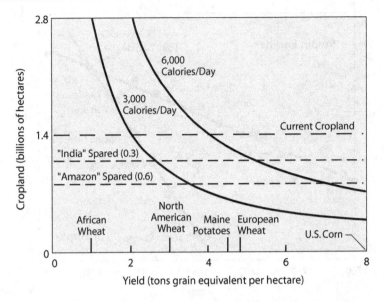

FIGURE 12–5. The sparing for nature of a reference area of 2.8 billion hectares of cropland by farmers raising yields for ten billion people consuming 3,000 or 6,000 calories daily.

larger than the Amazon basin—even with the calorie intake of today's American as the diet (see Fig. 12–5).

Water

With its vast oceans, the earth suffers no overall shortage of water. However, supplies of domestic fresh water for drinking, cooking, and washing, as well as fresh water for agricultural and industrial users, vary greatly in availability and cost. Also, quality standards vary greatly with intended usage. Total per capita water withdrawals in the US quadrupled between 1900 and 1970.[2] However, since 1975, per capita water use has fallen appreciably, at an annual rate of 1.3%.[3] Absolute US water withdrawals peaked about 1980. Total industrial water withdrawals plateaued a decade earlier than total US withdrawals and dropped by one-third, more steeply than the total. Notably, industrial withdrawals per unit of gross national product (GNP) have dropped steadily since 1940, from 14 gallons per constant dollar to 3 gallons in 1990. Not only intake but discharge per unit of production are perhaps one-fifth of what they were 50 years earlier.

Law and economics as well as technology have favored frugal water use. Various legislation encouraged the reduction of discharges, recycling, and conservation, as well as shifts in relative prices. Water withdrawals for all users in the industrialized countries span a 10-fold range, with the US and Canada at the highest end.[4] In the long run, with much higher thermodynamic efficiency for processes, removing impurities to recycle water will require small amounts of energy.

Materials

The intensity of use of diverse primary materials has also plummeted during the twentieth century.[2] Lumber, steel, lead (Pb), and copper (Cu) have lost relative importance, while plastics and aluminum (Al) have expanded. Many products, for example, cars, computers, and beverage cans, have become lighter and often smaller. Although the soaring numbers of products and objects, accelerated by economic growth, raised municipal waste in the US annually by approximately 1.6% per person in the past couple of decades, trash per unit of GNP dropped slightly. Over time new materials replace old, and theoretically each replacement should improve material properties per unit, thus lowering the intensity of use. Furthermore, as countries develop, the intensity of use of a given material or system declines as each country arrives at a similar level of development.

Since 1990, recycling has accounted for over half the metals consumed in the US, up from less than 30% in the mid-1960s.[5] There is a need to make waste minimization a property of the industrial system, even when it is not completely a property of an individual process, plant, or industry. Advancing information networks can help by offering inexpensive ways to link otherwise unconnected buyers and sellers to create new markets or waste exchanges.

The preceding projections of secular changes affecting environmental quality suggest that the benefits of increasingly efficient sources of energy, usage of land and water resources, and reduced use of raw materials can lead to lesser impacts of anthropological activities on the natural environment. With a stabilization of population size; a more hydrogen-, wind-, and solar-based energy system; further increases in agricultural productivity; more efficient utilization of water supplies; and a reduced dependence on natural raw materials, it should be possible to preserve more of our remaining natural ecosystems and ensure the benefits of the services that they supply to all of the earth's inhabitants.

COSTS VS. BENEFITS OF ENVIRONMENTAL MANAGEMENT

The environmental future is determined by our individual and collective willingness to bear the costs of pollution prevention, cleanup and/or isolation of previously discarded wastes, protection and creation of park lands and nature preserves, and enforcement of statutory regulations and international treaties and commitments related to environmental protection. In economic terms, willingness to pay is expressed in hard currency, as determined from appropriately designed survey data of a stratified sample of the public. In practice, the elicitation of results can be expected to vary with the ways in which the questions are constructed and with the extent and reliability of the background information presented to, or previously available to, the members of the public who respond to the survey's questions. For example, by including information on both the direct and indirect effects of expenditures for environmental protection, including intangible and/or nonmonetizable benefits, an individual's willingness to pay could vary substantially.

At the national level in the US, the willingness to commit public funds for environmental protection and to impose control costs on individuals and state and local governmental agencies has been tempered. For example, the Clean Air Act (CAA) of 1970 mandated that primary standards for criteria air pollutants be set to protect the public health with an adequate margin of safety and without consideration of costs. However, by executive order, every proposed federal regulation submitted to the Office of Management and Budget for review prior to promulgation must now be accompanied by a regulatory impact analysis, which includes a cost-benefit analysis. Furthermore, the amendments of 1990 required the EPA to report on the benefits and costs of the CAA retrospectively from 1970 to 1990 and to prospectively analyze the benefits and costs of the additional requirements in the 1990 amendments out to 2000 and 2010.

The findings of this analysis are discussed next so as to provide an example, using air quality, of an exercise that evaluates the benefits versus the costs of environmental management. In order to estimate the benefits and costs of the CAA, differences in economic, human health, and environmental outcomes under two alternative scenarios were examined: a control scenario and a no-control scenario.[6] The control scenario was based largely on historical data. The hypothetical no-control scenario assumed that no air pollution controls had been established. Each scenario was evaluated in terms of economics, emissions, air quality, physical effects, economic valuation, and uncertainty models.

While compliance with the provisions of the CAA incurred higher costs for many goods and services, which were borne by manufacturers, consumers, and taxpayers, emissions were substantially lower by 1990 under the control scenario than under no-control (e.g. SO_2 by 40%; NO_x by 30%; volatile organic carbons [VOCs] by 45%; CO by 50%; and particulate matter [PM] by 75%). These reductions were achieved during a period in which population grew by 22.3% and the national economy grew by 70%.

These reductions in air pollutant emissions translated into significantly improved air quality throughout the US. For SO_2, NO_x, and CO, the improvements under the control scenario were assumed to be proportional to the estimated reduction in emissions. Reductions in ground-level ozone (O_3) were achieved through reductions in emissions of its precursor pollutants, particularly VOCs and NO_x. These lower concentrations under the control scenario yielded human health/welfare and ecological benefits. For some benefit categories, quantitative functions were available from the scientific literature, allowing estimation of the reduction in incidence of adverse effects. Examples of these categories include the mortality and morbidity effects of a number of pollutants, visibility impairment, and effects on yields for some agricultural products.

Table 12–1 shows examples of the health impact in 1990 due to control, reflecting reductions estimated for the entire US population living in the 48 contiguous states. The table was based on epidemiological findings about correlations between pollution and observed health effects to estimate changes in the number of health effects that would occur if pollution levels change. A range is presented along with the mean estimate for each effect, reflecting uncertainties that have been quantified in the underlying health effects literature.

Adverse human health effects of the "criteria pollutants" dominated the quantitative effects estimates in part because although there were important residual uncertainties, evidence of physical consequences was greatest for these pollutants. Implementation of the CAA yielded other benefits that could not be quantitated or monetized, including all benefits accruing from reductions in hazardous air pollutants (also referred to as air toxics); reductions in damage to cultural resources, buildings, and other materials; reductions in adverse effects on wetland, forest, and aquatic ecosystems; and a variety of additional human health and welfare effects of criteria pollutants.

Existing scientific research suggests that reductions in both hazardous air pollutants and criteria air pollutants yielded widespread improvements in the functioning and quality of aquatic and terrestrial ecosystems. In addition to any intrinsic value to be attributed to these ecological systems, human welfare is enhanced through improvements in a variety of ecological services. For example, protection of fresh-water ecosystems achieved through reductions in deposition of acidic air pollutants can improve commercial and recreational fishing. Other potential ecological benefits of reduced acid deposition include improved wildlife viewing, maintenance of biodiversity, and nutrient cycling. Increased growth and productivity of US forests may have resulted from reductions in ground-level O_3. More vigorous forest ecosystems, in turn, yield a variety of benefits, including increased timber production; improved forest aesthetics for people enjoying outdoor activities such as hunting, fishing, and camping; and improvements in ecological services such as nutrient cycling and temporary sequestration of global warming gases. These improvements in ecological structure and function were not quantified in this assessment.

In order to compare or aggregate benefits across endpoints, the benefits had to be monetized. Assigning a monetary value to avoided incidences of each effect permits a summation, in terms of dollars, of monetized benefits realized, allowing the summation of the benefits to be compared to the cost of implementing the CAA. For this analysis, unit valuation estimates were derived from the economics literature and reported in dollars per case or, in some cases, episode or symptom-day, avoided for health effects, and dollars per unit of avoided damage for human

Table 12–1. Pollutant Health Benefits—Estimated Distributions of 1990 Incidences of Avoided Health Effects (in Thousands of Incidences Reduced) for 48 State Populations[a]

ENDPOINT	POLLUTANT(S)	AFFECTED POPULATION	ANNUAL EFFECTS AVOIDED[b] (THOUSANDS)			UNIT
			5TH PERCENTILE	MEAN	95TH PERCENTILE	
Premature mortality	PM[c]	30 and over	112	184	257	Cases
Chronic bronchitis	PM	All	498	674	886	Cases
Hospital admissions						
All respiratory	PM and O₃	All	75	89	103	Cases
Chronic obstructive pulmonary disease and pneumonia	PM and O₃	Over 65	52	62	72	Cases
Ischemic heart disease	PM	Over 65	7	19	31	Cases
Congestive heart failure	PM and CO	65 and over	28	39	50	Cases
Other respiratory-related ailments						
Shortness of breath, days	PM	Children	14,800	68,000	133,000	Days
Acute bronchitis	PM	Children	0	8700	21,600	Cases
Upper and lower respiratory symptoms	PM	Children	5400	9500	13,400	Cases
Any of 19 acute symptoms	PM and O₃	18–65	15,400	130,000	244,000	Cases
Asthma attacks	PM and O₃	Asthmatics	170	850	1520	Cases

Increase in respiratory illness						
Any symptom	NO_2	All	4840	9800	14,000	Cases
	SO_2	Asthmatics	26	264	706	Cases
Restricted activity and work loss days						
Minor restricted activity days	PM and O_3	18–65	107,000	125,000	143,000	Days
Work loss days	PM	18–65	19,400	22,600	25,600	Days

[a]The following additional human welfare effects were quantified directly in economic terms: household soiling damage, visibility impairment, decreased worker productivity, and agricultural yield changes.

[b]The 5th and 95th percentile outcomes represented the lower and upper bounds, respectively, of the 90% credible interval for each effect as estimated by uncertainty modeling. The mean is the arithmetic average of all estimates derived by the uncertainty modeling.

[c]In this analysis, PM was used as a proxy pollutant for all non-Pb criteria pollutants that may contribute to premature mortality.

welfare effects. Similar to estimates of physical effects provided by health studies, each of the monetary values of benefits applied in this analysis were expressed in terms of a mean value and a range around the mean. This range reflected the extent of the uncertainty in the economic valuation literature associated with a given effect. The mean values of these ranges are shown in Table 12–2.[7]

The total monetized benefits of the CAA realized during the period from 1970 to 1990 ranged from $5.6 trillion to $49.4 trillion, with a central estimate of $22.2 trillion. By comparison, the value of direct compliance expenditures over the same period equaled approximately $0.5 trillion. Subtracting costs from benefits results in net, direct, monetized benefits ranging from $5.1 trillion to $48.9 trillion, with a central estimate of $21.7 trillion, for the 1970 to 1990 period. While the central estimate in benefits may have been a significant underestimate, due to the exclusion of large numbers of benefits from the monetized benefit estimate (e.g., all air toxics effects, ecosystem effects, numerous human health effects), even the lower-bound estimate of monetized benefits substantially exceeded the costs of implementing the historical CAA. Monetized benefits consistently and substantially exceeded costs throughout the 1970 to 1990 period.

The previous discussion indicated that the benefits of the CAA and associated control programs substantially exceeded the costs during the 1970 to 1990 period. Even considering the large number of important uncertainties permeating each step of the analysis, it is extremely unlikely that the converse could have been true. The study highlighted important areas of uncertainty associated with many of the monetized benefits included in the quantitative analysis and listed benefit categories that could not be quantified or monetized given the then current state of the science at the time the study was performed. The results of the retrospective study also provided useful lessons with respect to the value and the limitations of cost-benefit analysis as a tool for evaluating environmental programs.

Other air pollution benefit and cost studies in recent years have produced findings that were generally consistent with those reported by EPA on the CAA. In the US, these included studies in the state of California and the city of Houston, Texas. A study of benefits and costs of air pollution in Canada was completed under the auspices of the Royal Society of Canada.[8]

CONCLUSIONS

A number of potentially serious drivers of our environmental future have not been discussed in any detail in this chapter, including global climate change and the major population displacements and ecological disruptions that it may cause; sudden catastrophic events, such as an asteroid impact, major earthquakes, or volcanic eruptions in heavily populated areas; nuclear or biological warfare on more than a limited, local scale; or a major pandemic of disease not amenable to public

TABLE 12–2. Central Estimates of Economic Value per Unit of Avoided Effect[a]

ENDPOINT	POLLUTANT	VALUATION[b]
Mortality	PM and Pb	$4,800,000 per case[c]
Chronic bronchitis	PM	$260,000 per case
Hospital admissions		
Ischemic heart disease	PM	$10,300 per case
Congestive heart failure	PM	$8300 per case
Chronic obstructive pulmonary disease	PM and O_3	$8100 per case
Pneumonia	PM and O_3	$7900 per case
All respiratory	PM and O_3	$6100 per case
Respiratory illness and symptoms		
Acute bronchitis	PM	$45 per case
Acute asthma	PM and O_3	$32 per case
Acute respiratory symptoms	PM, O_3, NO_2, SO_2	$18 per case
Upper respiratory symptoms	PM	$19 per case
Lower respiratory symptoms	PM	$12 per case
Shortness of breath	PM	$5.30 per day
Work loss days	PM	$83 per day
Mild restricted activity days	PM and O_3	$38 per day
Welfare benefits		
Visibility	DeciView	$14 per unit change in DeciView
Household soiling	PM	$2.50 per household per PM-10 change
Decreased worker productivity	O_3	$1[f]
Agriculture (net surplus)	O_3	Change in economic surplus

[a]In 1990 dollars.
[b]Mean estimate.
[c]Alternative results, based on assigning a value of $293,000 for each life-year lost are presented.
[d]Strokes are comprised of atherothrombotic brain infarctions and cerebrovascular accidents; both are estimated to have the same monetary value.
[e]The different valuations for stroke cases reflect differences in lost earnings between males and females.
[f]Decreased productivity valued as change in daily wages: $1 per worker per 10% decrease in O_3 and PM.

health controls. Rather, this chapter has discussed a number of factors than can broadly affect our environmental future and that are more amenable to reasonably informed speculation.

Application of foresight tools to new environmental challenges can effectively and efficiently deal with them before they become crises demanding immediate actions that can be expected to be either ineffective or inefficient. Our ever-advancing technological capabilities can provide a means of improving the standard of living of an expanding population while simultaneously enhancing environmental quality. Recognizing these possibilities and consciously planning to utilize them is the challenge. Finally, benefits and cost studies provide a framework for demonstrating that improvements in environmental quality do not necessarily impose net burdens on society and that the costs of regulations can be viewed more as productive investments than lost opportunities for social and economic improvements. These analyses provide a basis for an outreach program to the public and to legislative bodies that could lead to a more positive reception to investments in environmental cleanup and protection that can accelerate our progress toward environmental and public health improvements.

An understanding of the progress that has been made, the challenges that remain, and the tools that are provided by the environmental health science and engineering communities can help society in effectively addressing the unfinished business of traditional pollution problems, as well as new, more global challenges as they may arise.

REFERENCES

1. EPA-SAB. *Beyond the Horizon: Using Foresight to Predict the Environmental Future.* EPA-SAB-EC-95-007. US Environmental Protection Agency, Washington, DC 20460, Jan. 1995.
2. Ausubel, J.H. The liberation of the environment. *Daedalus 125*(3):1–7, 1996.
3. USGS. US Geological Survey, *Estimated Use of Water in the United States in 1990*, Circular 1081. Washington, DC: US Government Printing Office, 1993.
4. OECD. Organization for Economic Cooperation and Development, *The State of the Environment.* Paris: OECD, 1991.
5. Wernick, I.K. and Ausubel, J.H. National materials metrics for industrial ecology. *Resources Policy 21*(3):189–98, 1995.
6. EPA. *The Benefits and Costs of the Clean Air Act, 1970 to 1990.* US Environmental Protection Agency, Washington, DC 20460, Oct. 1997.
7. EPA. *The Benefits and Costs of the Clean Air Act, 1990–2010.* EPA-410-R-99-001. US Environmental Protection Agency, Washington, DC 20460, Nov. 1999.
8. RSC. The Royal Society of Canada Expert Panel Review of the Socio-Economic Modes and Related Components Supporting the Development of Canada-Wide Standards for Particulate Matter and Ozone. Royal Society of Canada, Ottawa, 2001.

Supplementary Bibliography

This bibliography lists selected sources that the reader may consult for more detailed information on the various topics discussed in the text.

General

Baird, C., Cann, M. *Environmental Chemistry*. 5th ed. 2012. W.H. Freeman and Company, New York, NY.
Hemond, H.F., Fechner, E.J. *Chemical Fate and Transport in the Environment*. 3rd ed. 2015. Elsevier, Waltham, MA.
Newman, M.C. *Fundamentals of Ecotoxicology: The Science of Pollution*. 4th ed. 2015. CRC Press, Boca Raton, FL.

The Atmosphere

Friedlander, S.K. *Smoke, Dust, and Haze: Fundamentals of Aerosol Dynamics*. 2000. Oxford Univ. Press, New York, NY.
Hinds, W.C. *Aerosol Technology: Properties, Behavior and Measurement of Airborne Particles*. 2nd ed. 1999. Wiley, New York, NY.
Seinfeld, J.H., Pandis, S.N. *Atmospheric Chemistry and Physics*. 3rd ed. 2016. Wiley, New York, NY.

Sternberg, S.P.K. *Air Pollution: Engineering, Science, and Policy.* 2015. College Publishing, Glen Allen, VA.

Vallero, D. *Fundamentals of Air Pollution.* 2014. Academic Press, Waltham, MA.

The Hydrosphere

Laws, E.A. *Aquatic Pollution.* 2000. Wiley, New York, NY.

Nathanson, J.A., Schneider, R.A. *Basic Environmental Technology: Water Supply, Waste Management and Pollution Control.* 6th ed. 2014. Pearson, Boston.

Nikinmaa, M. *Introduction to Aquatic Toxicology.* 2014. Academic Press, Waltham, MA.

The Lithosphere

Bleam, W. *Soil and Environmental Chemistry.* 2nd ed. 2017. Academic Press, San Diego CA.

Brevik, E.C., Burgessd, L.C. *Soils and Human Health.* 2012. CRC Press, Boca Raton, FL.

Food

Knechtges, P.L. *Food Safety: Theory and Practice.* 2011. Jones and Bartlett, Burlington, MA.

Omage, S.T. *Food and Nutritional Toxicology.* 2004. CRC Press, Boca Raton, FL.

Health Effects

Friis, Robert H. *Epidemiology 101.* 2nd ed. 2017. Jones and Bartlett, Mississauga, Ontario, Canada.

Gordis, Leon. *Epidemiology.* 5th ed. 2014. Elsevier Saunders, Philadelphia.

Hayes, A.W., Kruger, C.L. *Hayes' Principles and Methods of Toxicology.* 6th ed. 2014. CRC Press, Boca Raton, FL.

Klaassen, Curtis D., Watkins, John B. III. *Casarett and Doull's Essentials of Toxicology.* 3rd ed. 2015. McGraw-Hill Medical, New York, NY.

Lippmann, M. (Ed.) *Environmental Toxicants: Human Exposures and Their Health Effects.* 3rd ed. 2009. Wiley-Interscience, New York, NY.

Mercurio, S.D. *Understanding Toxicology: A Biological Approach.* 2017. Jones and Bartlett, Burlington, MA.

Richards, I.S., Bourgeois, M. 2014. *Principles and Practice of Toxicology in Public Health.* 2nd ed. Jones and Bartlett, Burlington, MA.

Rose, V.E., Cohrssen, B. (Ed.) 2010. *Patty's Industrial Hygiene.* 6th ed. Wiley-Interscience, New York, NY.

Risk Assessment

Greim, H., Snyder, R. *Toxicology and Risk Assessment: A Comprehensive Introduction.*
2008. Wiley, West Sussex, UK.

Nielsen, E., Ostergaard, G, Larsen, J.C. *Toxicological Risk Assessment of Chemicals.*
2008. Informa, New York, NY.

Simon, T. *Environmental Risk Assessment: A Toxicological Approach.* 2014. CRC Press,
Boca Raton, FL.

Environmental Sampling and Analysis

ACGIH. *Air Sampling Instruments.* 9th ed. 2000. American Conference of Governmental
Industrial Hygienists, Cincinnati, OH.

ASTM. *Annual Book of ASTM Standards*: Part 31, Water, and Part 26, Atmospheric
Analysis. 2016. American Society for Testing and Materials, West Gonshohocken, PA.

Patty, F., Clayton, G., and Clayton, F. *Industrial Hygiene and Toxicology.* 6th ed2010. .
Wiley, New York, NY.

US EPA. *Guidelines for Exposure Assessment.* 1992. EPA/600/Z-92/001. Risk
Assessment Forum. Office of Research and Development, Washington, DC, May 29.

Zhang, C. *Fundamentals of Environmental Sampling and Analysis.* 2007. Wiley,
Hoboken, NJ.

Index